9th International Conference on Applied Physics and Mathematics (ICAPM 2019)

Journal of Physics: Conference Series Volume 1298

Bangkok, Thailand
21 - 23 January 2019

ISBN: 978-1-5108-9575-1
ISSN: 1742-6588

Printed from e-media with permission by:

Curran Associates, Inc.
57 Morehouse Lane
Red Hook, NY 12571

Some format issues inherent in the e-media version may also appear in this print version.

Institute of Physics
Dirac House, Temple Back
Bristol BS1 6BE UK

Phone: 44 1 17 929 7481
Fax: 44 1 17 920 0979

techtracking@iop.org

Additional copies of this publication are available from:

Curran Associates, Inc.
57 Morehouse Lane
Red Hook, NY 12571 USA
Phone: 845-758-0400
Fax: 845-758-2633
Email: curran@proceedings.com
Web: www.proceedings.com

TABLE OF CONTENTS

MATHEMATICAL THEORY AND CALCULATION

MATHEMATICAL MODELING AND METHODS

ALGORITHM DESIGN AND OPTIMIZATION

IMAGE PROCESSING AND DATA CALCULATION

ASTROPHYSICS

MATERIAL ANALYSIS AND FLUID MECHANICS

POWER AND CONTROL ENGINEERING

IOP Conf. Series: Journal of Physics: Conf. Series **1298** (2019) 011001 doi:10.1088/1742-6596/1298/1/011001

Preface

Welcome to 2019 9th International Conference on Applied Physics and Mathematics (ICAPM 2019), which will be held in Chulalongkorn University, Bangkok, Thailand during January 21-23, 2019. Chulalongkorn University founded in March 1917 is Thailand's first institution of higher learning. Chulalongkorn University has 20 faculties, 23 colleges and research institutes, over 8,000 faculty members, and over 37,000 students including undergraduates, postgraduates, and certificate program students. Chulalongkorn University is one of the world's great universities. We are delighted to come to prestigious Chulalongkorn University to meet with so many of our distinguished collaborators and scientists to explore new knowledge and ideas.

ICAPM Conference is one important event organized by American Society for Research (ASR). It has been successfully held in 8 consecutive years since 2011. In the past two years, ICAPM 2017 and ICAPM 2018 Conference Proceedings have been archived into IOP Journal of Physics: Conference Series, Volumes 814 (2017) and 1039 (2018) indexed by ISI Web of Science (Conference Proceedings Citation Index (CPCI), Scopus, Engineering Information (EI), etc. ICAPM 2019 aims to provide a high level international forum which focus on interdisciplinary and multidisciplinary research concerning material analysis and preparation, machinery and manufacturing engineering, fluorescent materials and biological materials, mechatronics and control engineering, metal materials and alloys, applied mathematics and computational modeling, quantum physics and astrophysics, and computer and information technology. These are all fascinating topics related to future needs. 28 Selected and important results will also publish in the peer-reviewed conference journal.

We express our thanks and appreciation to our keynote speakers Prof. Yoshihiko Uematsu from Gifu University in Japan, Assoc. Prof. Ratchatin Chancharoen from Chulalongkorn University in Thailand, Prof. Naoyuki Ishimura from Chuo University in Japan, Prof. Jong-Shenq Guo from Tamkang University in Taiwan, Assoc. Prof. Vadim Rashitovich Gasiyarov from South Ural State University in Russia, and all the participants and committee members for all support to ICAPM 2019.

January 2019

Prof. Po-Liang Liu
National Chung-Hsing University, Taiwan

IOP Conf. Series: Journal of Physics: Conf. Series **1298** (2019) 011002 doi:10.1088/1742-6596/1298/1/011002

Conference Committee

Conference Chairs

Prof. Naoyuki Ishimura, Chuo University, Japan

Prof. Po-Liang Liu, National Chung-Hsing University, Taiwan

Program Chairs

Prof. Jong-Shenq Guo, Tamkang University, Taiwan

Assoc. Prof. Kuo-Chih Chuang, Zhejiang University, China

Technical Committee

Prof. J F Lin, Far East University, Taiwan

Prof. Ratinun Luampon, Rajamangala University of Technology, Thailand

Prof. Alexandra B. Santos-Putungan, University of the Philippines Los Banos, Philippines

Prof. Darwin Barayang Putungan, University of the Philippines Los Banos, Philippines

Prof. Pramod Maurya, NIT Durgapur, India

Prof. Chih-Ru Hsiao, Soochow University, Taiwan

Prof. Wen-Lin Chiou, Fu-Jen University, Taiwan

Prof. HIRANMOY GARAI, National Institute of Technology Durgapur, India

Asst. Prof. Kayvan Moradipour, Technical and Vocational University, Iran

Prof. Praveen Agarwal, Anand Intl. College of Engg, India

Prof. Dr. Abdur Rashid, Gomal University, Pakistan

Prof. Haydar Akca, College of Arts and Sciences, Abu Dhabi University, UAE

Prof. Xiaodi Li, School of Mathematical Sciences, Shandong Normal University, China

Prof. Roberto Montemanni, University of Applied Sciences of Southern Switzerland, Switzerland

Prof. Ilona Tregub, Financial University under the Government of the Russian Federation, Russia

Prof. Samir Hadid, Ajman University, United Arab Emirates

Prof. Vijay Vir Singh, Yusuf Maitama Sule University, Kano, Nigeria, Nigeria

Prof. Nik Mohd Asri Nik Long, Universiti Putra Malaysia, Malaysia

Prof.Joginder Dhiman, Himachal Pradesh University, India

Asst. Prof. Autcha Araveeporn, King Mongkut's Institute of Technology Ladkrabang, Thailand

Assoc. Prof.Pankaj Biswas, Shri Mata Vaishno Devi University, India

Assoc. Prof. Johan Matheus Tuwankotta, Bandung Institute of Technology, Indonesia

Assoc. Prof. Rene Batac, Department of Physics, De La Salle University, Philippines

Assoc. Prof. Wael Zaki, Khalifa University, UAE

Assoc. Prof. Reza Mohammadi, University of Neyshabur, IR

Assoc. Prof. Lv Meibo, Northwestern Polytechnical University, China

Assoc. Prof. Ge Zhi Lei, Northwestern Polytechnical University, China

Assoc. Prof. Dr. Abdul Ghaffar Jaffar, Balochistan University of Information Technology, Pakistan

Assoc. Prof. Zhilei Ge, Northwestern Polytechnical University, China

Dr. Vamsi Krishna Narla, School of Technology, GITAM (Deemed to be University), Hyderbad, India

Dr. Siti Suzilliana Putri Mohamed Isa, Universiti Putra Malaysia, Malaysia

IOP Conf. Series: Journal of Physics: Conf. Series **1298** (2019) 011002 doi:10.1088/1742-6596/1298/1/011002

Dr. Ibrahim Rose Irnawaty, UNIVERSITI SAINS ISLAM MALAYSIA, MALAYSIA
Dr. Zailan Siri, University of Malaya, Malaysia
Dr. Khushboo Verma, UNIVERSITY OF LUCKNOW, India
Dr. Wooi-Nee Tan, Multimedia University, Malaysia
Asst. Prof. Dr. Yan-Gu Lin, National Synchrotron Radiation Research Center, Taiwan
Asst. Prof. Dr. Mohanad Alkhasawneh, Qatar University, Qatar
Asst. Prof. Yong-Yuk Won, Dept. of Electronic Engineering, Myongji University, South Korea
Asst. Prof. Thamir Abdul Ameer Hassan, Al -Karkh University of Science Iraq, Iraq
Asst. Prof. Abdalla Mansur, Al Ain University of Science and Technology, United Arab Emirates
Asst. Prof. Chukiat Chaiboonsri, Chiang Mai University, Thailand
Asst. Prof. Analabha Roy, The University of Burdwan, India
Asst. Prof. Jai Prakash Verma, Netaji Subhash Chandra Bose Girls Government Degree College aliganj, India
Dr. Lizhi Xing, College of Economics and Management, Beijing University of Technology, China
Dr. Kamni, shri mata vaishno devi university, India
Dr. Keeratikan Piriyakul, KMUTNB, Thailand
Dr. Nur Nafhatun Md Shariff, Universiti Teknologi MARA, Malaysia
Dr. Wang Pei, Northwestern Polytechnical University, China
Dr. Balqees Aldabbagh, university of technology, Iraq
Dr. Mitali Konwar, Moran college, India
Dr. Suraya Kamil, Universiti Teknologi MARA, Malaysia
Dr. Zety Sharizat Hamidi, Universiti Teknologi MARA, Malaysia
Dr. Darmesah Gabda, Universiti Malaysia Sabah, Malaysia
Dr. Suntaree Unhapipat, Mahidol University, Thailand

IOP Conf. Series: Journal of Physics: Conf. Series **1298** (2019) 011003 doi:10.1088/1742-6596/1298/1/011003

Peer review statement

All papers published in this volume of *Journal of Physics: Conference Series* have been peer reviewed through processes administered by the proceedings Editors. Reviews were conducted by expert referees to the professional and scientific standards expected of a proceedings journal published by IOP Publishing.

IOP Conf. Series: Journal of Physics: Conf. Series **1298** (2019) 012001 doi:10.1088/1742-6596/1298/1/012001

A Class of Weighted Point Schemes for the Grünwald Implicit Finite Difference Solution of Time-Fractional Parabolic Equations Using KSOR method

F A Muhiddin[1], J Sulaiman[2] and A Sunarto[3]

[1] Faculty of Computer and Mathematical Sciences, Universiti Teknologi MARA Cawangan Sabah, Malaysia.
[1,2] Faculty of Science and Natural Resources, Universiti Malaysia Sabah, Malaysia.
[3] Faculty of Islamic Economics and Business, IAIN Bengkulu, Indonesia.

E-mail: [1]fatih170@sabah.uitm.edu.my, [2]jumat@ums.edu.my, [3]andang99@gmail.com

Abstract. In this study, system of Grünwald implicit approximation equations has been developed through the discretization of one-dimensional linear time-fractional parabolic equations using the Grünwald fractional derivative operator and second-order implicit finite difference scheme. The aim of this paper is to examine the effectiveness of Kaudd Successive Over-Relaxation (KSOR) iterative method, which is one of the weighted point iterative schemes for solving the proposed time-fractional parabolic equations by considering the Grünwald implicit approximation equation. To investigate the effectiveness of the proposed iterative method, numerical experiments and comparison are made in terms of number of iterations, execution time, and maximum absolute error. Based on numerical results, the accuracy of Grünwald implicit solution obtained by proposed iterative method is in excellent agreement, and it can be concluded that the proposed KSOR iterative method requires less number of iterations and execution time as compared to the existing point iterative method.

1. Introduction

For the past few decades, the interest on the application of fractional order partial differential equations (FPDEs) in diverse fields such as biomedics, engineering, finance and natural sciences fields [1-3] have grown among researchers for its potential in describing the non-Markovian random walks problems [4] which is also stated by [5] as an effective tool to describe real world phenomena such as diffusion process. In this paper, we consider the one-dimensional inhomogeneous time-fractional parabolic equations (TFPE) of order $0<\alpha<1$ as

$$\frac{\partial^\alpha U(x,t)}{\partial t^\alpha}+p(x)\frac{\partial U(x,t)}{\partial x}+q(x)\frac{\partial^2 U(x,t)}{\partial x^2}=f(x,t), \qquad t>0, \quad x\in\Omega, \quad 0<\alpha\le 1, \qquad (1)$$

subject to the initial condition

$$U(x,0)=f(x), \qquad a\le x\le b,$$

and the boundary conditions

$$U(a,t)=g_1(t);\ U(b,t)=g_2(t) \qquad 0\le t\le T,$$

where $p(x)$ and $q(x)$ are real parameters and $f(x,t)$ is the source term.

To construct the approximation equation of problem (1), we used the implicit finite difference scheme with Grünwald fractional derivative order α, which defined as [6,7]

$$D_G^{\alpha} f(t) = \frac{1}{(\Delta t)^{\alpha}} \lim_{N\to\infty} \sum_{k=0}^{j} g_{\alpha,k} f(t - k\Delta t), \qquad 0 < \alpha < 1 \tag{2}$$

where the Grünwald weights are $g_{\alpha,k} = \dfrac{\Gamma(k-\alpha)}{\Gamma(-\alpha)\Gamma(k+1)}$.

Numerous numerical approaches which dealing with TFPE could be found in the literatures. Several recent researches have used the reduced spline (RS) method based on a proper orthogonal decomposition (POD) technique [8], Crank-Nicholson scheme with finite element method [9] and alternating segment explicit-implicit/ implicit-explicit parallel difference method [10]. Meanwhile, some earlier researchers have applied the finite difference method based on implicit scheme [11-17] and compact difference scheme [18], which are mostly described based on Caputo sense.

Similar to the integer order PDEs, the discretization of TFPE will also generate a large sparse linear systems. Where in this case, iterative method is famously used by researchers for its reputation in reducing the time consumption of the direct method. Gauss-Seidel (GS) iterative method is one of the most used classical iterative methods. Yet, the convergence rate of this method is still relatively slow. Hence, previous researchers such as [19-20] have devoted their studies to improve the convergence rate of large sparse linear systems. Prior to that, [21-22] has introduced and discussed another approach called the Successive Over-Relaxation (SOR) iterative method.

Just recently, a new concept known as Kaudd Successive Over-Relaxation (KSOR) iterative method have been introduced by [23-24], which have sparked the interest of this paper. So far, only some researches [15-17] could be found in solving TFPE iteratively and none of the work related to solving TFPE using KSOR iterative method could be found in the literatures. Figure 1 shows the finite grid network that is used to formulate the approximate equation, where the GS and KSOR iterative methods are applied onto each interior node point until the convergence test can be reached.

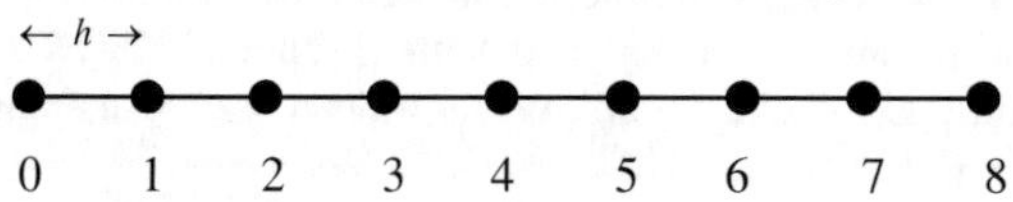

Figure 1. The distribution of uniform node points for the solution domain at m=8.

2. Grünwald finite difference approximation

In this section, the time-fractional parabolic equation (1) is discretized using the Grünwald fractional derivative as stated in equation (2) and implicit difference scheme. To derive numerical approximations based on Grünwald derivative, first let the mesh point $x_i = \alpha + ih$, where $i = 0,1,\ldots,N$ and $h = \Delta x = (b-a)/N$ denotes the uniform step-size. Hence, the discrete equation of problem (1) is given by

$$\frac{1}{(\Delta t)^{\alpha}} \sum_{k=0}^{j} g_{\alpha,k} U_{i,j-k} + \frac{p(x)}{2\Delta x}\left(U_{i+1,j} - U_{i-1,j}\right) + \frac{q(x)}{(\Delta x)^2}\left(U_{i+1,j} - 2U_{i,j} + U_{i-1,j}\right) = f_{i,j} \tag{3}$$

Then, by letting

$$G_k = \frac{g_{\alpha,k}}{(\Delta t)^{\alpha}}, \qquad a_i = \frac{p(x)}{2\Delta x}, \qquad b_i = \frac{q(x)}{(\Delta x)^2},$$

the simplified approximation equation can be stated in the form

$$\alpha_i U_{i-1,j} + \beta_i U_{i,j} + \gamma_i U_{i+1,j} = F_{i,j} \tag{4}$$

where,

$$\alpha_i = b_i - a_i, \qquad \beta_i = G_0 - 2b_i, \qquad \gamma_i = a_i + b_i,$$

and where, at time level $j=1$, $F_{i,j} = f_{i,j} - G_0$ and at time level $j = 2,3,\cdots,M$, $F_{i,j} = f_{i,j} - G_k \sum_{k=1}^{j} g_{\alpha,k} U_{i,j-k}$.

Meanwhile, the illustration of their respective computational molecules when time level $j=1$ and $j=2$ as shown in Figures 2 and 3.

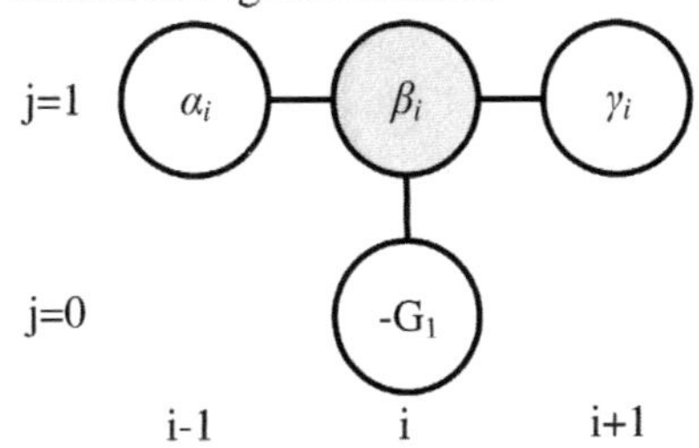

Figure 2. The computational molecule for Grünwald time-fractional parabolic approximation equation at time level j=1.

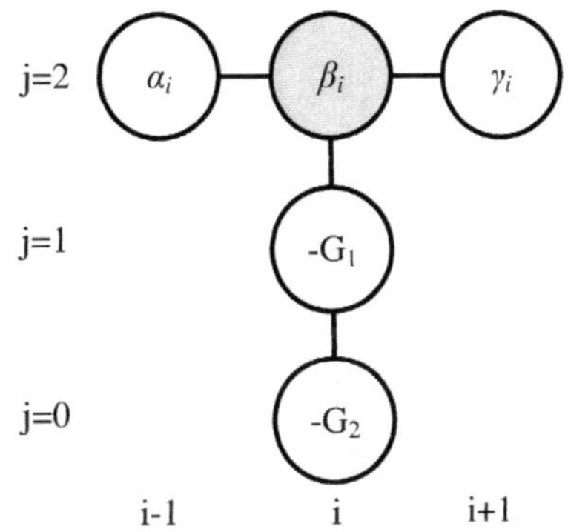

Figure 3. The computational molecule for Grünwald time-fractional parabolic approximation equation at time level j=2.

Based on Figure 2, it shows that the computational molecules for Grünwald time-fractional parabolic approximation equation is similar to the regular computational molecules of implicit finite difference approximation at time level $j=1$. Meanwhile, at time level $j = 2,3,\ldots,M$, a series of points will appear as a tail-like as shown in Figure 3.

Hence, equation (4) will lead to large sparse linear system in matrix form as

$$A\underline{U}_j = \underline{F}_j \tag{5}$$

where

$$A = \begin{bmatrix} \beta_1 & \gamma_1 & & & & \\ \alpha_2 & \beta_2 & \gamma_2 & & & \\ & \alpha_3 & \beta_3 & \gamma_3 & & \\ & & \ddots & \ddots & \ddots & \\ & & & \alpha_{m-2} & \beta_{m-2} & \gamma_{m-2} \\ & & & & \alpha_{m-1} & \beta_{m-1} \end{bmatrix}_{(m-1)\times(m-1)}, \underline{U} = \begin{bmatrix} U_{1,j} \\ U_{2,j} \\ \vdots \\ U_{m-2,j} \\ U_{m-1,j} \end{bmatrix}, \quad \underline{F} = \begin{bmatrix} F_{1,j} - \alpha U_{0,j} \\ F_{2,j} \\ \vdots \\ F_{m-2,j} \\ F_{m-1,j} - \gamma_{m-1} U_{m,j} \end{bmatrix},$$

and

$$\alpha_i = b_i - a_i, \qquad \beta_i = G_0 - 2b_i, \qquad \gamma_i = a_i + b_i.$$

3. Derivation of KSOR iteration scheme

Due to the performance of KSOR iterative method as discussed in [23-24], we investigate the effectiveness of KSOR iterative method in solving problem (1) based on Grünwald derivative and implicit finite difference scheme. To do so, the linear system in equation (5) are solved iteratively by using GS iterative method as a control method. First, let the coefficient matrix A be decomposed as

$$A = D + L + V \tag{6}$$

where D, V and L are diagonal, lower triangular and upper triangular matrices respectively.

To formulate the KSOR iterative method, let us recall the SOR iterative method stated in general form as [21-22,25]

$$\underline{U}_j^{(k+1)} = (1-\omega)\underline{U}_j^{(k)} + \omega(D+L)^{-1}\left(\underline{F}_j - V\underline{U}_j^{(k)}\right) \tag{7}$$

noted that, the SOR iterative method in equation (7) can be reduced to GS iterative method when $\omega = 1$.

By using the decomposition matrix in equation (6), the KSOR method can be written in general form as [23-24]

$$\underline{U}_j^{(k+1)} = \left(((1+\omega)D - \omega L)^{-1}(D+\omega V)\right)\underline{U}_j^{(k)} + ((1+\omega)D - \omega L)^{-1}\left(\omega \underline{F}_j\right) \tag{8}$$

where the relaxation parameter ω is extended to $R-[-2,0]$ for KSOR iterative method, compared to $\omega \in (1,2)$ for SOR iterative method.

The implementation of KSOR iterative method to solve problem (1) is as described in Algorithm 1.

Algorithm 1: KSOR scheme

i. Initialize $\underline{U}_0$ and $\varepsilon \leftarrow 10^{-10}$
ii. Assign the optimal value of ω,
iii. For $j = 1,2,\cdots,M$, perform:
   a. Set initial value, $\underline{U}_j^{(0)} \leftarrow 0$
   b. Calculate the approximate value of $\underline{U}_j^{(k+1)}$ iteratively
   $$\underline{U}_j^{(k+1)} \leftarrow \left(((1+\omega)D - \omega L)^{-1}(D+\omega V)\right)\underline{U}_j^{(k)} + ((1+\omega)D - \omega L)^{-1}\left(\omega \underline{F}_j\right)$$
   c. Perform the convergence test, $\left|U_{i,j}^{(k+1)} - U_{i,j}^{(k)}\right| \le \varepsilon = 10^{-10}$. If yes, go to step (d). Otherwise repeat step (b).
   d. Check time level, $j = M$. If yes, go to step (iv). Otherwise, repeat step (a).
iv. Display approximate solutions.

4. Numerical experiments

In this section, we tested one-dimensional inhomogeneous TFPEs for α=0.333, α=0.666 and α=0.999 in order to verify the performance of the KSOR iterative method over the GS iterative method. Comparison was made based on three criteria, which are the iteration numbers, time of execution and maximum absolute error. In the implementation of the iterative methods, the convergence test considered the tolerance error $\varepsilon = 10^{-10}$.

Example 1: Consider the following one-dimensional linear inhomogeneous fractional Burger's equation [26]

$$\frac{\partial^\alpha U(x,t)}{\partial t^\alpha} + \frac{\partial U(x,t)}{\partial x} - \frac{\partial^2 U(x,t)}{\partial x^2} = \frac{2t^{2-\alpha}}{\Gamma(3-\alpha)} + 2x - 2.$$

with p(x) =1 and q(x) = -1, and subject to the initial condition $U(x,0) = x^2$.

The exact solution is $U(x,t) = x^2 + t^2$.

Example 2: Consider the following one-dimensional inhomogeneous time-fractional parabolic equation [27]

$$\frac{\partial^{\alpha}U(x,t)}{\partial t^{\alpha}}-\frac{\partial^{2}U(x,t)}{\partial x^{2}}=f(x,t),$$

with p(x) =0 and q(x) = -1, while $f(x,t)=\frac{2t^{2-\alpha}}{\Gamma(3-\alpha)}t^{2-\alpha}\sin(2\pi x)+4\pi^{2}t^{2}\sin(2\pi x)$,

and subject to the initial condition $U(x,0)=0$. The exact solution is given by $U(x,t)=t^{2}\sin(2\pi x)$.

Example 3: Consider the following one-dimensional linear inhomogeneous time-fractional parabolic equation [28]

$$\frac{\partial^{\alpha}U(x,t)}{\partial t^{\alpha}}-\frac{\partial^{2}U(x,t)}{\partial x^{2}}=f(x,t), \qquad t>0, x\in R, 0<\alpha\leq 1,$$

with p(x) =0 and q(x) = -1, while $f(x,t)=\frac{2e^{x}t^{2-\alpha}}{\Gamma(3-\alpha)}-t^{2}e^{x}$,

subject to the initial condition $U(x,0)=0$. The exact solution is given by $U(x,t)=t^{2}e^{x}$.

Then, the numerical results of both GS and KSOR iterative methods for examples (1) to (3) are recorded as in Tables 1 to 3. Overall, the percentage of reductions of all examples at α=0.333, 0.666, 0.999 are respectively between 93.89-99.80% for the iteration numbers and 13.45-99.35% for the time of execution.

Table 1. Comparison of Number Iterations (k), execution time (t) (in seconds) and maximum absolute error for the iterative methods using example 1 at α=0.333, 0.666, 0.999.

M	Method	α=0.333			α=0.666			α=0.999		
		k	t	Max Error	k	t	Max Error	k	t	Max Error
128	GS	18325	7.89	2.5972e-02	8957	5.36	1.3065e-02	2772	3.69	1.2478e-03
	KSOR	404	3.2	2.5972e-02	283	3.11	1.3065e-02	166	3.14	1.2480e-03
256	GS	67139	41.9	2.5973e-02	32944	23.62	1.3066e-02	10244	11.42	1.2473e-03
	KSOR	813	6.67	2.5972e-02	569	6.47	1.3065e-02	331	6.33	1.2480e-03
512	GS	243922	276.4	2.5975e-02	120271	142.56	1.3067e-02	37649	53.24	1.2454e-03
	KSOR	1621	14.51	2.5972e-02	1136	13.63	1.3065e-02	659	13.08	1.2480e-03
1024	GS	877165	1914.74	2.5981e-02	435083	965.24	1.3074e-02	137338	323.96	1.2376e-03
	KSOR	3246	34.07	2.5972e-02	2269	30.86	1.3065e-02	1311	27.7	1.2480e-03
2048	GS	3114564	13356.94	2.6008e-02	1556326	6730.67	1.3102e-02	496352	2216.78	1.2066e-03
	KSOR	6365	87.08	2.5972e-02	4519	75.36	1.3065e-02	2618	63.69	1.2480e-03

Table 2. Comparison of Number Iterations (k), execution time (t) (in seconds) and maximum absolute error for the iterative methods using example 2 at α=0.333, 0.666, 0.999.

M	Method	α=0.333			α=0.666			α=0.999		
		k	t	Max Error	k	t	Max Error	k	t	Max Error
128	GS	14148	6.4	3.4745e-04	7120	4.72	5.2947e-04	2358	3.57	4.5604e-04
	KSOR	385	3.23	3.4734e-04	264	3.13	5.2935e-04	144	3.09	4.5594e-04
256	GS	47194	28	2.0203e-04	24125	17.5	3.8529e-04	8332	10.12	3.1294e-04
	KSOR	764	6.66	2.0160e-04	513	6.4	3.8483e-04	291	6.2	3.1267e-04
512	GS	151187	149.13	1.6691e-04	79109	86.99	3.5049e-04	29799	42.17	2.7708e-04
	KSOR	1467	13.99	1.6516e-04	1025	13.61	3.4871e-04	576	12.74	2.7687e-04
1024	GS	454367	812.72	1.6304e-04	251077	509.06	3.4511e-04	107228	240.64	2.6674e-04
	KSOR	2906	31.95	1.5605e-04	2049	30.49	3.3969e-04	1125	27.04	2.6791e-04
2048	GS	1241856	4591.56	1.7665e-04	839911	3310.98	3.3991e-04	384794	1578.31	2.5732e-04
	KSOR	5708	80.73	1.5380e-04	4056	71.23	3.3742e-04	2155	60.45	2.6568e-04

Table 3. Comparison of Number Iterations (k), execution time (t) (in seconds) and maximum absolute error for the iterative methods using example 3 at α=0.333, 0.666, 0.999.

M	Method	α=0.333			α=0.666			α=0.999		
		k	t	Max Error	k	t	Max Error	k	t	Max Error
128	GS	18947	7.94	1.1755e-03	9174	5.54	2.5778e-03	2824	3.78	2.2069e-03
	KSOR	423	3.16	1.1757e-03	295	3.13	2.5780e-03	172	3.11	2.2070e-03
256	GS	69499	40.98	1.1744e-03	33775	23.09	2.5767e-03	10436	11.31	2.2058e-03
	KSOR	853	6.69	1.1750e-03	592	6.44	2.5773e-03	342	6.27	2.2064e-03
512	GS	252889	267.84	1.1724e-03	123467	138.85	2.5746e-03	38383	51.68	2.2036e-03
	KSOR	1707	14.43	1.1748e-03	1181	13.68	2.5772e-03	681	12.89	2.2063e-03
1024	GS	911238	1836.1	1.1650e-03	447430	922.83	2.5670e-03	140195	309.11	2.1956e-03
	KSOR	3392	33.8	1.1748e-03	2353	30.74	2.5771e-03	1356	27.92	2.2062e-03
2048	GS	3243868	12617	1.1357e-03	1604114	6334.88	2.5367e-03	507555	2083.86	2.1641e-03
	KSOR	6774	88.57	1.1747e-03	4680	75.59	2.5771e-03	2708	63.53	2.2062e-03

5. Conclusion

In this paper, we proposed the Grünwald implicit finite difference scheme and KSOR iterative method to solve inhomogeneous time-fractional parabolic equations. By taking GS iterative method as the control method, the numerical results of three examples show that the KSOR iterative method is able to substantially improve the iteration numbers and time of execution of the GS iterative method in solving one-dimensional inhomogeneous time-fractional parabolic equations. This concluded that, the KSOR iterative method is superior than the GS iterative method.

6. Acknowledgments

The authors would like to express sincere gratitude to Universiti Malaysia Sabah for funding this research under UMSGreat research grant for postgraduate student: GUG0221-1/2018.

7. References

[1] Xu Q and Xu Y 2018 *Commun. Nonlinear Sci. Numer. Simulat.* **64** 135-148
[2] Edeki S, Ugbebor O and Owoloko E 2017 *Int. J. Appl. Math.* **47**(4)
[3] Paradisi P 2015 *Commun. Appl. Ind. Math.* **e-530** 1-25
[4] Gaspar F J and Rodrigo C 2017 *SIAM J. Sci. Comput.* **39**(4) A1201–24
[5] Sontakke B R and Shelke A S 2017 *Global J. Pure Appl. Math.* **13**(8) 4333-45
[6] Podlubny I 1999 *Fractional Differential Equations* (San Diego: Academic Press)
[7] Zahra W K and Elkholy S M 2013 *Electr. J. Math. Anal. Appl.* **1**(2) 230-41
[8] Ghaffari R and Ghoreishi F 2019 *Appl. Numer. Maths.* **137** 62–79
[9] Gunzburger M and Wang J 2019 *Int. J. Numer. Anal. Mod.* **16**(2) 225-39
[10] Wu L, Zhao Y and Yang X 2018 *J. Appl. Math. Phys.* **6**(5) 1017-33
[11] Chen C M, Liu F, Turner I and Anh V 2007 *J. Comput. Phys.* **227** 886-97
[12] Karatay I, Bayramoğlu Ş R and Şahin A 2011 *Appl. Numer. Math.* **61**(12) 1281-88
[13] Mishra T N and Rai K N 2015 *Int. J. Appl. Math. Res.* **4**(1) 135
[14] Murio D A 2008 *Comput. Math. Appl.* **56**(4) 1138-45
[15] Baharuddin, Sunarto A and Dalle J 2017 *J. Eng. Appl. Sci.* **12** 3220-24
[16] Sunarto A, Sulaiman J and Saudi A 2016 *Global J. Pure Appl. Math.* **12**(4) 3469-479
[17] Sunarto A, Sulaiman J and Saudi A 2014 *Journal of Physics: Conference Series* **495** 012032
[18] Gao G and Sun Z 2011 *J. Comput. Phys.* **230**(3) 586-95
[19] Hackbusch W 1994 *Iterative Solution of Large Sparse Systems of Equations* (New York: Springer-Verlag)
[20] Saad Y 2003 *Iterative methods for sparse linear systems* 2nd ed (Philadelphia: SIAM)
[21] Young D M 1971 *Iterative solution of large linear systems* (London: Academic Press)
[22] Young D M 1972 *J. Approx. Theory* **5** 137-48
[23] Youssef I K 2012 *J. Math. Stat.* **8** 176-84
[24] Youssef I K and Ibrahim R A 2013 *Life Sci. J.* **10**(2) 304-12
[25] Muhiddin F A and Sulaiman J 2017 *Journal of Physics: Conference Series* **890** 012065
[26] Uddin M and Haq S 2011 *Commun. Nonlin. Sci. Numer. Simul.* **16**(11) 4208-14
[27] Jiang Y and Ma J 2011 *J. Comput. Appl. Math.* **235**(11) 3285-90
[28] Ma Y 2014 *J. Appl. Math. Bioinformatics* **4**(2) 125-45

IOP Conf. Series: Journal of Physics: Conf. Series **1298** (2019) 012002 doi:10.1088/1742-6596/1298/1/012002

Newton Method with AOR Iteration for Finding Large Scale Unconstrained Minimizer with Tridiagonal Hessian Matrices

K Ghazali[a,1], J Sulaiman[a,2], Y Dasril[b,3], and D Gabda[a,4]

[a]Mathematics with Economics Programme, Universiti Malaysia Sabah, 88400 Kota Kinabalu, Sabah, Malaysia
[b]Faculty of Electronic and Computer Engineering, Universiti Teknikal Malaysia Melaka, 76100 Melaka, Malaysia.

[1]khadizah@ums.edu.my, [2]jumat@ums.edu.my, [3]yosza@utem.edu.my, [4]darmesah@ums.edu.my

Abstract. Finding the large scale unconstrained minimizer using Newton method has required the calculation of large and complicated linear systems results from solving the Newton direction. Therefore, in this paper, we propose a method for solving large scale unconstrained optimization problems with tridiagonal Hessian matrices to reduce the complexity of calculating Newton direction. Our proposed method was a combination of Newton method and Accelerated Over Relaxation (AOR) iterative method. To evaluate the performance of the proposed method, combination of Newton method with Gauss-Seidel iteration and Newton method with Successive Over Relaxation (SOR) iteration were used as reference method. Finally, the numerical experiment illustrated that the proposed method produce results that are more efficient compared to the reference methods with less execution time and minimum number of iterations.

1. Introduction

Unconstrained optimization problems shows an important role in many areas of mathematics, engineering, medical, computational economics and management. As the scale of unconstrained optimization problems is getting large, we need faster optimization methods that not only work well in theory but also work well in practice by using computing skill. Therefore, developing an efficient methods to solve large scale unconstrained optimization problems is a very important and active research area.

Generally, unconstrained optimization problems can be solved using either direct search methods or gradient descent methods. For this paper, we choose gradient descent methods for solving large scale unconstrained optimization problems. There are various existing gradient descent methods that have been used by previous researcher such as steepest descent method [1], Newton's method [2], conjugate gradient method [3], Quasi-Newton method [4] and Levenberg-Marquardt's algorithm [5]. Out of these methods, Newton's method as one of the most well-known classical methods of solution for unconstrained optimization problems to be discussed in this paper. This is due to the fact that Newton method has an excellent rate of convergence namely quadratic [6] and it also represents the most reliable method for solving large problem [7]. However, this method requires an initial guess to be selected closer to the optimal solution in order to guarantee its convergence. Another problem is that the Newton's method can be very costly in calculating and storing the inverse of the Hessian especially for large scale problems.

Thus in this paper, motivated by how to avoid the computational complexity of the inverse Hessian, we propose a new approach for finding large scale unconstrained minimizer. The new approach use a combination of Newton method with AOR iterative method and namely as Newton-AOR method. This combination uses AOR iterative method as inner iteration in finding the Newton direction then Newton's method is uses as the outer iteration to estimate the solution of problem. The AOR iterative method has been introduced by Hadjidimos [8] using two parameter generalization of the SOR method. This AOR method classified as one of the numerical methods that has an advantage of the efficient point iteration for solving any linear systems [9-11]. To evaluate the performance of the Newton-AOR method, this paper also considers a combination of Newton method with Gauss-Seidel iteration and Newton method with SOR iteration as reference method and they are named as Newton-GS method and Newton-SOR method respectively.

To investigate the capability of Newton-AOR method, let us consider a large scale unconstrained optimization problem being expressed as

$$\min_{\underline{x}\in\mathbb{R}^n} f(\underline{x}) \tag{1}$$

where $f:\mathbb{R}^n \to \mathbb{R}$ is the objective function and at least twice continuously differentiable. In the process of estimating the solution using Newton's method, a Newton direction is required which can be obtained by solving the linear system resulting from problem (1). Solving this linear system for higher dimension directly becomes computationally too expensive for calculating and storing the Hessian matrix. Thus, using AOR iterative method to discard the disadvantages of Newton's method can be reduced the computational cost of Newton Direction.

2. Approaches to Newton Scheme with a Tridiagonal Hessian Matrix

In this section, the Newton iteration for solving large scale unconstrained optimization was formulated for minimizing the quadratic approximation to the objective function, $f(\underline{x})$ in problem (1) at the current point $\underline{x}^{(k)}$ using the first three terms of Taylor series expansion as:

$$f(\underline{x}) \approx f(\underline{x}^{(k)}) + \left[\nabla \underline{f}(\underline{x}^{(k)})\right]^T (\underline{x}-\underline{x}^{(k)}) + \frac{1}{2}(\underline{x}-\underline{x}^{(k)})^T \nabla^2 f(\underline{x}^{(k)})(\underline{x}-\underline{x}^{(k)}), \tag{2}$$

where $\nabla \underline{f}(\underline{x}^{(k)})$ represent the gradient of the first partial derivatives of $f(\underline{x})$ and $\nabla^2 f(\underline{x}^{(k)}) = \mathbf{H}(\underline{x}^{(k)})$ represent as the Hessian matrix of second partial derivatives of $f(\underline{x})$. This quadratics approximation (2) will achieves its minimum value by differentiating it with respect to $\underline{x}$ and equating the resulting expression to zero as;

$$\nabla \underline{f}(\underline{x}^{(k)}) + \mathbf{H}(\underline{x}^{(k)})(\underline{x}-\underline{x}^{(k)}) = 0. \tag{3}$$

Thus, we can simplify equation (3) and setting $\underline{d}^{(k)} = \underline{x}^{(k+1)} - \underline{x}^{(k)}$, then we can formally define the Newton direction by:

$$\underline{d}^{(k)} = -\left[\mathbf{H}(\underline{x}^{(k)})\right]^{-1} \nabla \underline{f}(\underline{x}^{(k)}), \tag{4}$$

where $\left[\mathbf{H}(\underline{x}^{(k)})\right]^{-1}$ is the inverse of the Hessian matrix $\mathbf{H}(\underline{x}^{(k)})$. This the Newton direction (4) can be obtained by solving the Newton equation;

$$\mathbf{H}(\underline{x}^{(k)})\underline{d}^{(k)} = -\nabla \underline{f}(\underline{x}^{(k)}). \tag{5}$$

The Newton direction (4) is a descent direction as long as $\mathbf{H}(\underline{x}^{(k)})$ is positive definite. Since $\mathbf{H}(\underline{x}^{(k)})$ is positive definite, then its inverse must exist and is positive definite. Notice that, we are interested in solving problem (1) for a large scale unconstrained optimization. Thus, the Hessian matrix, $\mathbf{H}(\underline{x}^{(k)})$ involved in this study is a large sparse matrix. As a particularly interesting case, Hessian of a tridiagonal matrix of order n can be considered [12];

$$\mathbf{H}\left(\underline{\mathrm{x}}^{(k)}\right)=\left[h_{i,j}\right]=\begin{bmatrix} b_1 & c_1 & 0 & 0 & \cdots & 0 \\ a_2 & b_2 & c_2 & 0 & \cdots & 0 \\ 0 & a_3 & b_3 & c_3 & \ddots & \vdots \\ 0 & 0 & \ddots & \ddots & \ddots & 0 \\ \vdots & \vdots & \ddots & a_{n-1} & b_{n-1} & c_{n-1} \\ 0 & 0 & 0 & 0 & a_n & b_n \end{bmatrix}, \tag{6}$$

with $h_{i,i} > 0 \; \forall \; i$, $h_{i,j} < 0$ for $i \neq j$, $h_{i,j} = 0 \; \forall \; |i-j| > 1$ and $i, j = 1, 2, \ldots, n$. Since matrix (6) is symmetric, hence we have

$$a_i = c_{i-1}, \; i = 2, 3, \ldots, n. \tag{7}$$

Furthermore, this tridiagonal matrix $\mathbf{H}\left(\underline{\mathrm{x}}^{(k)}\right)$ has a diagonal entries that satisfies the inequality

$$|b_i| \geq |a_i| + |c_i|, \; \forall \; i = 01,2, \ldots, n, \tag{8}$$

so that the condition of positive definite can be fulfilled.

3. Formulation of proposed Iterative Method

As mention in the previous section, the Hessian of a tridiagonal matrix, $\mathbf{H}\left(\underline{\mathrm{x}}^{(k)}\right)$ is large and sparse. Therefore, finding the inverse of $\mathbf{H}\left(\underline{\mathrm{x}}^{(k)}\right)$ in high dimensions can be very expensive arithmetic operation. Hence, solving equation (5) by using direct method as in [13,14] will lead to a solution involving very trivial work. To overcome this problem, we propose an iterative method as in [15-17] for solving linear system of equation (5). Let the linear system (5) be rewritten in general form as

$$\mathbf{H}\underline{d} = \underline{f} \tag{9}$$

where, $\mathbf{H}$ is as in the form of matrix (6), $\underline{d}^T = [d_1 \; d_2 \; d_3 \ldots d_n]$ and $\underline{f}^T = [f_1 \; f_2 \; f_3 \ldots f_n]$. To derive the formulation of proposed iterative method, we decompose the real coefficient matrix $\mathbf{H}$ of the linear system (9) as;

$$\mathbf{H} = \mathbf{D} - \mathbf{L} - \mathbf{U} \tag{10}$$

in which $\mathbf{D}$ is the nonzero diagonal part, $\mathbf{L}$ is strictly lower triangular part and $\mathbf{U}$ is strictly upper part, of $\mathbf{H}$. By applying the decomposition in equation (10) into linear system (5), the iterative formulation of the SOR method can be stated in vector form as [18];

$$\underline{d}^{(k+1)} = (\mathbf{D} - \omega\mathbf{L})^{-1}(\omega\mathbf{U} + (\mathbf{1} - \omega)\mathbf{D})\underline{d}^{(k)} + \omega(\mathbf{D} - \omega\mathbf{L})^{-1}\underline{f} \tag{11}$$

For the implementation of SOR point iterations, each component $d_i^{(k+1)}$ can be computed as;

$$d_i^{(k+1)} = (1-\omega)d_{i-1}^{(k+1)} + \frac{\omega}{a_{i,i}}\left(f_i - \sum_{j=1}^{i-1} a_{i,j}d_j^{(k+1)} - \sum_{j=i+1}^{n} a_{i,j}d_j^{(k)}\right), i = 1,2, \ldots, n \tag{12}$$

where ω represents as a relaxation parameter with the optimal value in the range of $[1,2)$ and selected based on the smallest number of inner iterations. In any case, the optimal value is given by [18];

$$\omega_{OPT} = \frac{2}{1 + (1 - \rho^2)^{1/2}}$$

with ρ denotes as the spectral radius of the iteration matrix. In analogy to what was done for SOR iterations and by considering the implementation of two different relaxation parameters, which are r as an acceleration parameter and ω as an overrelaxation parameter. Therefore, we derived the formulation of AOR method [18] with the iterative scheme is given as;

$$d_i^{(k+1)} = (1-\omega)d_{i-1}^{(k+1)} + \frac{\omega}{a_{i,i}}\left(f_i - \sum_{j=1}^{i-1} a_{i,j}d_j^{(k)} - \sum_{j=i+1}^{n} a_{i,j}d_j^{(k)}\right) - \frac{r}{a_{i,i}}\sum_{j=1}^{i-1} a_{i,j}\left(d_j^{(k+1)} - d_j^{(k)}\right), i = 1,2, \ldots, n \tag{13}$$

By using the formulation of AOR iterative method to calculate the Newton direction (4) in Newton equation (5), we proposed the reliable algorithm of Newton-AOR scheme for solving problem (1), as stated in Algorithm 1. Note that for $\omega = r = 1$, equation (13) is reduced to the GS method and if $\omega = r$, then equation (13) coincides with SOR method. Hereby, to evaluate the performance of our proposed iterative method, the Newton-GS and Newton-SOR iterative methods are used as the reference method.

Algorithm 1 : Newton-AOR Scheme

i. Initialize

Set up the objective function : $f(\underline{x}), f(\underline{x}^*) \leftarrow \mathbb{R}, \underline{x}^{(0)} \leftarrow \mathbb{R}^n, \varepsilon_1 \leftarrow 10^{-6}, \varepsilon_2 \leftarrow 10^{-8}$, $n \leftarrow \{1000, 5000, 10000, 20000, 30000\}$

ii. Assign the optimal value of ω and r

iii. For $j = 1,2, \ldots, n$, implement

a. Set $\underline{d}^{(0)} \leftarrow 0$

b. Calculate $f(\underline{x}^{(k)})$

c. For $i = 1,2, \ldots, n$, calculate iteratively ; solve equation (9) by using equation (13)

d. Check the convergence test, $\|\underline{d}^{(k+1)} - \underline{d}^{(k)}\| < \varepsilon_2$. If yes, go to step (e). Otherwise go back to step (b)

e. For $i = 1,2, \ldots, n$, calculate; $\underline{x}^{(k+1)} \leftarrow \underline{x}^{(k)} + \underline{d}^{(k)}$

f. Check the convergence test, $\left\|\nabla \underline{f}(\underline{x}^{(k)})\right\| \leq \varepsilon_1$. If yes, go to (iii). Otherwise go back to step (a)

iv. Display approximate solutions

4. Numerical Experiments

Next we tested our algorithm using three different test functions taken from [19-21] and we implemented the algorithm in C language to solve problem (1). For each test function we considered five numerical experiments with dimensions varied from 1000 to 30000 variables as listed in Table 1 and tested with three different initial points, $\underline{x}^{(0)}$ which is randomly selected but closer to the solution point, $\underline{x}^*$. Therefore for three test functions, we have 45 test cases. In our numerical experiments, the termination criterion are $\|\underline{d}^{(k+1)} - \underline{d}^{(k)}\| < 10^{-8}$ for inner iteration and $\|\nabla f(\underline{x})\| < 10^{-6}$ for outer iteration. For the sake of comparison, we used combination Newton method with other two similar subroutines (GS and SOR point iterations). Thus, we compare the efficiency of our proposed method with Newton-GS iteration and Newton-SOR iteration subject to the number of inner iterations and the execution time. Now we consider three test functions (were generally used for unconstrained optimization test problem) with its Hessian of the Newton direction (4) that leads a tridiagonal matrix as follows;

Example 1: Dixon and Price Function [19]

$$f(\underline{x}) = (x_1 - 1)^2 + \sum_{i=2}^{n} i(2x_i^2 - x_{i-1})^2 \tag{14}$$

This function has a global minimum, $f^* = 0$ at $x_i^* = 2^{-\left[\frac{2^i-2}{2^i}\right]}$, for $i = 1, 2, \ldots, n$. The used starting points, $\underline{x}^{(0)}$ were

(a) $\underline{x}^{(0)} = (0.6, 0.6, \ldots, 0.6)$, **(b)** $\underline{x}^{(0)} = (1.0, 1.0, \ldots, 1.0)$ and **(c)** $\underline{x}^{(0)} = (0.6, 1.0, \ldots, 0.6, 1.0)$

Example 2: NONSCOMP Function [20]

$$f(\underline{x}) = (x_1 - 1)^2 + \sum_{i=2}^{n} 4\left(x_i - x_{i-1}^2\right)^2 \tag{15}$$

This function has a global minimum, $f^* = 0$ at $x_i^* = 1$, for $i = 1, 2, \ldots, n$. The used starting points, $\underline{x}^{(0)}$ were:

(a) $\underline{x}^{(0)} = (1.5, 1.5, \ldots, 1.5)$, **(b)** $\underline{x}^{(0)} = (3.0, 3.0, \ldots, 3.0)$ and **(c)** $\underline{x}^{(0)} = (1.5, 1.0, \ldots, 1.5, 1.0)$

Example 3: Cube Function [21]

$$f(\underline{x}) = (x_1 - 1)^2 + \sum_{i=1}^{n-1} 100\left(x_{i+1} - x_i^3\right)^2 \tag{16}$$

This function has a global minimum, $f^* = 0$ at $x_i^* = 1$, for $i = 1, 2, \ldots, n$. The used starting points, $\underline{x}^{(0)}$ were:

IOP Conf. Series: Journal of Physics: Conf. Series **1298** (2019) 012002 doi:10.1088/1742-6596/1298/1/012002

(a) $\underline{x}^{(0)} = (-1.2, -1.2, \ldots, -1.2)$, **(b)** $\underline{x}^{(0)} = (0.9, 0.9, \ldots, 0.9)$ and **(c)** $\underline{x}^{(0)} = (1.2, 1.0, \ldots, 1.2, 1.0)$

The numerical results for 45 test cases stated in Table 1. The efficiency comparison results for the execution time (seconds) and the number of iteration are presented in Table 1. From Table 1 also, it is obvious that the number of inner iteration for our proposed method reduce well compared from the referenced method. Whereas, Table 2 shows that the decrement percentage of the number of inner iterations and the execution time for Newton-AOR method is superior that the Newton-SOR method compared to Newton-GS method.

Table 1. Comparison of number of inner iteration and execution time (second) for Newton-GS, Newton-SOR and Newton-AOR method.

Example	Method	Order of Hessian matrix, n				
		1000	5000	10000	20000	30000
		Number of inner iterations (Number of outer iterations) / Execution time				
1 (a)	Newton-GS	193(20)/0.01	199(26)/0.05	202(29)/0.10	204(31)/0.20	206(33)/0.31
	Newton-SOR	100(11)/0.00	103(14)/0.03	104(15)/0.06	105(16)/0.12	106(17)/0.20
	Newton-AOR	91(10)/0.00	94(13)/0.02	95(14)/0.05	96(15)/0.10	97(16)/0.16
1 (b)	Newton-GS	282(16)/0.01	288(22)/0.05	290(24)/0.11	293(27)/0.23	294(28)/0.34
	Newton-SOR	147(8)/0.01	150(11)/0.05	152(13)/0.07	153(14)/0.13	154(15)/0.19
	Newton-AOR	139(7)/0.00	142(10)/0.03	143(11)/0.06	144(12)/0.12	145(13)/0.17
1 (c)	Newton-GS	276(18)/0.02	282(24)/0.06	284(26)/0.11	287(29)/0.23	288(30)/0.35
	Newton-SOR	157(9)/0.01	160(12)/0.03	162(14)/0.07	163(15)/0.14	164(16)/0.21
	Newton-AOR	138(8)/0.00	140(10)/0.03	142(12)/0.06	143(13)/0.12	144(14)/0.19
2 (a)	Newton-GS	208448(10)/4.39	208450(12)/21.33	208451(13)/42.57	208452(14)/84.69	208452(14)/127.50
	Newton-SOR	8523(6)/0.20	8632(6)/1.00	8632(6)/1.99	8632(6)/3.87	8632(6)/5.92
	Newton-AOR	5640(6)/0.14	5640(6)/0.65	5640(6)/1.27	5640(6)/2.61	5640(6)/3.96
2 (b)	Newton-GS	10460805(12)/207.53	10460807(14)/1075.94	10460808(15)/2147.06	10460809(16)/4269.45	10460809(16)/8301.51
	Newton-SOR	76137(322)/1.79	86359(337)/10.02	90525(202)/20.86	90525(202)/41.77	90525(202)/62.40
	Newton-AOR	58982(108)/1.39	58982(108)/6.87	58982(108)/13.60	58983(109)/26.89	58983(109)/40.48
2 (c)	Newton-GS	209275(10)/4.30	209277(12)/21.92	209278(13)/43.64	209279(14)/86.68	209282(15)/130.42
	Newton-SOR	8437(6)/0.20	8437(6)/1.00	8437(6)/2.16	8437(6)/4.34	8437(6)/6.46
	Newton-AOR	5463(6)/0.13	5463(7)/0.61	5463(7)/1.27	5463(7)/2.51	5463(7)/3.83
3 (a)	Newton-GS	341(44)/0.05	341(44)/0.11	341(44)/0.19	341(44)/0.38	341(44)/0.56
	Newton-SOR	324(44)/0.03	324(44)/0.10	324(44)/0.18	324(44)/0.36	324(44)/0.55
	Newton-AOR	319(45)/0.02	319(45)/0.09	319(45)/0.17	319(45)/0.35	319(45)/0.53
3 (b)	Newton-GS	414(47)/0.04	414(47)/0.13	414(47)/0.22	414(47)/0.42	414(47)/0.65
	Newton-SOR	393(44)/0.03	393(44)/0.11	393(44)/0.21	393(44)/0.41	393(44)/0.62
	Newton-AOR	378(48)/0.02	378(48)/0.10	378(48)/0.20	378(48)/0.39	378(48)/0.60
3 (c)	Newton-GS	385(45)/0.04	385(45)/0.13	385(45)/0.22	385(45)/0.39	385(45)/0.57
	Newton-SOR	340(45)/0.03	340(45)/0.10	340(45)/0.20	340(45)/0.38	340(45)/0.56
	Newton-AOR	334(46)/0.02	334(46)/0.09	334(46)/0.19	334(46)/0.36	334(46)/0.55

Table 2. Decrement percentage of the number of inner iterations and the execution time compared to Newton-GS method.

$\underline{x}^{(0)}$	Example / Method	Number of inner Iterations (%)			Execution Time (%)		
		1	2	3	1	2	3
(a)	Newton-SOR	48.19 – 48.54	95.86 – 95.91	4.99	35.48 – 100.00	95.31 – 95.44	1.79 – 40.00
	Newton-AOR	52.76 – 52.97	97.29	6.45	48.39 – 100.00	96.81 – 97.02	5.36 – 60.00
(b)	Newton-SOR	47.59 – 47.92	99.12 – 99.27	5.07	0.00 – 44.12	99.02 – 99.25	2.38 – 25.00
	Newton-AOR	50.68 – 50.85	99.44	8.70	40.00 – 100.00	99.33 – 99.51	7.14 – 50.00
(c)	Newton-SOR	42.96 – 43.26	95.97	11.69	36.36 – 50.00	94.99 – 95.44	1.75 – 25.00
	Newton-AOR	50.00 – 50.35	97.39	13.25	45.45 – 100.00	96.98 – 97.22	3.51 – 50.00

5. Conclusion

As presented in this paper, the combination of Newton method with AOR point iterative method has speed up the process for solving large scale unconstrained optimization problems with a tridiagonal Hessian matrix. This can be seen through the execution time and the number of iterations given in Table 1 as a result of our implementation proposed algorithm. In addition to that, the numerical results stated in Table 2 clearly has showed that our proposed method has reduced the number of inner iterations better than the reference algorithm with less execution time (second). Therefore, it can be concluded that our proposed iterative method (Newton-AOR) is able to show substantial improvement in the number of inner iterations and execution time compared to the Newton-GS and Newton-SOR point iterative methods. In future, this work will investigate the efficiency of the combination of Newton method with block iterative method as in [16, 22].

6. Acknowledgments

The authors are grateful for the fund received from Universiti Malaysia Sabah upon publication of this paper (GUG0160-2/2017).

7. References

[1] Napitupulu H, Sukono, Mohd B I, Hidayat Y and Suplan S 2018 Steepest descent method implementation on unconstrained optimization problem using C++ program. *Materials Sci. and Eng.* **332** 012024.

[2] Babaie-Kafaki S 2016 Emrouznejad A (eds) Big Data Optimization: Recent Developments and Challenges. Studies in Big Data, **18**. Springer DOI 10.1007/978-3-319-30265-2_17

[3] Moyi A U and Leong W J 2016 A sufficient descent three-term conjugate gradient method via symmetric rank-one update for largescale optimization *Optimization* **65** (1) 121-143.

[4] Aderibigbe F M, Adebayo K J and Dele-Rotimi A O 2015 *America Journal of Applied Mathematics* **3**(2) 47-50.

[5] Andreas F and Pradyumn K S 2008 *Operations Research Letters* **36** 643–646.

[6] Taheri S, Mammadov M and Seifollahi S 2015 *Optimization*, **64**(2) 249-263.

[7] Nocedal J 1992 Theory of Algorithms for Unconstrained Optimization. *Acta Numerica* **1**, 199-242.

[8] Hadhidimos A 1978 Accelerated Overrelaxation Method. *Mathematics of Computation*, **32**(141), 149–157.

[9] Yousif W S and Martins M M 2008 Explicit De-couple Group AOR method for solving elliptic partial differential equations. *Neural Parallel and Scientific Computations* **16**(4) 531-542.

[10] Akhir M K M and Sulaiman J 2017 Accelerated over relaxation iterative method using triangle element approximation for solving 2D Helmholtz Equations. *J. Phys: Conf. Ser.* **890** 012060. DOI:10.1088/1742-6596/890/1/012060

[11] Basran N A, Eng J H, Saudi A, and Sulaiman J 2018 Numerical analysis of AOR iteration for anisotropic diffusion equation in image blurring. *AIP Conf. Proc.*, 020040. https://doi.org/10.1063/1.5054239

[12] Li H-B, Huang T-Z, Lui X-P and Li H 2010 On the inverses of general tridiagonal matrices. *Linear Algebra and its Applications* **433**. 965-983.

[13] Killingbeck J P and Grosjean A 2010 A Gauss elimination method for resonances. *J Math Chem* **47**. 1027.

[14] Yunfei Liu, Jun Lv and Xiaowei Gao 2016 "The application of simultaneous elimination and backsubstitution method (SEBSM) in finite element method". *Engineering Computations*, **33**(8). 2339-2355. https://doi.org/10.1108/EC-10-2015-0287

[15] Sulaiman J, Hasan M K, Othman M and Karim S A A 2014 Fourth-order solutions of nonlinear two-point boundary value problems by Newton-HSSOR iteration. *AIP Conf. Proc.* **1602**. 69-75.

[16] Ghazali K, Sulaiman J, Dasril Y and Gabda D 2019 Application of Newton-4EGSOR Iteration for Solving Large Scale Unconstrained Optimization Problems with a Tridiagonal Hessian Matrix. In: Alfred R., Lim Y., Ibrahim A., Anthony P. (eds) Computational Science and Technology. Lecture Notes in Electrical Engineering, **481**. Springer, Singapore. https://doi.org/10.1007/978-981-13-2622-6_39

[17] Sulaiman J, Hasan M K, Othman M and Karim S A A 2015 Application of Block Iterative Methods with Newton Scheme for Fisher's Equation by Using Implicit Finite Difference. *Jurnal Kalam.* **8**(1) 039-46

[18] Young D M 1954 Iterative methods for solving partial difference equations of elliptic type. *Trans. Amer. Math. Soc.* **76**. 92-111 https://doi.org/10.1090/S0002-9947-1954-0059635-7

[19] Laguna, M. and Marti, R 2005 Experimental Testing of Advanced Scatter Search Designs for Global Optimization of Multimodal Functions *Journal of Global Optimization* **33**(2). 235-55

[20] Andrei N 2008. An unconstrained optimization test function collection, *Adv. Modeling and Opt.*, **10**(1): 147-161.

[21] Jamil M and Yang X 2013 A literature survey of benchmark functions for global optimization problems. *Int. J. Mathematical Modelling and Numerical Opt.* **4**(2). 150-194.

[22] Ghazali K, Sulaiman J, Dasril Y and Gabda D 2018 Newton method with explicit group iteration for solving large scale unconstrained optimization problems, *IOP Conf. Ser.: J. Phys. Conf. Ser.* **1132** 012056.

IOP Conf. Series: Journal of Physics: Conf. Series **1298** (2019) 012003 doi:10.1088/1742-6596/1298/1/012003

The Frontier of Estimator in Agriculture Economics: Application for Rice Products in 7 Top Countries in Asia

Chukiat Chaiboonsri[1] and Satawat Wannapan[2]

[1,2] Puey Ungphakorn Excellence Centre of Econometrics
Faculty of Economics, Chiang Mai University,Thailand

[1] chukiat1973@gmail.com [2] lionz1988@gmail.com

Abstract. The purpose to econometrically calculate the comparing estimators in parametric estimations is successfully done in this paper. The two powerful estimators such as the James-Stein estimation (JSE) and Maximum Likelihood estimation (MLE) were employed to solve the most efficiently parametric forecasting. Experimentally, the yearly time-series data of rice exporting products during 2006 to 2016 was collected and observed from 7 Asian countries, which are the major exporters in the world, such as Thailand, India, Pakistan, Vietnam, Myanmar, China, and Cambodia. Methodologically, the comparison between the two estimators was based on the linear regression to provide residual terms. Empirically, the results of the estimator comparison implied that the JSE can be substantially replaced the counterpart when doing in the multiple-factors estimation, and error terms estimated were compared for seeking the best estimator to predict trends of rice products. Consequently, it is sensible to introduce the J-S estimator can be one of interesting tools in the modern econometric researches.

1. Introduction

In modern econometrics, the big problems that many quantitative researchers have been trying to computationally solve are the parametric estimation and prediction. Interestingly, to precisely calculate our parameters, is the maximum likelihood estimator really the best estimator for data predictive estimating in the recent moment? Additionally, there has been long time since a hundred years ago that quantitatively econometric papers consider only two dimensions, which refer to one dependent variable related to one independent factor (an unknown-mean variable), others are fixed as constant following assumptions. However, the huge mistake is that we are losing the ability of truth expressions by fixing and assuming as the traditional way. Can we put more dimensions of exogenous variables and make a precise estimation without any assumption simultaneously? This question leads the authors to employ the mathematically experimental study by using the JSE (James-Stein's estimator) comparing with the maximum likelihood estimator (MLE). Consequently, the empirical result of this paper will be the challenge one that can extend the frontier of econometric estimations and forecasts in the current moment of modern econometrics.

According to the field of agricultural economics, especially rice exporting productions, the trends of this fundamental food have been econometrically predicted for seeking the precise occasions to bargain on the future market and to manage rice supplies in the real market. Obviously, there are many internal and external factors affecting the fluctuations of the trends, and we should not consider them separately. Consequently, the purpose of the paper is aimed to improve the econometric estimation by comparing two estimators between the J-S estimator and MLE estimator. The hypothesis of the result is to believe that the J-S estimation will be the new tool that can overcome the problem of outliners capturing in the MLE estimation. The paper employs the sampled data of agriculture goods, which are rice products in top 7 countries in Asia. The yearly data is collected during 2006 to 2016. In the term of literature, it is undeniable that mathematical results or even econometrical outcomes are strikingly contrary to generally held belief even though an obviously valid proof is given. James Stein's estimator, which can be alternatively called "Stein's idea", made this issue dramatically in 1961. This is the beginning of the story called "shrinkage estimation", which deliberate biases are introduce to improve all performance and its modern implementations. Because of the J-S estimator has not been the talk of the town in econometrics yet. This seems the empirical application of J-S testing is still rare in current moment. From literature, for example, the original paper considered into the estimation with quadratic loss by Stein [3], the discussed paper called "Stein's Paradox in Statistics" by Efron and Morris [2], the paper of Brown [1] regarding the admissibility of invariant estimators in more location parameters, the applied time-series paper with James-Stein estimators by Senda and Taniguchi [5], and the paper of quantum physics using James-Stein estimator by Stander [6].

2. Methodology

2.1 Simulation Method for Estimator Comparison

Before the beginning of real sampled estimations, the comparison between the MLE and JSE is investigated by simulated computations. In this paper, the bootstrap was the computational method relied on the relationship between samples and (unknown) populations by a comparable linkage between the samples at hand and appropriately designed (observable) resamples [7]. Let we compute the bootstrap percentile of the function $y = f(q_i)$, where q_i is a motif statistic. The samples, $q_1, \ldots, q_n$, are assumed to be identically independence distribution (i.i.d) and the function $P(q \mid I)$, stands for the parametric density of choices [8]. Thus, we can set $y = f(q_i)$ for each *I*, the bootstrap percentile is given by

$$P_{BS}(f, q) = \frac{|y_i \mid y < y|}{n}. \tag{1}$$

2.2 The James-Stein Estimator

The James–Stein estimator (JSE) is an estimator that provides an estimation of the mean of a multivariate normal distribution, and it dominates the classical Maximum-Likelihood estimator (MLE). The JSE is crucial to stand on their limitations as follows; 1) the JSE is a biased estimator, essentially, it trades bias for risk; 2) the James-Stein estimator is based on a minimax estimator, which is an estimator with the property that its largest risk is no greater than that of any other estimator. If the mean of a multivariate normal distribution is minimax, any estimator that dominates the MLE must also be minimax; 3) the JSE is defined as a minimax, which is an estimator for which no other estimator exists that dominates it. James and Stein showed that the MLE of the mean of a multivariate normal distribution is not an admissible estimator if the dimension exceeds two [4].

According to this paper, the J-S estimator is employed to solve the linear regression problem. The JSE readily extended to deal with a nonidentity covariance matrix, and this leads to two crucial concepts, namely, the "effective dimension" and "shifting the origin". We conduct $X \in R^P$ and denote that a normally distributed random vector with mean μ as well as positive covariance matrix Ω with a normal distribution, $X \sim N(\mu, \Omega)$. Mathematically, P is defined as $P = \left(\Omega^{-1/2}\right)$ and existed since Ω is positive and $PX \sim N(P\mu, I)$. Consequently, the mean equation can be expressed as follows;

$$\hat{\mu}^{JS} = \left(1 - \frac{p-2}{X'\Omega^{-1}X}\right)X. \tag{2}$$

IOP Conf. Series: Journal of Physics: Conf. Series **1298** (2019) 012003 doi:10.1088/1742-6596/1298/1/012003

This estimator belongs to the class of *spherically symmetric* estimator because of $\hat{\mu} = h(X'\Omega^{-1}X)X$, where $h: R \to R$ which belongs to any *Borel-measurable* function. Theoretically, if $tr[\Omega]/\lambda_{\max}[\Omega] \le 2$ and there is no spherically symmetric estimator that dominates the Maximum Likelihood estimator (MLE), this can be called as the *effective dimension* which is a real-valued quantity. To aware of the restriction on Ω existing and to proof the effectively dimensional condition is crucial for observing the problem for a different angle. Considering a diagonal covariance matrix, $\Omega = diag\{\lambda_1,....\lambda_P\}$. The squared error can be written as $\|\hat{\mu} - \mu\|^2 = \sum_{j=1}^{P} \lambda_, \left((\hat{P\mu})_j - (P\mu)_j\right)^2$, where $P = \Omega^{-(1/2)}$. Because the JSE of $P\mu$ in general does not take a small risk for every individual element, it is evident that if one of the λ_j is relatively large, there is no longer possible to compensate for the possibility of introducing a larger MSE into this particular element [4].

The next technique is *shifting the origin*. It is well known that the risk of the JSE, which decreases as $\|\mu\|^2 \to 0$. Accordingly, the risk of the JSE is decreased by shifting the origin to $\bar{\mu}$, replacing μ by $(\hat{\mu} + \bar{\mu})$, and estimating $\hat{\mu}$. For the parametric estimator by the JSE, the spherically symmetric JSE for the regression parameter is based on the JSE. The differences are that we include the matrix D in the regression model and the term $\left(\min\{(p-2)2(p^*-2)\}\right)^+$. Consequently, the James-Stein estimator for x in the regression is

$$\hat{x}^{JS} = \left(1 - \sigma^2 \frac{\left(\min\{(p-2),2(p^*-2)\}\right)^+}{\hat{x}^{ML'}(C'RC)\hat{x}^{ML}}\right)^+ \hat{x}^{ML}, \tag{3}$$

where R is defined as $R = (DD')^{-1}$. C is expressed as $C \in R^{n \times p}$, and $(C'RC)^{-1}$ is explained as the covariance matrix. p is the dimension of x, n the dimension of z, $\hat{x}^{ML}$ is the effective dimension, which is written as

$$P^* = \frac{tr\left\{(C'RC)^{-1}\right\}}{\lambda_{\max}\left\{(C'RC)^{-1}\right\}}. \tag{4}$$

If $n > p$ and σ^2 is unknown, σ^2 is derived as follows;

$$\sigma^2 = \frac{\left\|D^{-1}(z - C\hat{x}^{ML})\right\|^2}{n - p + 2}. \tag{5}$$

For the MLE and JSE, they are taken to rely on the following properties:

1) For all $x \in R^P$, $J^{JS}(x) \le J^{ML}(x)$, where $J^{ML} = E\left[\left\|\hat{x}^{ML} - x\right\|^2\right]$ and $J^{JS} = E\left[\left\|\hat{x}^{JS} - x\right\|^2\right]$ are the risk of $\hat{x}^{ML}$ and $\hat{x}^{JS}$, respectively.
2) If $p^* > 2$, $J^{JS} < J^{ML}$, with an upper bound on $J^{JS}(x)$ decreasing as x converging to the origin; $\left\|\Omega^{(1/2)}x\right\|^2 \to 0$.
3) If $p^* \le 2$, $\hat{x}^{JS} = \hat{x}^{ML}$.

3. Empirical Results

3.1. The Simulation Comparison between MLE Estimator and J-S Estimator

3.1.1 Case 1: changing in observations and factor dimensions. Before the real data goes to the computation, first of all, the simulated approach by the bootstrap sampling is employed to display the difference between the two estimators. Experimentally, 20 and 50 bootstrapping observations are generated with the difference of factor dimensions. For the former calculation, results of the comparison are shown in the table 1, and it is pictured in the figure 1, which shows the comparison of estimated errors.

Empirically, speaking to the case of 20 samples with 2 factor dimensions, the result evidently expresses that the JSE and MLE represent the similar values of errors. However, the difference is more obvious when the dimensions are increased as multiple factors. The result confirms that the JSE provides the efficient estimator rather than the MLE in the case of 6 factor dimensions. The errors estimated by the JSE are lower than the counterpart, which are 0.4268 and 1.0775, respectively. The estimated outputs are clear when we focus on the figure 1. The mean squared error lose estimated by the JSE is nearly zero when the dimensions are more than six elements. Conversely, the errors provided by the MLE are constant when the dimensions are increased. This states the MLE cannot efficiently capture robustness in the situation contained multiple variables. To proof that the JSE is better than the MLE surely, the second simulation analysis is employed to complete this issue. 50 bootstrapped observations are generated. The results are represented in the table 2.

Table 1. The simulation of 20 observations with 2 and 6 factor dimensions

Factor dimensions (p)	James-Stein (Square error)	MLE (Square error)
p = 2, theta_norm = 1	1.1281 (0.1551)	1.1281 (0.1551)
p = 6, theta_norm = 1	0.4268 (0.0637)	1.0775 (0.0959)

Table 2. The simulation of 50 observations with 2 and 6 factor dimensions

Factor dimensions (p)	James-Stein (Square error)	MLE (Square error)
p = 2, theta_norm = 1	1.1281 (0.1551)	1.1281 (0.1551)
p = 6, theta_norm = 1	0.4268 (0.0637)	1.0775 (0.0959)

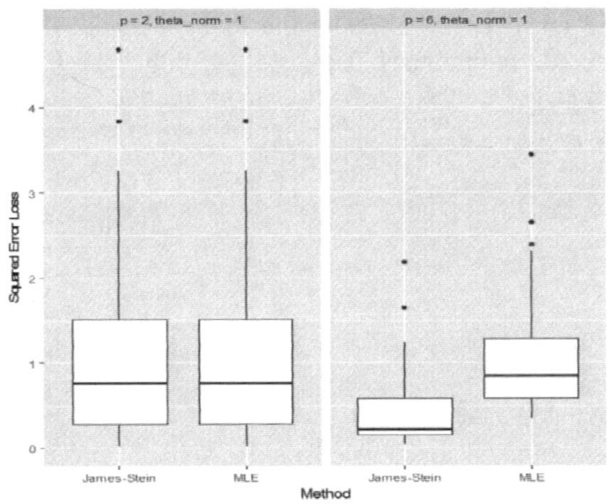

Figure 1. The simulation of 20 observations with 2 and 6 factor dimensions by the JSE and MLE

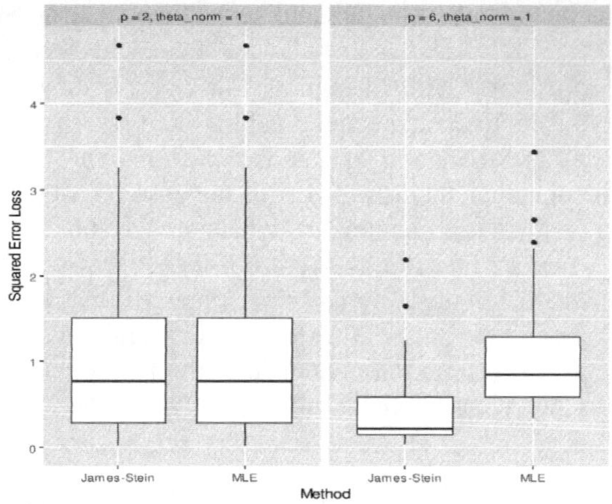

Figure 2. The simulation of 50 observations with 2 and 6 factor dimensions by the JSE and MLE

Evidently, from the table 2, the errors estimated from the J-S estimation are lower than the MLE when the factor dimensions are increased, even though it is not a huge increment in simulated observations. This strongly confirms that the J-S estimator is quite efficient to substitute the MLE, especially in the case of multi-dimensions analyses. Furthermore, the results are crystal clear when we consider into the figure 2.

3.1.2 Case 2: changing only in factor dimensions. This section is to employ 100 bootstrapping observations with differences of factor dimensions during 1 to 16 elements. Empirically, the results numerically detail that error terms estimated from the JSE are obviously lower that the counterpart when the dimensions are raised continuously. This can imply that the JSE is better than the MLE in the case of multiple independent variables, which is more realistic and closed to the true value in parametric estimations. These results are graphically represented as the figure 3.

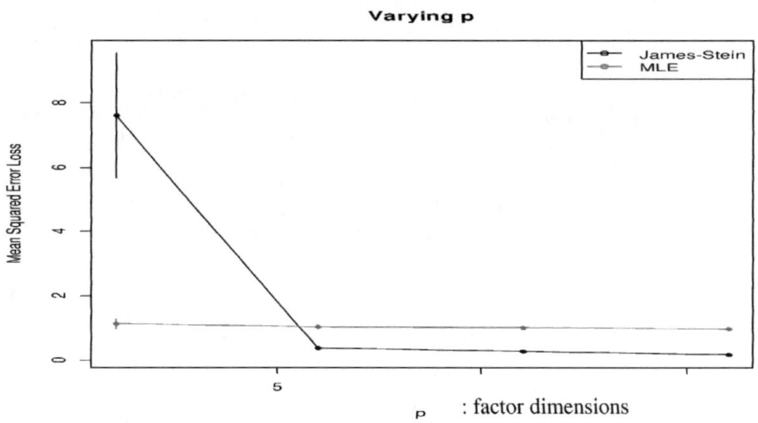

Figure 3. The simulation of 100 observations with changing factor dimensions by the JSE and MLE

3.2 Applications for Measuring Rice Exporting in 7 Top Countries in Asia

3.2.1 Data description of rice exporting in 7 top countries in Asia by the James-Stein Estimator. Descriptively, the symmetrical distributions which regard to the divided proportion between the estimated mean and median are pictured as the figure 4. Technically, the median-mean distribution should close to one theoretically, which explains the estimator is perfectly precise, and it can capture outliners completely. Empirically, the symmetric distributions estimated by the JSE are mostly higher than the counterpart provided by the MLE, which are the consumptions in Singapore, Malaysia, Indonesia, and Philippines from the J-S estimation, except for Thailand's consumption.

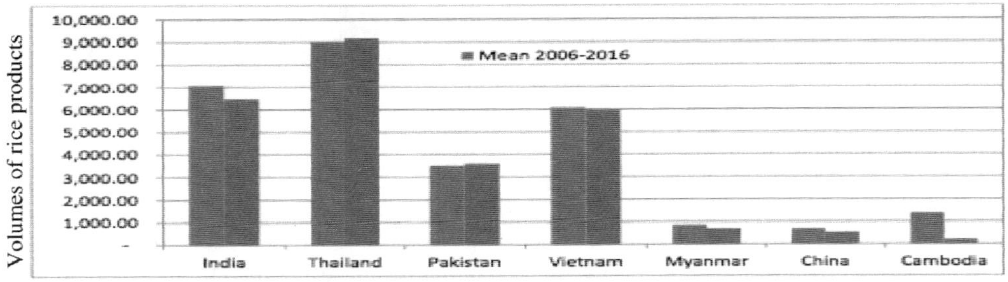

Figure 4. The presentation of estimated mean and median from the JSE estimation

3.2.2 *The Comparison measurement between the MLE and J-S estimator.* In the final section, the simulation data of rice exporting indexes in 7 top countries in Asia was processed in the linear regression, which is employed the MLE and JSE to provide error terms. In this case, the estimated residuals were used to indicate the precise estimation. The results displayed in the figure 5 expresses that the J-S estimator can be more fitted for the estimation rather than the MLE, which are 4 of 7 cases of parametric estimations such as Vietnam, Thailand, Pakistan, and India, respectively. On the other hand, the MLE can be fitted for the rice exporting products in Myanmar, China, and Cambodia, respectively.

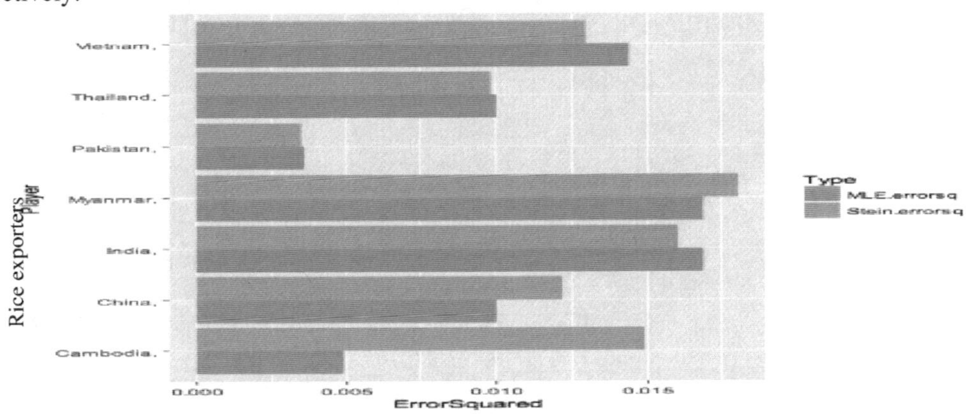

Figure 5. The comparison of estimated errors from the JSE and MLE

4. Conclusion

The clarification in the comparison of statistical estimators which are the James-Stein estimator (JSE) and Maximum Likelihood estimator (MLE) to apply in the agricultural indicator was successfully finished in this paper. Experimentally, it is obvious that the univariate factor, which is the data of rice exporting products in 7 Asian countries, is the main issue to do the parametric estimation. However, the problem leaded to produce this paper is the precision of the estimation, especially when the dimensions which can be referred to multiple-factor analyses are increased more than two elements. Consequently, the comparison between two estimators will provide the better way for improving the precision of econometrical forecasting.

Empirically, the results of the estimator comparison were clear that the JSE is the better estimator than the counterpart when doing in the multiple-factors estimation. Additionally, the simulation approach called *bootstrapping* was employed to extend the investigation in time-series data robustness. The results were obvious that the JSE can efficiently capture outliners in simulated data with high numbers of factor dimensions rather than the MLE. In the application for agricultural data (rice exporting products in top Asian countries), it is evident that the efficiency of parametric forecasting between the JSE and MLE based on the univariated linear regression can be strongly substitutable. This implies that the JSE can perfectly replace the MLE and completely overcome the problems of robustness, multicolinearity, even restricted linear assumptions. Accordingly, the JSE is potentially the frontier of econometric estimations and forecasts in the current moment of modern econometrics.

5. References

[1] Brown, L. D. (1966), "On the admissibility of invariant estimators of one or more location parameters", Annals of Mathematical Statistics, 37 (5): 1087–1136,doi:10.1214/aoms/1177699259.
[2] Efron, B. and Morris, C. (1977), "Stein's Paradox in Statistics." Scientific American. 236 (5): 119–127.
[3] James, W.; Stein, C. (1961), "Estimation with quadratic loss", Proc. Fourth Berkeley Symp. Math. Statist. Prob., 1, pp. 361–379.
[4] Manton, J. H., Krisnamurthy, V., and Vincent Poor, H. (1998). James–Stein state filtering algorithms. IEEE TRANSACTIONS ON SIGNAL PROCESSING, 46(9), pp. 2431-2447.
[5] Senda, M. and Taniguchi, M. (2006). James–Stein estimators for time series regression models. Journal of Multivariate Analysis, 97(9), pp. 1984-1996.
[6] Stander, M. (2017), Using Stein's estimator to correct the bound on the entropic uncertainty principle for more than two measurements. March 2, 2017. Available at: https://arxiv.org/abs/1702.02440.
[7] Vinod, H.D. (2013). *Maximum entropy bootstrap simulations for variance estimation*. Available at SSRN: https://ssrn.com/abstract=2295723.
[8] O'Neil, P.K. and Erill, I. (2016). Parametric bootstrapping for biological sequence motifs BMC. *Bioinformatics*, 17.

IOP Conf. Series: Journal of Physics: Conf. Series **1298** (2019) 012004 doi:10.1088/1742-6596/1298/1/012004

Differential equations learning from spatial-time series data by the fast iterative shrinkage-thresholding algorithm

Pumipat Tongudom[1], **Montri Maleewong**[1]

[1]Department of Mathematics, Faculty of Science, Kasetsart University, Bangkok, Thailand

Email:pumipat.to@ku.th

Abstract. The numerical method of data-driven modeling is presented in this work. It can be used to find the corresponding governing equations in terms of differential equations from some provided spatial and time series data. The problem can be reduced to solve by sparse linear regression technique that can be referred to one of the powerful technique in machine learning algorithm. Here we propose to apply the Fast Iterative Shrinkage-Thresholding Algorithm (FISTA) and then study the capability of its predictions. The environmental problem involves advection and diffusion effects are demonstrated to show the accuracy and the efficiency for finding all of the coefficients in the partial differential equation corresponding to the input data.

1. Introduction

From the past decade until now, the data storage devices and the hardware have become much cheaper.

The power of computing increases very fast. So, creating mathematical model has become more convenient and easier than before by using the collected data and efficient algorithms. In the past, differential equations are derived from the principles of physics or various conservative laws. These principles have been developed to describe natural phenomena. However, the problems in real situations such as neuroscience, environmental system, propagation of disease, or etc. have much complexities. It is very difficult to find underlying mathematical models to describe the physical phenomena.

In the present, there have been various numerical methods that can create differential equations from spatial and time series data, such as artificial neural network [1], nonlinear regression [2], sparse identification of nonlinear dynamical systems [3], and the PDE-FIND with the sparse linear regression [4]. The main objective of our present study is to apply the PDE-FIND to discover the governing equations from the collected data. The proposed PDE-FIND algorithm is modified using the FISTA algorithm which gives faster and better convergence rate than using the ISTA [5] and the TWIST [6]. In addition, we have demonstrated how to apply the proposed method for finding various parameters in the considering differential equations. This method can be applied to solve environmental problems involving advection and diffusion effects.

2. Theory and related works

2.1. Differential equation learning

In order to find the governing equations or mathematical models from the collected spatial and time series data. Generally, we assume that the differential equations are in the following form.

$$u_t = N\left(u, v, uv, u_x, u_{xx}, \ldots, x, t, \mu\right), \tag{1}$$

where u_t is the derivative of an unknown function in time, $N(□)$ is an unknown operator, μ is the parameter in our method. The main goal is to find $N(□)$ implying that the mathematical model can be found from the given data.

The right-hand side of (1) can be expressed in a vector form as

$$U_t = \Theta(U, Q)\xi, \tag{2}$$

where $\Theta(U,Q)$ is the library of candidate terms in (1) that contains various derivative terms of $u(x,t)$, Q represents any additional input terms, and ξ is the sparse vector identifying the derivative terms in the corresponding differential equations. Overall steps can be summarized in Figure 1. The challenging point is the optimization process to find numerically the sparse vector. The FISTA and ISTA methods are applied in this work. The final output will be the governing equations or PDEs that can represent the dynamics of collected data.

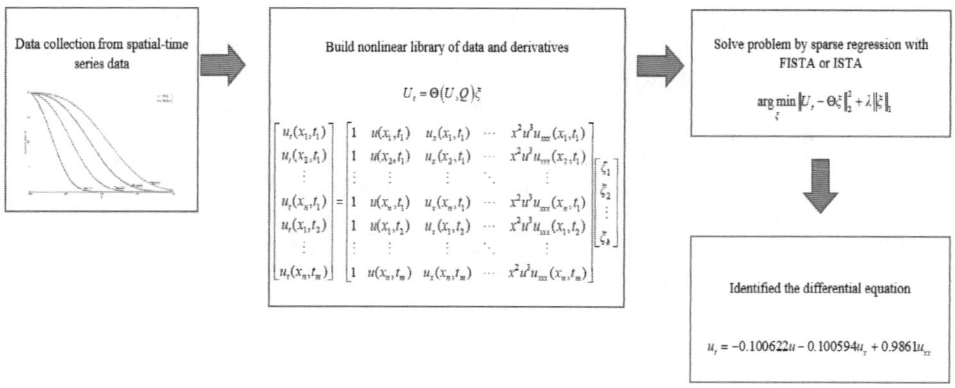

Figure 1. Overall steps in the differential equation learning for identifying mathematical model from the spatial and time series data.

2.2. *Building libraries of candidate terms*

The first step for building libraries of candidate terms starts from collecting all spatial and times series data, then rearrange them into a column vector $U \in □^{n\times m}$ of n spatial locations and m time data points. We collect U, Q in a library as

$$\Theta(U,Q) \in □^{n\times m\times d},$$

where d is the number of candidate terms. The derivatives of all data with respect to time t and spatial x are calculated using the finite difference method. All of the candidate terms and the partial derivatives are represented in terms of $\Theta(U,Q)$ as

$$\Theta(U,Q) = \begin{bmatrix} 1 & U & U^2 & \ldots & Q & \ldots & U_x & UU_x & \ldots & Q^2U^3U_{xx} \end{bmatrix}. \tag{3}$$

Each column of $\Theta(U,Q)$ contains candidate terms across of $n\times m$ space-time grid points. For example, if we have 128 spatial data points, 100 time data points, and 20 candidate terms, then $\Theta(U,Q) \in □^{128\times 100\times 20}$.

2.3. Sparse Linear Regression

After we have already built the library and the corresponding candidate terms, we need to find the sparse vector ξ for identifying the final candidate terms in the governing equation. In this work, the sparse linear regression method with the concept of the regularized least square can be applied as follows.

$$\xi = \arg\min_{\xi} \left\| U_t - \Theta\xi \right\|_2^2 + \lambda \left\| \xi \right\|_1, \tag{4}$$

where $\lambda\left\|\xi\right\|_1$ is the penalty term, $\left\|\xi\right\|_1$ is the l_1-penalty term , λ is a tuning parameter that controls the strength of the l_1-penalty and ξ is the iterative sparse vector. More details can be found in [4].

2.3.1. Iterative shrinkage-thresholding algorithm (ISTA)

The iterative shrinkage-thresholding algorithm is one of the most powerful methods for solving equation (4) where each iteration involves matrix-vector multiplication involving Θ and Θ^T that are called the shrinkage or soft-threshold step.
The general form of ISTA is that

$$\xi_{k+1} = \psi_\alpha\left(\xi_k - t\Theta^T\left(\Theta\xi_k - U_t\right)\right), \tag{5}$$

where t is a time step size, and k is an iteration counter. The shrinkage operator ψ_α is defined by

$$\psi_\alpha\left(\xi_k\right) = sign\left(\xi_k\right)\left(\left|\xi_k\right| - \alpha\right)^+. \tag{6}$$

More details for calculating (5) and (6) can be found in [5, 8]. The advantage of ISTA method is that it is simple and accurate. However, the ISTA method may converge slowly for some problems. Short summary of the ISTA method is shown in algorithm 1.

2.3.2. Fast iterative shrinkage-thresholding algorithm (FISTA)

In some problems, the ISTA method may converge slowly. Then, in this work, we modify the PDE-FIND algorithm by applying the fast iterative shrinkage-thresholding algorithm (FISTA) in [7], and related work in [9]. The different concept between the ISTA and the FISTA is that the shrinkage operator $\psi_\alpha(\square)$ is not calculated directly, but rather for a vector z_k which uses a linear combination of the first two previous vectors $\{\xi_{k-1}, \xi_{k-2}\}$. More additional computations are also performed using these formulae,

$$z_1 = \xi, t_1 = 1 \text{ and } t_{k+1} = \frac{1 + \left(1 + 4t_k\right)^{1/2}}{2}. \tag{7}$$

It can be proved from [7] that the convergence rates of the ISTA and the FISTA are $O(\frac{1}{k})$ and $O(\frac{1}{k^2})$, respectively, where k is the iteration number. Short summary of the FISTA method is shown in algorithm 2.

Algorithm 1: ISTA$\left(\Theta, U_t, \lambda, iters\right)$	**Algorithm 2:** FISTA$\left(\Theta, U_t, \lambda, iters\right)$

1. Set $\xi = [0]_{d\times 1}$ 2. Set a Lipschitz constant $L = \left\|\Theta^T\Theta\right\|$ 3. for $k = 1,..,iters$: $\xi_{k+1} = \psi_\alpha\left(\xi_k - t\Theta^T\left(\Theta\xi_k - U_t\right)\right)$ 4. If converge, return iterative sparse vector, else "error message"	1. Set $\xi = [0]_{d\times 1}$ 2. Set a Lipschitz constant $L = \left\|\Theta^T\Theta\right\|$ 3. Set $z_1 = \xi, t_1 = 1$ 4. for $k = 1,..,iters$: 4.1. $\xi_{old} = \xi$ 4.2. $\xi = \psi_\alpha\left(z_k - \frac{1}{L}\Theta^T\left(\Theta z_k - U_t\right)\right)$ 4.3. $t_{k+1} = \frac{1+\left(1+4t_k\right)^{1/2}}{2}$ 4.4. $z_{k+1} = \xi + \left(\frac{t_k - 1}{t_{k+1}}\right)\left(\xi - \xi_{old}\right)$ 5. If converge, return ξ else "error message"

The summary of the whole PDE-FIND method is shown in algorithm 3. The collected data in both spatial and time series data are divided into two sets composing of training and testing with the ratio of 80:20. The most important step in algorithm 3 is step 7 that solves the sparse linear regression problem. In this work, we will show the efficiency of the PDE-FIND algorithm by comparing the convergence rate between FISTA and ISTA methods.

Algorithm 3: TrainData$\left(\Theta, U_t, \lambda, max_iters, iters\right)$ (modified from [3] and [4])

\# Split the data into training and testing sets 80:20

1. Set $\Theta \rightarrow \left[\Theta^{train}, \Theta^{test}\right]$
2. Set $U_t \rightarrow \left[U_t^{train}, U_t^{test}\right]$

\# Set an appropriate l_0 – penalty parameter.

3. Set $\lambda_0 = 10^{-3}$

\# Get a baseline predictor

4. $\xi_{best} = (\Theta^{train})^{-1} U_t^{train}$
5. $error_{best} = \left\|\Theta^{test} - \xi_{best} U_t^{test}\right\|_2^2 + \lambda_0 \left\|\xi_{best}\right\|_0$

\# Find the best predictor

6. for $i = 1,..,max\_iters$:
   \# Train and evaluate performance by FISTA
   6.1. $\xi = \text{FISTA}(\Theta^{train}, U_t^{train}, \lambda, iters)$
   6.2. $error = \left\|\Theta^{test} - \xi_{best} U_t^{test}\right\|_2^2 + \lambda_0 \left\|\xi_{best}\right\|_0$
   6.3. If $error \le error_{best}$:
      6.3.1. $error = error_{best}$
      6.3.2. $\xi_{best} = \xi$
7. Return ξ_{best}

3. Numerical experiment

To demonstrate the efficiency of the proposed differential equation learning method, we consider the environmental problem that involves advection, diffusion and reaction effects. The efficiency in terms

of the convergence rate are investigated.
In this section, testing is done by using the numerical data from one-dimensional advection-diffusion problem. The problem is solved by the forward-time central space method (FTCS). The exact solutions are available in some cases. In order to measure the accuracy and the rate of convergence using the proposed algorithm, we define the mean absolute percentage error (MAPE) by

$$MAPE = \frac{1}{n}\sum_{i=1}^{n}\left|\frac{\hat{w}_i - w_i}{w_i}\right| \times 100\% ,$$

where n is the number of candidate terms, $\hat{w}_i$ is the given exact coefficients in the PDE (if available), and w_i is the predicted numerical coefficients. For easy cases of boundary conditions, exact solutions can be found. It implies that the given data is exact without noise. But for the case of complex boundaries, the set of data are generated using the FTCS scheme. These numerical data can be regarded as data with noise. These two sets of data are used as input data in our method. Next, the accuracy can be investigated.

3.1. One-dimensional advection-diffusion problem

Consider the environmental problem involves advection and diffusion effects that can be represented by the PDE,

$$\frac{\partial C}{\partial t} = D\frac{\partial^2 C}{\partial x^2} - u\frac{\partial C}{\partial x} ,$$

where $C(x,t)$ is the concentration of pollution, u is the flow velocity in the x direction, and D is the diffusion coefficient. We will consider two cases of Dirichlet and Neumann conditions.

3.1.1. Dirichlet boundary condition

The boundary and initial conditions are given by

$$C(0,t)=0, \quad C(L,t)=100, C(x,t)=\frac{100x}{L}, \qquad 0 \le x \le L.$$

From [10], the exact solution is that

$$C(x,t) = 100\left[\frac{e^{\frac{Px}{L}}-1}{e^P-1} + \frac{4\pi e^{\frac{Px}{2L}}\sinh(P/2)}{e^P-1}\sum_{m=1}^{\infty}A_m + 2\pi e^{\frac{Px}{2L}}\sum_{m=1}^{\infty}B_m\right],$$

where P is the Peclet number defined by $P = \frac{uL}{D}$, the coefficients A_m, B_m are given by

$$A_m = (-1)^m\frac{m}{\beta_m}\sin\left(\frac{m\pi x}{L}\right)e^{-\lambda_m t}, \; B_m = \left[(-1)^{m+1}\frac{m}{\beta_m}\left(1+\frac{P}{\beta_m}\right)e^{\frac{-P}{2}} + \frac{mP}{\beta_m^2}\right]\sin\left(\frac{m\pi x}{L}\right)e^{-\lambda_m t},$$

where $\beta_m = \left(\frac{P}{2}\right)^2 + (m\pi)^2$ and $\lambda_m = \frac{u^2}{4D} + \frac{m^2\pi^2 D}{L^2} = \frac{D\beta_m}{L^2}$.

The PDE is dimensionless form, we set $u=-0.1$, $D=0.01$, and $L=2$. The numerical data are generated by the FTCS scheme with $\Delta x = 0.02$, and $\Delta t = 0.005$. The plot of pollution concentration is shown in Figure 2. The numerical results are accurate. These two sets of data will be used as input data in algorithm 3. The problem is solved by FISTA and ISTA methods.

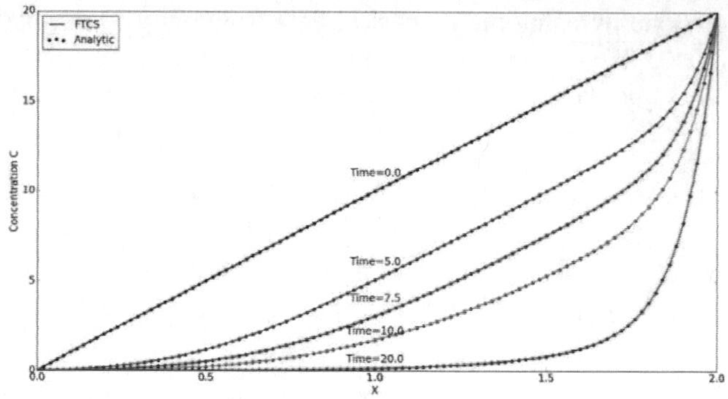

Figure 2. Variations of the pollution concentration in the time and spatial domains.

When the input data is numerical data, the PDE can be found as

$$C_t = -0.10562 C_x + 0.010114 C_{xx},$$

when the input data is exact with the same spacing in space and time, the PDE is that

$$C_t = -0.100284 C_x + 0.010118 C_{xx}.$$

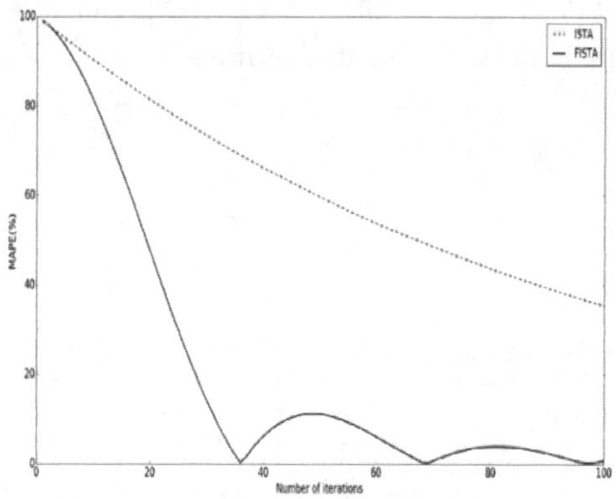

Figure 3. The number of iterations for ISTA and FISTA algorithms versus MAPE error.

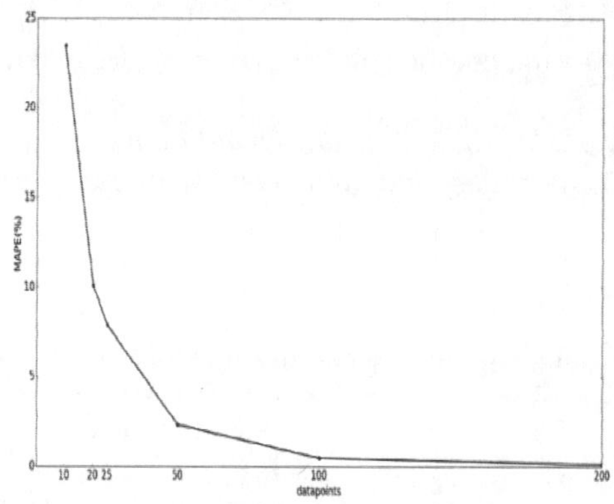

Figure 4. The MAPE versus the number of input data points.

The number of iterations resulted from the ISTA and the FISTA algorithms versus mean absolute percentage error (MAPE) are shown in Figure 3. The convergence rate of FISTA is faster than ISTA. The MAPE versus the number of input data points is shown in Figure 4. When the number of input data increases, the present algorithm can find the PDE better.

3.1.2. Mixed boundary condition
The boundary conditions are given by

$$C(0,t) = 20, \frac{\partial C(L,t)}{\partial x} = 0, \qquad t > 0.$$

To test the accuracy, the exact coefficients are given by $u = -0.1$, and $D = 0.01$. Here, we fix the pollution concentration at the left boundary while we leave naturally on downstream. Exact solution can be found from [10]. Input data from the FTCS and analytical solutions provides respectively the PDE,

$$C_t = -0.100041C_x + 0.009974C_{xx}, \text{ and } C_t = -0.100048C_x + 0.01C_{xx}.$$

They are in good agreement when we use two different sets of input data (noise and without noise). In term of convergence rate, FISTA method converges within 5 iterations while ISTA method converges within 10 iterations when the MAPE is set as 0.15%. Effects of increasing input data are shown in Table 1 where the number of time data points is fixed as 2000, then we vary the number of spatial data points. It shows that the errors decrease very fast when the number of data points increases.

Table 1. Effects of increasing input spatial data points

Spatial points n	PDE	MAPE(%)
10	$C_t = -0.0569C_x + 0.0046C_{xx}$	48.49
50	$C_t = -0.1004C_x + 0.0093C_{xx}$	3.50
200	$C_t = -0.1001C_x + 0.0100C_{xx}$	0.17

4. Conclusions

The differential equation learning algorithm is presented in work. Input data will be separated as the training and testing sets, and then solve iteratively the sparse linear regression problem. Previously, the ISTA algorithm is implemented to find the coefficients in the PDE, but in this work, we have modified and implemented the FISTA algorithm in order to find all of the coefficients in the considering PDE. The accuracy and convergence rate are also shown. The testing PDE involves advection and diffusion effects with various types of boundary conditions. The PDE-FIND with FISTA method can be applied to find the target PDE which implies that all of the coefficients in the PDE can be estimated at the same time.

5. References

[1] GonzAlez-Garciaa R, Rico-Martinez R and Kevrekidis I G 1998 Identification of distributed parameter systems: A neural net based approach *Computers chem. Engng* **22** 965

[2] Voss H, Kolodner P, Abel M and Kurths J 1999 Amplitude equations from spatiotemporal binary-fluid convection data *Phys. Rev. Lett.* **83** 3422

[3] Bruntona S L, Proctorb J L, and Kutzc J N 2016 Discovering governing equations from data by sparse identification of nonlinear dynamical systems *Proc. Natl. Acad. Sci. USA.* **113** 3932

[4] Rudy S H, Brunton S L, Proctor J L, Kutz J N 2017 Data-driven discovery of partial differential equations *Science Advances* e1602614

[5] Daubechies I, Defrise M, and Mol C D 2004 An iterative thresholding algorithm for linear inverse Problems with a sparsity constraint *Comm. Pure Appl. Math.* **57** 1413

[6] Bioucas-Dias J and Figueiredo M 2007 Two-step iterative shrinkage-thresholding algorithms for image restoration *IEEE Trans. Image Process* **16** 2992

[7] Beck A and Teboulle M 2009 A fast iterative shrinkage-thresholding algorithm for linear inverse problems *Siam J. Imaging Science* **2** 183

[8] Tibshirani R 1996 Regression shrinkage and selection via the Lasso *J.R. Statist. Soc.* B **58** 267

[9] Nesterov Y E 1983 A method of solving a convex programming problem with convergence rate $O(1/k^2)$ *Dokl. Akad. Nauk SSSR* **27** 543

[10] Thongmoon M and Mckibbin R 2006 A comparison of some numerical methods for the advection-diffusion equation *Res. Lett. Inf. Math. Sci.* **10** 49

Measurement of the cell density of microalgae by an optical method

Piyachat Wungmool, Nattha Rangsi, Tongchai Hormwantha, Malee Sutthiopad and Chaiya Luengviriya*

Department of Physics, Kasetsart University, 50 Phaholyothin Road, Jatujak, Bangkok 10900, Thailand.

* Email: fscicyl@ku.ac.th

Abstract. Microalgae are currently interesting in industry because it can generate many useful products, such as supplementary food, cosmetic and biomass energy. During the cultivation, the growth of microalgae is traditionally followed by cell counting with an optical microscope and a calculation of the cell density. To obtain a precise value of the cell density, the measurement should be performed by ones with good laboratory skills and it takes a long time. We develop a photometer to determine the cell density of microalgae. The photometer consists of a light dependent resistor as the sensor, a light emitting diode as the light source, and a microcontroller which controls all components as well as the cell density calculation using a calibration equation. To test the performance of our device, we cultivate marine microalgae *Dunaliella Salina* with different aerations and measure the cell density during 18-day cultivation period by using the photometer and the cell counting with an optical microscope. The results show that the cell density values obtained from both measurement methods increase with day of cultivation in a similar manner. Thus, the photometer is a good alternative device to follow the growth of microalgae. Furthermore, it is relatively inexpensive, quick and easy for the operator to use.

1. Introduction

Microalgae are photosynthetic microorganisms that transform sunlight into lipids as an energy reservoir. Lipids can be used as food, feedstock or substrate for biofuel production [1, 2]. Microalgae need a small area for cultivation (0.1–0.2 m^2/year/kg biodiesel), grow fast (10 g/m^2/day) and accommodate lipid contents (30 – 70%) higher than food crops [3-5].

Growth of a microalgae population can be divided into several major phases, i.e., lag phase, exponential phase, declining relative growth phase, stationary phase and algal cell death [6]. The phase of microalgae can be specified by considering their growth rate, i.e., the rate of change in the cell density per unit time. In addition, the growth rate is a key quantity considered in a comparison of different cultivation methods as well as to find the proper time for harvesting the microalgae.

Normally, the measurement of the cell density is performed by using a counting chamber under an optical microscope. The counting chamber consists of a glass microscope slide with grid of perpendicular lines. The grid has specific dimensions and the area covered by the lines is known so that it possible to count the number of cells in a specific volume of solution. This method needs good laboratory skills and it takes a long time of about 45 minutes to complete the measurement [7-8].

Recently, the arising of automated cell counting instruments based on image processing has provided the possibility of investigation of a large number of samples in a shorter time. However, they are constructed for specific cell types and may be not suitable for other types of cells due to technical limitations in their operation [8-12].

In this article, we present an optical method to measure the cell density of microalgae. We constructed a photometer and a calibration equation using the cell density measured by the traditional counting method. To test our photometer using different samples, we measured the cell density of microalgae *Dunaliella salina* which cultivated with four different methods of aeration and compared the results with those measured by counting cells under a microscope.

Dunaliella salina is a halophile green microalga and known for high accumulation of the important antioxidant beta-carotene under stress conditions including high light intensity, high NaCl concentration, suitable temperatures, or lack of nitrate [13-14]. Humans and animals can convert beta-carotene into vitamin A which are important for good vision, strong immunity, and general health [15–16].

2. Photometer

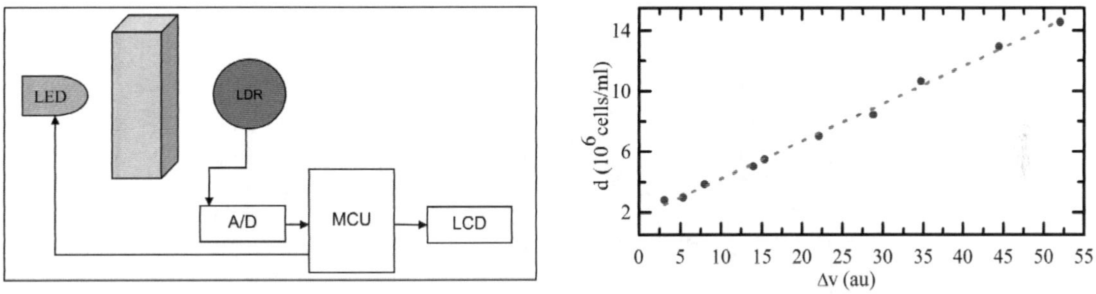

Figure 1. (left) Schematic of photometer and (right) the calibration curve of the photometer. Circles depict the cell density *d* measured by counting cells under a microscope vs the difference of digital voltage Δv obtained from the photometer. The dashed line is a linear fit of the data.

The circuit diagram of photometer is shown in figure 1(left). The microcontroller unit (MCU) supplies power to the light emitting diode (LED) which illuminates the sample. The light transmitted from the sample incidents the light dependent resistor (LDR). Then the voltage signals from LDR are converted (A/D) to digital signals with a resolution of 10 bits (1024 digital levels) corresponding to the signals of 0 - 5 volts. The microcontroller calculates the cell density of microalgae using the digital signals and the calibration equation and reports the results via an LCD monitor.

To develop the calibration equation of the cell density, 10 samples with different cell density of microalgae *Dunaliella Salina* were prepared by diluting a concentrated microalgae sample. Then the cells in each sample were counted under an optical microscope. The cell density *d* was in between 2.3 $\times 10^6$ cells ml^{-1} and 14.1$\times10^6$ cells ml^{-1}. The same samples were placed into the photometer and the digitized voltage v was measured. To avoid errors originated from the initial light intensity in different measurements, we measured the digitized voltage v_0 of the cultivation medium (without microalgae) as the blank measurement and used the difference of digital voltage $\Delta v = v - v_0$ to construct the calibration equation. As shown in figure 1(right), the difference of digital voltage Δv increases linearly with the cell density *d*. A linear regression results the calibration equation as $d = 0.2476\Delta v + 1.236$ with the R-squared value of 0.99606.

3. Monitoring of microalgae cell density

To test the performance of the photometer, microalgae *Dunaliella Salina* strain KU 11 [17] were cultured with different conditions. As shown in figure 2, the microalgae were cultivated in transparent glass chambers containing 50 litres culture medium with 2 M NaCl prepared as in [13] and illuminated by fluorescent lamps from top all times. From the left to the right of figure 3, different aerations were used: (a) without aeration, (b) circulation by water pump, (c) bubbles by aerator, and (d) both circulation by water pump and bubbles by aerator. At the beginning (day 1), the samples were very dilute and appeared transparent. Then, the cell density and the opaque of the chambers increased with time (e.g. day 10 and 18).

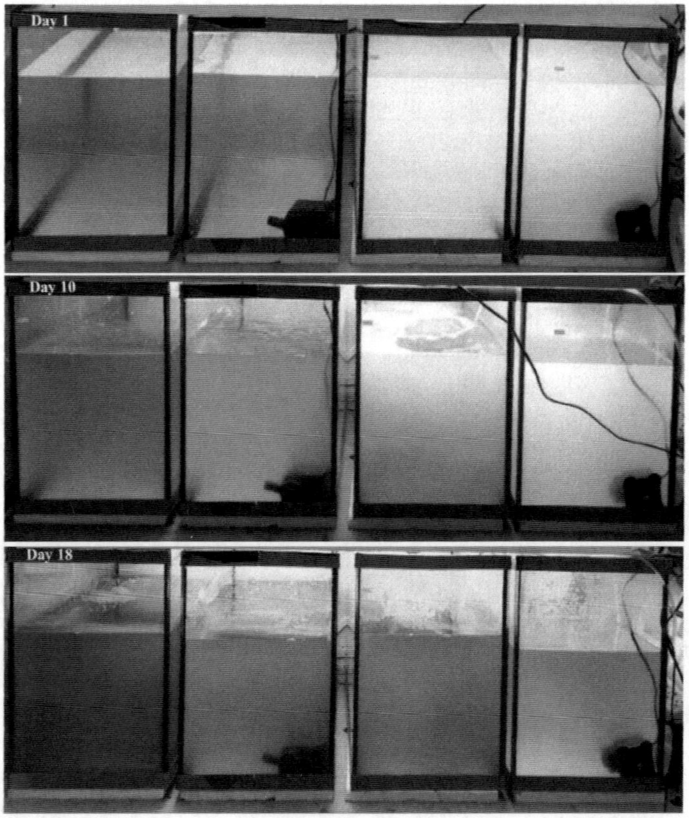

Figure 2. Cultivation of microalgae *Dunaliella Salina*. From left to right chambers, different aeration methods were used: without aeration, circulation by water pump, bubbles by aerator, and both circulation by water pump and bubbles by aerator. Images from top to bottom show the cultures at day 1, 10, and 18, respectively.

In each day of cultivation, we monitored the cell density of microalgae *Dunaliella Salina* in all chambers by using the photometer. The density was also measured by counting under a microscope for comparison. As shown in figure 3, the cell density was initially about 2×10^6 cells ml^{-1} in all chambers. In the course of time, both the cell densities as measured by the photometer and by counting under a microscope in each chamber increased with similar manners. At the last day of observation, the cell densities in chambers from left to right in figure 2 measured by the photometer were 11.9×10^6, 11.6×10^6, 8.2×10^6 and 5.4×10^6 cells ml^{-1}, respectively. The results imply that the growth of cells depends on the aeration.

4. Discussion and Conclusion

We have demonstrated an optical method to measure the cell density of microalgae. As examples, we cultivated microalgae *Dunaliella Salina* using different aerations. During the first two weeks of cultivation, the cell density monitored by our photometer were almost identical to that measured by counting under a microscope. In later days, the cell density obtained from the photometer were lower than that from the counting method. This difference might result from changes of cell properties such as color and cell size which affect the optical density as reported for silver nanoparticles [18]. Further studies should be conducted to improve the performance of the photometer.

Since the cell density monitored by our photometer increased in a similar trend to that measured by counting under a microscope, this device is an appropriate tool for observation of the growth of microalgae in a long duration of cultivation. We expect that the photometer will be useful for monitoring microalgae cultivation in either research laboratories or industries since this optical method is fast and low cost and it does not require laboratory skill to operate.

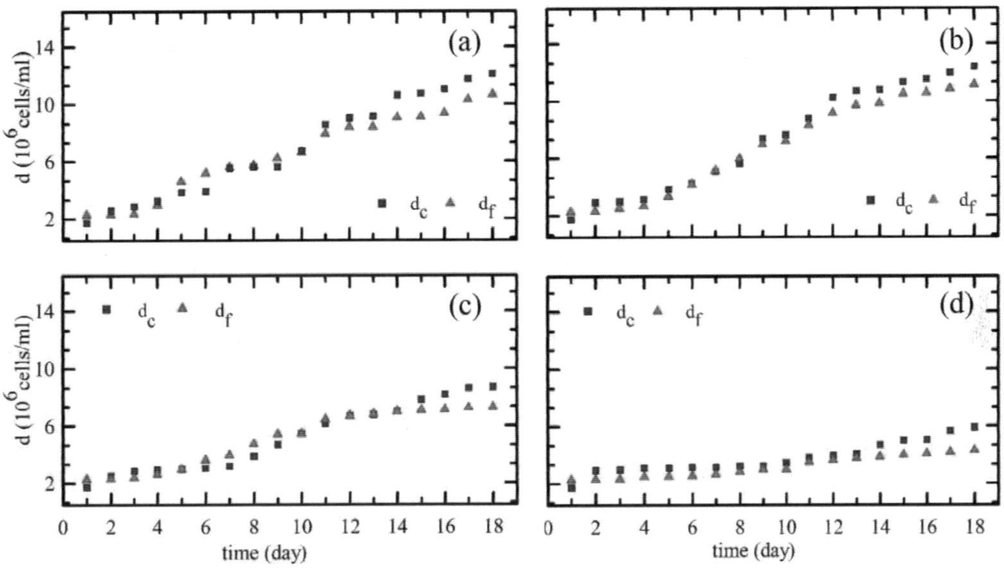

Figure 3. Cell density of microalgae *Dunaliella Salina* cultivated with different aeration methods: (a) without aeration, (b) circulation by water pump, (c) bubbles by aerator, and (d) both circulation by water pump and bubbles by aerator. d_f and d_c depict the cell density measured by the photometer and by counting under a microscope, respectively.

Acknowledgments
We would like to thank the Faculty of Science, the Research and Development Institute (KURDI), the Center for Advanced Studies of Industrial Technology, and the Graduate School, Kasetsart University; and Agricultural Research Development Agency (Public Organization) for financial support.

5. References

[1] Chisti Y 2007 *J. Biotechno. Adv.* **25** 3
[2] Mallick N 2002 *J. Biomet.* **15** 4
[3] Chen W, Zhang C, Song L, Sommerfeld M and Hu Q 2009 *J. Microbiol. Methods* **77** 1
[4] Lakaniemi A M, Hulatt C J, Thomas D N, Tuovinen O H and Puhakka J A 2011 *J. Biotechnol. Biofuels.* **4** 34
[5] Weyer K M, Bush D R, Darzins A and Willson B D 2010 *J. Bioenerg. Res.* **3** 204
[6] Price K and Farag I H 2016 *J. Eng. and Tech. Res.* **1** 8
[7] Nielsen L K, Smyth G K and Greenfield P F 1991 J. Biotech. Prog. 7: 560
[8] Louis K S, Siegel A C and Levy G A 2011 *J. Meth. Mol. Biol.* p 7
[9] Camacho-Fernandez C, Hervas D, Rivas-Sendra A, Marin M P and Segui-Simarro J M 2018 *J. Plant Methods* **14** 30
[10] Cadena-Herrera, Joshua E, Esparza-De L, Ramirez-Ibanez N D, Lopez-Morales C A, Perez N O, Flores-Ortiz L F and Medina-Rivero E 2015 *J. Biotechnology Reports* **9** 16
[11] Maruhashi F, Murakami S and Baba K 1994 *Automated monitoring of cell concentration and*

viability using image analysis system (Dordrecht: Springer)

[12] Tucker K G, Chalder S, Al-Rubeai M, Thomas C R 1994 *J. Enzyme Microb Technol* **16** 29

[13] Ben-Amotz A, and Avron M 1983 *J. Plant Physiol.* **72** 593

[14] Ben-Amotz A, and Avron M 1990 *J. Trends Biotechnol.* **8** 121

[15] Prieto A, Pedro Canavate J and Garcıa-Gonzalez M 2011 *J. Biotechnol.* **151** 180

[16] Nielsen L K, Smyth G K and Greenfield P F 1991 *J. Biotech. Prog*. **7** 560

[17] Wu Z, Dejtisakdi W, Kermanee P, Chunhong Ma, Arirob W, Sathasivam R and Juntawong N 2018 *J. Biotech. and Appl. Biochem.* **65** 4

[18] Paramelle D, Sadovoy A, Gorelik S, Free P, Hobley J and Fernig DG 2014 *J. Analyst*, **139** 4855

IOP Conf. Series: Journal of Physics: Conf. Series **1298** (2019) 012006 doi:10.1088/1742-6596/1298/1/012006

Delay-induced stability switches in an SIRS epidemic model with saturated incidence rate and temporary immunity

A M Pasion and J A Collera

Department of Mathematics and Computer Science, University of the Philippines Baguio, Governor Pack Road, Baguio City 2600, Philippines

E-mail: ampasion@up.edu.ph

Abstract. This work considers a time-delayed SIRS epidemic model with temporary immunity and nonlinear incidence rate, where the susceptible host population satisfies the logistic equation and the incidence rate is of saturated form with the susceptible. The time delay represents a period of temporary immunity where disease-recovered individuals return to the susceptible class after a fixed period of time. By analyzing the associated characteristic equation with delay-dependent coefficients and regarding the time lag as the bifurcation parameter, the local stability of the endemic equilibrium is investigated and sufficient conditions for the occurrence of stability switches via Hopf bifurcations are obtained. It is shown that the delay parameter can induce a finite number of stability switches before completely destabilizing the system. Numerical simulations are carried out to illustrate theoretical results.

1. Introduction

Delay differential equations (DDE) have been successfully utilized to model various infectious diseases. Incorporating time delays in epidemic models considers the fact that the transmission dynamic behavior of a disease at time t depends not only on the state of time t but also on the state of previous times. Delays have been used to account the effect of disease characteristics such as latent period, course of infection, period of immunity and disease awareness (see, for example, [1-19] and the references cited therein).

The basic elements of an SIRS epidemic model are the three epidemiological classes: susceptibles S, the number of individuals who are susceptible to infection; infectives I, the number of individuals who have the disease and able to spread the infection; and the removed class R, the number of individuals who have been infected and then removed with immunity. In 1992, Mena-Lorcat and Hethcote [20] considered five SIRS epidemiological models for populations of varying size, one of which is governed by the following system of ordinary differential equations (ODE)

$$\begin{cases} \dot{S}(t) &= \Lambda - \mu S(t) - \beta S(t)I(t) + \delta R(t) \\ \dot{I}(t) &= \beta S(t)I(t) - \mu I(t) - \alpha I(t) - \gamma I(t) \\ \dot{R}(t) &= \gamma I(t) - \mu R(t) - \delta R(t) \end{cases} \tag{1}$$

where Λ is the recruitment rate of the population, μ is the natural death rate due to causes unrelated to the infection, β is the transmission coefficient of the disease, α is the disease-related death rate, γ is the recovery rate of infected hosts, and δ is the rate at which recovered hosts lose their immunity and become susceptible again.

In model (1), the temporary immunity time is distributed exponentially as $e^{-\delta t}$ with mean time of $1/\delta$. This means that the period of temporary immunity for each individual ranges from zero (no immunity) to infinite (permanent immunity). If we assume that the duration of temporary immunity is a constant $\tau = 1/\delta$ and no individuals in the removed class will return to the susceptible class until time $t > \tau$, then we obtain the SIRS model of Brauer *et al.* [2] governed by the following ODE-DDE system

$$\begin{cases} \dot{S}(t) &= \Lambda - \mu S(t) - \beta S(t)I(t), \\ \dot{I}(t) &= \beta S(t)I(t) - (\mu + \alpha + \gamma)I(t), \\ \dot{R}(t) &= \gamma I(t) - \mu R(t), \\ & 0 \le t \le \tau, \end{cases} \quad \begin{cases} \dot{S}(t) &= \Lambda - \mu S(t) - \beta S(t)I(t) + \gamma e^{-\mu\tau} I(t-\tau), \\ \dot{I}(t) &= \beta S(t)I(t) - (\mu + \alpha + \gamma)I(t), \\ \dot{R}(t) &= \gamma I(t) - \mu R(t) - \gamma e^{-\mu\tau} I(t-\tau), \\ & t > \tau. \end{cases} \tag{2}$$

In the previous model, the immunity gained by experiencing the disease is only temporary, that is, the disease-recovered individuals who are still alive after the period of temporary immunity τ will return to the susceptible class. The delay-dependent coefficient $0 < e^{-\mu\tau} \le 1$ is the probability that the recovered individuals survive during the time period $t \le \tau$. Hence, the term $\gamma e^{-\mu\tau} I(t-\tau)$ reflects the fact that a recovered individual has survived from natural death in the removed class before becoming susceptible again to the disease. The authors of the research articles [1-4] incorporated the term $\gamma e^{-\mu\tau} I(t-\tau)$ in the study of their respective models.

The model problem of this article is a modification of the second model. From a practical point of view, we consider the nonlinear incidence rate $\beta SI/(1+\sigma S)$ introduced by May and Anderson [21] which is saturated with the susceptibles. The parameter σ is the saturation factor that measures the inhibitory effect from the behavioral changes of the susceptible individuals when their number increases. Several scholars [7-9, 12, 20] utilized the saturated incidence rate $\beta SI/(1+\sigma S)$ in their respective models. It is also assumed that in the absence of infection, the susceptible host population follows the logistic growth $rS(1 - S/K)$ with carrying capacity K and a specific growth rate r. We propose the generalized SIRS epidemiological model as follows

$$\begin{cases} \dot{S}(t) &= rS(t)\left(1 - \dfrac{S(t)}{K}\right) - \dfrac{\beta S(t)I(t)}{1+\sigma S(t)}, \\ \dot{I}(t) &= \dfrac{\beta S(t)I(t)}{1+\sigma S(t)} - (\mu + \alpha + \gamma)I(t), \\ \dot{R}(t) &= \gamma I(t) - \mu R(t), \\ & 0 \le t \le \tau, \end{cases} \tag{3}$$

$$\begin{cases} \dot{S}(t) &= rS(t)\left(1 - \dfrac{S(t)}{K}\right) - \dfrac{\beta S(t)I(t)}{1+\sigma S(t)} + \gamma e^{-\mu\tau} I(t-\tau), \\ \dot{I}(t) &= \dfrac{\beta S(t)I(t)}{1+\sigma S(t)} - (\mu + \alpha + \gamma)I(t), \\ \dot{R}(t) &= \gamma I(t) - \mu R(t) - \gamma e^{-\mu\tau} I(t-\tau), \\ & t > \tau. \end{cases} \tag{4}$$

We add here that if $\tau = 0$, then the model reduces to an SIS model and if $\tau = \infty$, we obtain the SIR models (ODE version) in [8, 9]. Furthermore, Xu *et al.* [4] studied a similar problem by considering the nonlinear incidence rate $\beta SI/(1+\nu I)$ which is saturated with the infectives [22]. The saturation level ν measures the inhibitory effect from the crowding effect of infected individuals. This time, in order to account the psychological behaviors of the members of the susceptible population, we analyze the dynamics of SIRS model (3)-(4).

2. Main results

The initial condition for model (3)-(4) takes the form

$$S(0) > 0, I(0) > 0, R(0) \geq 0. \tag{5}$$

For each initial condition (5), ordinary differential equation system (3) has a unique solution $(S(t), I(t), R(t))$ for $t \in [0, \tau]$, which is exactly the initial history function of delayed system (4). It is easy to show that solutions remain non-negative for all $t > 0$.

2.1. *Linearized analysis*

The dynamics of epidemic model (3)-(4) are mainly determined by the first two equations of delayed system (4)

$$\begin{cases} \dot{S}(t) = rS(t)\left(1 - \dfrac{S(t)}{K}\right) - \dfrac{\beta S(t)I(t)}{1+\sigma S(t)} + \gamma e^{-\mu\tau} I(t-\tau), \\ \dot{I}(t) = \dfrac{\beta S(t)I(t)}{1+\sigma S(t)} - (\mu+\alpha+\gamma)I(t) \end{cases} \tag{6}$$

because the first two equations in system (4) do not depend on the third equation, and therefore this equation can be omitted without loss of generality. We analyze subsystem (6) throughout the remainder of this section.

Theorem 1. *For the model system* (6), *there always exist disease-free equilibria* $E_0 = (0,0)$ *and* $E_1 = (K,0)$. *If*

$$R_0 := \frac{K[\beta - \sigma(\mu+\alpha+\gamma)]}{\mu+\alpha+\gamma} > 1, \tag{7}$$

there exists an endemic equilibrium $E_+ = (\bar{S}, \bar{I})$, *where*

$$(\bar{S}, \bar{I}) = \left(\frac{K}{R_0}, \frac{r\bar{S}^2}{K(\mu+\alpha+\gamma-\gamma e^{-\mu\tau})}(R_0 - 1)\right). \tag{8}$$

Let $X(t) = [S(t), I(t)]^T$. Then the linearized system corresponding to (6) around an equilibrium $E_* = (S^*, I^*)$ is

$$\dot{X}(t) = \begin{bmatrix} a_1 & a_2 \\ a_3 & a_4 \end{bmatrix} X(t) + \begin{bmatrix} 0 & b_1 \\ 0 & 0 \end{bmatrix} X(t-\tau) \tag{9}$$

where

$$a_1 = r - \frac{2rS^*}{K} - \frac{\beta I^*}{(1+\sigma S^*)^2}, a_2 = -\frac{\beta S^*}{1+\sigma S^*}, a_3 = \frac{\beta I^*}{(1+\sigma S^*)^2}, a_4 = \frac{\beta S^*}{1+\sigma S^*} - (\mu+\alpha+\gamma), b_1 = \gamma e^{-\mu\tau},$$

and with corresponding characteristic equation

$$D(\lambda, \tau) := \left| \lambda I - \left(\begin{bmatrix} a_1 & a_2 \\ a_3 & a_4 \end{bmatrix} + e^{-\lambda\tau} \begin{bmatrix} 0 & b_1 \\ 0 & 0 \end{bmatrix} \right) \right| = 0. \tag{10}$$

By the stability theory of delayed differential equations, the equilibrium E_* is locally asymptotically stable if all characteristic roots of (10) have negative real parts and is unstable if there is one characteristic root which has a positive real part [23, 24].

2.2. *Stability of* E_0 *and* E_1

Using (10), the characteristic equation at the disease-free equilibrium $E_0 = (0,0)$ reduces to $(\lambda - r)(\lambda + (\mu+\alpha+\gamma)) = 0$ whose roots are $r > 0$ and $-(\mu+\alpha+\gamma) < 0$. Thus, equilibrium E_0 is always an unstable saddle. On the other hand, by using the value of R_0 in (7) the characteristic equation of the linearized system around E_1 is obtained to be

$$(\lambda + r)\left(\lambda - \frac{\mu+\alpha+\gamma}{1+\sigma K}(R_0 - 1)\right) = 0.$$

From this, we see that the disease-free equilibrium E_1 is locally asymptotically stable when $R_0 < 1$ and unstable when $R_0 > 1$. Moreover, if $R_0 = 1$, then $\lambda = 0$ is a simple characteristic root. Thus, E_1 is linearly neutrally stable if $R_0 = 1$. The following theorem summarizes our results.

Theorem 2. *For model system* (6), *the disease-free equilibrium* $E_1 = (K, 0)$ *is locally asymptotically stable if* $R_0 < 1$, *linearly neutrally stable if* $R_0 = 1$ *and unstable if* $R_0 > 1$.

2.3. Stability and existence of Hopf bifurcation at E_+

We now investigate the stability of the endemic equilibrium $E_+ = (\bar{S}, \bar{I})$. Note that a necessary and sufficient condition for the existence of the endemic equilibrium E_+ is the condition $R_0 > 1$ must hold. So from here on out, we always assume that $R_0 > 1$. Using the values of $\bar{S}$, $\bar{I}$ and R_0 defined in (8)and (7), we derive the following expressions

$$R_0 = \frac{K}{\bar{S}}, \quad \frac{\beta\bar{S}}{1+\sigma\bar{S}} = \mu+\alpha+\gamma \quad \text{and} \quad \frac{\beta\bar{I}}{1+\sigma\bar{S}} = \frac{r(\mu+\alpha+\gamma)}{\mu+\alpha+\gamma-\gamma e^{-\mu\tau}}\left(1 - \frac{1}{R_0}\right).$$

Thus, the characteristic equation of the linear system around E_+ is obtained to be

$$\bar{D}(\lambda,\tau) := \underbrace{[\lambda^2 + p(\tau)\lambda + d(\tau)]}_{P(\lambda,\tau)} + \underbrace{[c(\tau)]}_{Q(\lambda,\tau)} e^{-\lambda\tau} = 0 \tag{11}$$

where

$$p(\tau) = a + b(\tau), \quad d(\tau) = (\mu+\alpha+\gamma)b(\tau), \quad c(\tau) = -\gamma b(\tau)e^{-\mu\tau},$$

$$b(\tau) = \frac{r(\mu+\alpha+\gamma)}{(1+\sigma\bar{S})(\mu+\alpha+\gamma-\gamma e^{-\mu\tau})}\left(1 - \frac{1}{R_0}\right), \quad a = r\left(\frac{2}{R_0} - 1\right).$$

For characteristic equation (11), observe that the delay τ does not only appear in terms with $e^{-\lambda\tau}$, but also in several places. The reason behind this is that our model system (6) involves delay-dependent parameters. Time-delayed systems with delay-independent parameters are simpler to study and most of the time, one can easily compute the critical delay values at which Hopf bifurcations occur (see, for example, [24]). Since biological models with delay-dependent parameters are now frequently arising in the literature, Beretta and Kuang [25] in 2002 contributed a geometric method to effectively study the difficult characteristic equations emerging from model systems involving delay dependent parameters. They combined graphical information with analytical work to show that the stability of a given equilibrium solution is simply determined by the graph of some functions of the time delay which can be expressed explicitly and thus can be easily depicted by popular graphing softwares. In practice though, one cannot compute explicitly these critical delay values at which Hopf bifurcations take place. We apply their technique here since it is applicable to our problem.

In order to apply the geometric criterion due to Beretta and Kuang [25], we need to verify the following properties for all $\tau \geq 0$:

(A1) $\lambda = 0$ is not a root of $\bar{D}(\lambda,\tau) = 0$, that is, $P(0,\tau) + Q(0,\tau) = d(\tau) + c(\tau) \neq 0$.

(A2) If $\lambda = i\omega, \omega \in \mathrm{R}$, then $P(i\omega,\tau) + Q(i\omega,\tau) \neq 0$.

(A3) $\limsup\{|Q(\lambda,\tau)/P(\lambda,\tau)| : |\lambda| \to \infty, Re\lambda \geq 0\} < 1$

(A4) $F(\omega,\tau) := |P(i\omega,\tau)|^2 - |Q(i\omega,\tau)|^2$ has at most a finite number of real zeros.

(A5) Each positive root $\omega(\tau)$ of $F(\omega,\tau) = 0$ is continuous and differentiable with respect to τ whenever it exists.

Note that $d(\tau) + c(\tau) = (\mu+\alpha+\gamma-\gamma e^{-\mu\tau})b(\tau) > 0$. Thus, $\lambda = 0$ is not a root of $D(\lambda,\tau)$. (A2) is obviously true because $P(i\omega,\tau) + Q(i\omega,\tau) = -\omega^2 + ip(\tau)\omega + d(\tau) + c(\tau) \neq 0$. From (11), we know that

$|Q(\lambda,\tau)/P(\lambda,\tau)|\to 0$ as $|\lambda|\to+\infty$. Thus, (A3) follows immediately. Solving for $F(\lambda,\omega)$, we obtain $F(\lambda,\omega)=\omega^4+(p^2(\tau)-2d(\tau))\omega^2+(d^2(\tau)-c^2(\tau))$. It is obvious that (A4) is satisfied and by the Implicit Function Theorem, (A5) is also satisfied. Therefore, all properties (A1)-(A5) are satisfied.

If $\tau=0$ is a characteristic root, then equation (11) reduces to $\lambda^2+p(0)\lambda+d(0)+c(0)=0$. Since $d(0)+c(0)$ is always positive, we know that from the Routh-Hurwitz criteria that all of its roots have negative real part if and only if $p(0)>0$. Hence, the equilibrium E_+ is locally asymptotically stable for $\tau=0$ if and only if $p(0)>0$. Assume now that $\lambda=i\omega\,(\omega>0)$ is a root of (11). Splitting the real and imaginary parts yields,

$$\sin\omega\tau=\frac{p(\tau)\omega}{c(\tau)},\quad \cos\omega\tau=\frac{\omega^2-d(\tau)}{c(\tau)}. \tag{12}$$

If we can find (τ,ω) satisfying (12), then (11) will have a pair of pure imaginary roots $\pm i\omega$ at τ. Squaring both sides in (12), adding, and rearranging gives

$$F(\omega,\tau):=\omega^4+(p^2(\tau)-2d(\tau))\omega^2+d^2(\tau)-c^2(\tau)=0. \tag{13}$$

Using the quadratic formula, the potential positive roots of equation (13) are obtained to be

$$\omega_\pm(\tau)=\left(\frac{1}{2}(-p^2(\tau)+2d(\tau))\pm\frac{1}{2}\sqrt{(-p^2(\tau)+2d(\tau))^2-4(d^2(\tau)-c^2(\tau))}\right)^{1/2}. \tag{14}$$

It is easy to see that the coefficient $d^2(\tau)-c^2(\tau)$ of equation (13) is always positive for all $\tau\ge 0$ from their values defined in (11). Because of this, we can only consider the following two cases for the solutions of (13). If $-p^2(\tau)+2d(\tau)<0$ for all $\tau\ge 0$, then there are no positive $\omega(\tau)$ solutions of equation (13). On the other hand, if $(-p^2(\tau)+2d(\tau))^2-4(d^2(\tau)-c^2(\tau))>0$ and $-p^2(\tau)+2d(\tau)>0$ for all τ in some interval, then (13) has two distinct positive roots $\omega_+(\tau)$ and $\omega_-(\tau)$ whose values are given in (14). Assume now that R_c, R_k, τ_c and τ_k have values defined by

$$R_c=1+\frac{1+\sigma\bar{S}}{2+\sigma\bar{S}},\quad R_k=1+\frac{(\mu+\alpha)(1+\sigma\bar{S})}{(\mu+\alpha+\gamma)+(\mu+\alpha)(1+\sigma\bar{S})},$$

$$\tau_c=-\frac{1}{2\mu}\ln\left(\frac{4r(\mu+\alpha+\gamma)^2}{\gamma^2(\mu+\alpha)(1+\sigma\bar{S})}\left(1-\frac{1}{R_0}\right)\right)\quad\text{where}\quad\frac{4r(\mu+\alpha+\gamma)^2}{\gamma^2(\mu+\alpha)(1+\sigma\bar{S})}\left(1-\frac{1}{R_0}\right)<1,$$

$$\tau_k=-\frac{1}{2\mu}\ln\left(\frac{4r(\mu+\alpha+\gamma)(1+\sigma\bar{S})\left(\frac{2}{R_0}-1\right)^2}{\gamma^2\left(1-\frac{1}{R_0}\right)}\right)\quad\text{where}\quad\frac{4r(\mu+\alpha+\gamma)(1+\sigma\bar{S})\left(\frac{2}{R_0}-1\right)^2}{\gamma^2\left(1-\frac{1}{R_0}\right)}<1.$$

Let us consider the following hypotheses:

(H1) $R_c\le R_0\le 2$ and $\tau\in[0,\tau_c]$;

(H2) $1<R_0\le R_k$ and $\tau\in[0,\tau_k]$.

Define $I:=[0,\bar{\tau}]$ where $\bar{\tau}=\tau_c$ if (H1) holds true or $\bar{\tau}=\tau_k$ if (H2) holds true. If (H1) or (H2) is satisfied, then we can show using the techniques established in [26] that both $(-p^2(\tau)+2d(\tau))^2-4(d^2(\tau)-c^2(\tau))$ and $-p^2(\tau)+2d(\tau)$ are positive for all $\tau\in I$. We thus have the following result.

Lemma 3. *If either (H1) or (H2) is satisfied, then both $\omega_+(\tau)$ and $\omega_-(\tau)$ are positive on I. Furthermore, the endemic equilibrium E_+ is locally asymptotically stable at $\tau=0$ whenever (H1) or (H2) holds.*

Now, for any $\tau\in I$, we can define the angles $\theta_\pm(\tau)\in(0,2\pi)$ as the solutions of (12), that is,

$$\sin\theta_{\pm}(\tau)=\frac{p(\tau)\omega_{\pm}(\tau)}{c(\tau)},\quad \cos\theta_{\pm}(\tau)=\frac{\omega_{\pm}^2(\tau)-d(\tau)}{c(\tau)}.$$

We note here that the plus or minus signs are used consistently on both sides of any equations consisting them. Indeed, we can define the functions $\theta_{\pm}:I\to \mathrm{R}$ by

$$\theta_{\pm}(\tau)=\begin{cases}\pi+\arctan\left(\dfrac{p(\tau)\omega_{\pm}(\tau)}{\omega_{\pm}^2(\tau)-d(\tau)}\right) & \text{if}\quad \omega_{\pm}^2(\tau)<d(\tau),\\[2ex] 2\pi+\arctan\left(\dfrac{p(\tau)\omega_{\pm}(\tau)}{\omega_{\pm}^2(\tau)-d(\tau)}\right) & \text{if}\quad \omega_{\pm}^2(\tau)>d(\tau).\end{cases}\tag{15}$$

The relation between the arguments $\theta_{\pm}(\tau)$ and $\omega_{\pm}(\tau)\tau$ for $\tau\in I$ must be

$$\omega_{\pm}(\tau)\tau=\theta_{\pm}(\tau)+2n\pi,\quad n=0,1,2,\ldots.$$

Hence, we can introduce the sequence of maps $S_n^{\pm}:I\to \mathrm{R}$ defined by

$$S_n^{+}(\tau)=\tau-\frac{\theta_{+}(\tau)+2n\pi}{\omega_{+}(\tau)}\quad\text{and}\quad S_n^{-}(\tau)=\tau-\frac{\theta_{-}(\tau)+2n\pi}{\omega_{-}(\tau)}.\tag{16}$$

It is clear that the functions are continuous and differentiable on I by the assumptions. Thus, we give the following theorem which is due to Beretta and Kuang [25].

Theorem 4. *If there exists an integer $n\geq 0$ such that $S_n^{+}(\tau)=0$ or $S_n^{-}(\tau)=0$ at some $\tau=\tau^*$, then characteristic equation* (11) *has a pair of simple pure imaginary roots $\pm i\omega(\tau^*)$ where $\omega(\tau^*)=\omega_{+}(\tau^*)$ if $S_n^{+}(\tau^*)=0$ and $\omega(\tau^*)=\omega_{-}(\tau^*)$ if $S_n^{-}(\tau^*)=0$. If $\omega(\tau^*)=\omega_{+}(\tau^*)$, this pair of simple conjugate pure imaginary roots crosses the imaginary axis from left to right if $\delta_{+}(\tau^*)>0$ and crosses the imaginary axis from right to left if $\delta_{+}(\tau^*)<0$, where*

$$\delta_{+}(\tau^*):=sign\left\{\left.\frac{dRe\lambda}{d\tau}\right|_{\lambda=i\omega_{+}(\tau^*)}\right\}=sign\left\{\left.\frac{dS_n^{+}(\tau)}{d\tau}\right|_{\tau=\tau^*}\right\}.$$

On the other hand, if $\omega(\tau^*)=\omega_{-}(\tau^*)$, this pair of simple conjugate pure imaginary roots crosses the imaginary axis from left to right if $\delta_{-}(\tau^*)>0$ and crosses the imaginary axis from right to left if $\delta_{-}(\tau^*)<0$, where

$$\delta_{-}(\tau^*):=sign\left\{\left.\frac{dRe\lambda}{d\tau}\right|_{\lambda=i\omega_{-}(\tau^*)}\right\}=-sign\left\{\left.\frac{dS_n^{-}(\tau)}{d\tau}\right|_{\tau=\tau^*}\right\}.$$

Corollary 5. *If such τ^* in Theorem 4 exists with $\delta_{\pm}(\tau^*)\neq 0$ (depending on whether $\omega(\tau^*)=\omega_{\pm}(\tau^*)$), then model system* (6) *undergoes Hopf bifurcation at the endemic equilibrium E_{+} when $\tau=\tau^*$.*

3. Numerical simulations

For numerical simulations, let us consider the delayed SIRS model (3)-(4) with the following set of parameters $(r,K,\beta,\sigma,\mu,\alpha,\gamma)=(0.1,100,0.02,0.005,0.001,0.465,0.5)$. For such parameters, Theorem 4 can be used since we have $1<R_c=1.569<1.570=R_0<2$ and $\tau\in I=[0,\tau_c]=[0,62]$. In other words, hypothesis (H1) holds and stability switches for the equilibrium E_{+} may depend on all real roots of $S_n^{+}(\tau)$ and $S_n^{-}(\tau)$ in $I=[0,62]$. Figure 1 shows the graphs of $S_n^{+}(\tau)$ and $S_n^{-}(\tau)$ on I. The functions $S_0^{+}(\tau)$ and $S_0^{-}(\tau)$ both have zeros given by $\tau_0^{+}=15.32$ and $\tau_0^{-}=32.48$, respectively. Figure 1 also shows that $S_1^{+}(\tau)$ has a zero $\tau_1^{+}=39.73$. From Theorem 4, we have $\delta_{+}(\tau_0^{+})=+1$,

$\delta_-(\tau_0^-) = -1$ and $\delta_+(\tau_1^+) = +1$. Thus, Hopf bifurcations occur at these values of τ. The occurrence of stability switches (via Hopf bifurcations) at τ_0^+, τ_0^- and τ_1^+ are illustrated in Figure 2 using DDE-Biftool, which is a numerical continuation and bifurcation analysis tool developed by Engelborghs *et al.* [27, 28]. The green and red represents the stable and unstable parts of the branches, respectively as τ varies.

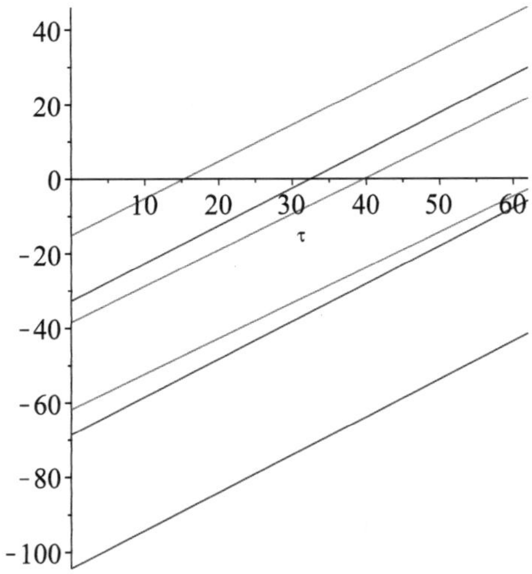

Figure 1. The functions $S_n^+(\tau)$ (red) and $S_n^-(\tau)$ (blue) are plotted on I.

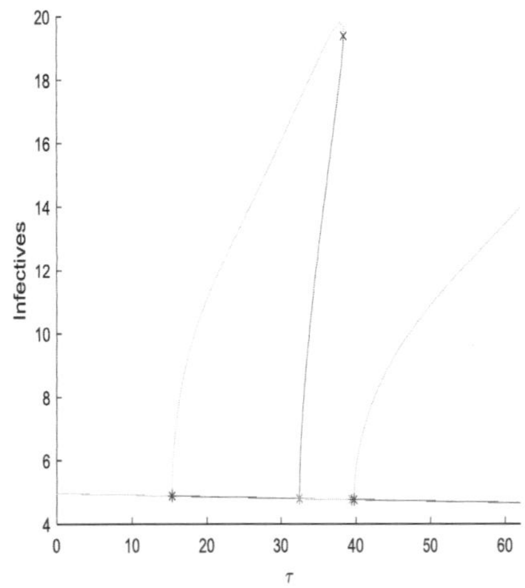

Figure 2. Stability switching and emergence of periodic solutions at Hopf bifurcation points (*).

4. Conclusion

In this paper, we formulated and studied a delayed SIRS epidemic model with a fixed period of temporary immunity and nonlinear incidence rate which is saturated with the susceptible host population. We investigated the local stability of the endemic equilibrium from the point of view of stability switches via Hopf bifurcations by regarding the time lag τ as bifurcation parameter. Under certain restrictions on the non-delay parameters, we obtained explicit intervals at which critical delay values exist. We have shown that stability switching via Hopf bifurcation occurs at these critical values of τ. Lastly, we supported our theoretical results with some numerical examples.

Acknowledgments

The authors acknowledge the support of the Department of Science and Technology (DOST) through its Accelerated Science and Technology Human Resource Development Program (ASTHRDP) and the University of the Philippines Baguio through its Research Dissemination Grant (UPB-RDG).

5. References

[1] Kyrychko Y N and Blyuss K B 2005 *Nonlinear Anal. Real World Appl.* **6** 495–507
[2] Brauer F, den Driessche P V and Wang L 2008 *Math. Biosci.* **215** 1–10
[3] Jiang Z and Wei J 2008 *Chaos Solitons Fractals* **35** 609–619
[4] Xu R, Ma Z and Wang Z 2010 *Comput. Math. Appl.* **59** 3211–3221
[5] Zhang F, Li Z and Zhang F 2008 *Appl. Math. Comput.* **199** 285–291
[6] Yang Y and Xiao D 2010 *Discrete Contin. Dyn. Syst. Ser. B* **13** 195–211
[7] Zhang T and Teng Z 2008 *Appl. Math. Model.* **32** 1403–1416
[8] Zhang J Z, Jin Z, Liu Q X and Zhang Z Y 2009 *Discrete Dyn. Nat. Soc.* **2008**
[9] Rihan F A and Anwar M N 2012 *Int. J. Differ. Equ.* **2012**

[10] Liu L 2015 *Adv. Differ. Equ.* **2015** 329
[11] Liu J 2016 *J. Appl. Math. Comput.* **50** 557–576
[12] Xu R, Zhang S and Zhang F 2016 *Math. Methods Appl. Sci.* **39** 3294–3308
[13] Xia W, Kundu S and Maitra S 2018 *Adv. Differ. Equ.* **2018** 336
[14] Kumar A and Nilam 2018 *Int. J. Comput. Methods* 1850055
[15] Li Y, Liu X, Wang L and Zhang X 2018 *Int. J. Biomath.* 1850091
[16] Agaba G O, Kyrychko Y N and Blyuss K 2017 *Ecol. Complex.* **31** 50–56
[17] Zhao H and Zhao M 2017 *J. Biol. Dyn.* **11** 8–24
[18] Basir F A, Ray S and Venturino E 2018 *Appl. Math. Comput.* **337** 372–385
[19] Beretta E and Breda D 2011 *Math. Biosci. Eng.* **8** 931–952
[20] Mena-Lorcat J and Hethcote H W 1992 *J. Math. Biol.* **30** 693–716
[21] Anderson R M and May R M 1978 *J. Anim. Ecol.* 219–247
[22] Capasso V and Serio G 1978 *Math. Biosci.* **42** 43–61
[23] Hale J K and Lunel S M V 1993 *Introduction to Functional Differential Equations. Applied Mathematical Sciences* vol 99 (Springer-Verlag, New York)
[24] Kuang Y 1993 *Delay Differential Equations with Applications in Population Dynamics. Mathematics in Science and Engineering* vol 191 (Academic Press, Boston)
[25] Beretta E and Kuang Y 2002 *SIAM J. Math. Anal.* **33** 1144–1165
[26] Fan G, Min-oo M and Wolkowicz G 2009 *Can. Appl. Math. Q.* **17**
[27] Engelborghs K, Luzyanina T and Roose D 2002 *ACM Trans. Math. Softw.* **28** 1–21
[28] Sieber J, Engelborghs K, Luzyanina T, Samaey G and Roose D 2016 *DDE-BIFTOOL v.3.1.1 Manual-Bifurcation analysis of delay differential equations*

IOP Conf. Series: Journal of Physics: Conf. Series **1298** (2019) 012007 doi:10.1088/1742-6596/1298/1/012007

Comparing the sandpile model with targeted triggering and the Olami-Feder-Christensen model as models of seismicity using recurrence network analysis

D C Biton and R C Batac

Department of Physics, College of Science, De La Salle University,
2401 Taft Avenue, 1004 Manila, Philippines

dionessa_biton@dlsu.edu.ph

Abstract. Slowly driven sandpile models has found applications in modelling earthquakes due to the observed power law statistics in its magnitude distributions, like the behaviour of earthquakes. Adding a probability to target the most susceptible site in the grid, the sandpile model recovers even the spatio-temporal statistics of earthquake events. In this work, we compare the sandpile model with targeted triggering to the Olami-Feder-Christensen (OFC) model: a standard earthquake model that also exhibits self-organized criticality. The sandpile model captures the magnitude distributions of earthquake events at a value of targeted triggering probability p = [0.004,0.007]. The triggering probability value p = 1.0, showing that the most susceptible site is always triggered, follows the magnitude distribution of the OFC model. A comparison was done by constructing a record-breaking recurrence network for the events. Spatial and magnitude criteria set the temporally directed links between events across the entire record. Both the models recover power-law exponents comparable to those previously obtained for earthquake data, which is 1.0 for recurrence distance and recurrence time distributions, and 2.1 for the in-degree distributions for the farthest recurrence criteria. The sandpile model with targeted triggering exhibits a behaviour in between a slowly driven sandpile and the OFC model.

1. Introduction

Earthquake modelling had attracted a lot of attention in the study of dynamical systems due to its self-organized critical behaviour [1-6]. A full understanding of earthquakes is essential since it may cause a lot of destruction in properties and lives [7-9].

Aside from earthquakes, many other systems in nature exhibit self-organized criticality (SOC), wherein the components move to a critical state independent of their initial conditions. Self-organized critical systems exhibit long-tailed distributions of some physical quantities [4]. One defining property of earthquake events are the observed power law distributions in the magnitude of the energy release in each event. This was defined as a universal law by Gutenberg and Richter, stating that the probability distribution of energy follows the scaling as shown in equation (1) [10]. This power law behavior in the distribution of the energy released during an earthquake is important to note since it indicates that earthquakes are SOC [11].

$$p(E) \sim E^{-5/3} \tag{1}$$

An example of a self-organized critical system is the Bak Tang Wiesenfeld (BTW) sandpile model which is a 2-dimensional grid that receives a slow external perturbation and grows in height as if grains of sand in a pile. The small perturbation can cause avalanche events which may be bigger in size than the external trigger. The sandpile model has been used to model seismicity because the distribution of avalanche sizes captures the magnitude distributions of earthquakes defined in the GR Law [1].

Aside from the magnitude distributions, earthquakes are also observed to cluster in space and time [12]. Correlated earthquakes occur in proximity with one another, such that after a large earthquake, a series of aftershocks occur near the original mainshock. This spatial and temporal clustering of events is not easily reproduced in the original sandpile model. A modification on this was done by adding a parameter that targets the most susceptible site in the pile. The interevent statistics from the modified model capture the space and time distributions of earthquakes [2].

Another earthquake model is the Olami-Feder-Christensen which is associated to the two-dimensional Burridge-Knopoff spring block model. It simulates an earthquake fault system through the transfer of energy along a system of blocks. The blocks are connected by a spring, which moves back to a stable position when subjected to an external force [3]. This model not only follow the Gutenberg-Richter law [13], it also exhibits the clustering in space and time of events [11]. To investigate the clustering in space and time of correlated events, Davidsen et al. used a spatio-temporal structure of seismicity from an idea of the causal relationship of events beyond the simplistic idea that each event has at most one correlated event. They constructed a network where the nodes are the events and a temporally directed link connects the two events if the latter event is spatially closest to the previous event compared to all events that occurred in between the two [14]. This method scans through the whole record, wherein the latest spatial distance record will be broken if an event nearer than the current record occurred. The network grows every time a new event arrives since the links of previous records are retained in the network. Extending the idea that correlated events cluster in space, a previous work by Tarun et al. added a farthest distance criterion for connection, wherein connection is made between to events if they are farthest in space compared to all events in between [5].

Magnitudes of earthquakes are important aspects to fully understand earthquake characteristics. Previous studies showed that using the magnitude as a metric resulted to observations regarding the clustering of correlated earthquakes [15]. Aside from using the spatial criteria in constructing the network introduced by Davidsen et al., magnitude was incorporated by Janer et al. by segregating the UP connections, if the link is from an event of lower magnitude to an event of higher magnitude, from the DOWN connections, if the link is connecting to an event of lower magnitude [6].

In our work, we constructed a temporally directed network using the spatial near and far connections and the magnitude up and down connections for avalanche events from a sandpile model with targeted triggering and an Olami-Feder-Christensen (OFC) model. The statistics from the network was used to compare the two models with the basis on the observed behaviour in earthquakes. For our analyses, we note the power law behaviour observed in earthquake magnitudes and the spatiotemporal correlations in earthquake events as a basis for comparison with the models similar to what was done in previous works [2, 5, 11].

2. Methodology

A sandpile model with targeted triggering and an Olami-Feder-Christensen (OFC) model was constructed and compared using recurrence network analysis.

2.1. Sandpile Model with Targeted Triggering

Sandpile is a cellular automata model where an external triggering causes unstable sites in the grid form avalanches. The 2-dimensional grid follows the von-Neumann neighbourhood, $n = \{(x, y), (x \pm \Delta x, y), (x, y \pm \Delta y)\}$, where $\Delta x = \Delta y = 1$, such that the cell at (x, y) have four nearest neighbours in its right, left, top, and bottom. The state of each cell evolves in every discrete time step $\Delta t = 1$.

Compared to the Bak-Tang-Wiesenfeld (BTW) sandpile model wherein the possible states are discrete, $s = \{0, 1, 2, 3, 4\}$ and the unstable state is at $\sigma_{max} = 4$, the sandpile model used has a continuous set of states $s = \{\sigma | \sigma \in [0, \sigma_{max})\}$, where $\sigma_{max} = 1$. The update rule f is due to the external triggering $\nu = 10^{-3}\sigma_{max}$. For each update, an external trigger ν was added to the state σ_{xy} of a selected neighbourhood centered at (x, y). This rule, shown as f_1 in equation (2), allows the system to receive a continuous but slow driving.

$$f_1 : \sigma_{xy}^t + \nu \quad (2)$$

A conditional rule f_2 represents the redistribution done to the system when the triggering results to an unstable site with a state greater than or equal to the unstable state $\sigma_{xy} \geq \sigma_{max}$. The unstable state σ^* is equally redistributed to the state σ_{nn} of its four neighbours as shown in equation (3). The site with an unstable state will then have a state $\sigma_{xy} = 0$ after the redistribution. The redistribution rule causes the cascading nature of the model since local redistributions from the original neighbourhood may make other cells unstable, thus continuing the redistribution across other neighbourhoods outside of the initially collapsing site. Compared to other sandpile models, the driving rule uses a targeted triggering probability p which is a probability that the driving term ν will be directed to the most susceptible site. This represents the high chance of rupturing of most susceptible locations.

$$f_2 : \sigma_{nn} = \sigma_{nn} + \sigma^*(4)^{-1} \quad (3)$$

In our work, we used a sandpile grid with size $L = 1024$ with initial random states $\sigma \in (0, \sigma_{max})$. The driving was done for 10^7 iteration times where the first 10^6 iterations were discarded to remove the transient effects. The recorded locations were from the location of the initial toppling. The magnitude of the avalanche event from an initial external driving was measured from the number of sites that was affected by the toppling event. The minimum area of avalanche A is 5 cells. A threshold magnitude was also applied wherein only the avalanche events greater than the threshold $A_{th} = 10$ is considered. This removes the very small single neighbourhood avalanches representing that in an actual catalogue, very weak earthquakes are not completely captured.

2.2. Olami-Feder-Christensen model

The Olami-Feder-Christensen (OFC) model is another model for seismicity which is a simplification of the Burridge-Knopoff spring block model, which considers a network or springs and blocks where blocks move to their balanced positions after being submitted to a force which made them unstable.

All states were initialized to have random values between 0 and the threshold force F_{th}. If any site (x, y) has a force greater than the threshold force $F_{xy} \geq F_{th}$, the force F_{xy} on the site is redistributed to the neighbours and the force on the site will be zero as shown in equations 4 and 5. The redistribution process is repeated until the earthquake is fully evolved. For the driving, compared to the sandpile model, a global perturbation was done such that the external driving was added to all the sites. Same threshold area $A_{th} = 10$ was used for the events recorded from the OFC model. The value for the dissipation parameter used is $\alpha = 1.5$.

$$F_{nn} \rightarrow F_{nn} + \alpha F_{xy} \quad (4)$$

$$F_{xy} \rightarrow 0 \quad (5)$$

2.3. Recurrence Network Construction

A temporally directed network was constructed using the spatial coordinates and magnitudes of the avalanche events to characterize the spatio-temporal clustering of the system. A previous work by Davidsen et al. [14] used a generalized theory of records in finding the probable causal relationships among earthquake events.

Two spatial criteria were used for the network construction, the nearest and the farthest criteria. The nearest criteria connect the events if the later event is nearer in space to the earlier event compared to all intervening events. This is demonstrated in figure 1 (a) where a sample network for 5 events was constructed. In the figure, each circle or node represents an event. The numbers correspond to the interval at which the event occurred and the relative spaces among the nodes represents the relative distances between events. The magnitudes of events are shown through the radius of the circles. When the second event arrived, a connection was made from the first event to the second event, since interevent connections are always done. In this specific network, the record connection of event 1 is the connection to event 2. When the third event arrived, the second event connected to the third event since the two are consecutive events. The spatial criterion is then applied when a link was made from event 1 and event 3. Since event 3 is nearer in space to event 1 compared to the connection of event 1 to event 2, the record connection was broken due to the nearest distance criterion. The new record of event 1 is a connection to event 3. Same rule was observed when creating connections to event 4 and event 5. For the farthest distance connection as shown in figure 1 (b), a similar rule with the nearest criterion of connection was applied. However, to create a connection, the farthest event compared to other intervening events, connects with a previous event. All interevent connections are made.

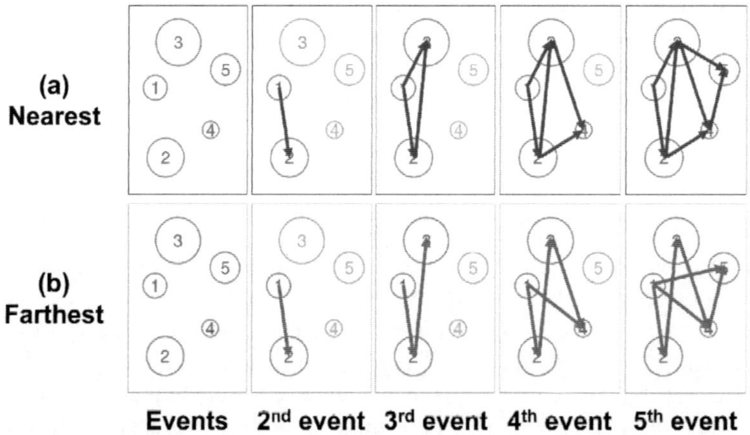

Figure 1. Network construction for a sample of 5 events using the (a) nearest criterion of connection and (b) the farthest criterion of connection. The numbers in the events correspond to each temporal coordinate, the relative distances correspond to the spatial aspect, and radii of the circles in each node represent the magnitudes of the events.

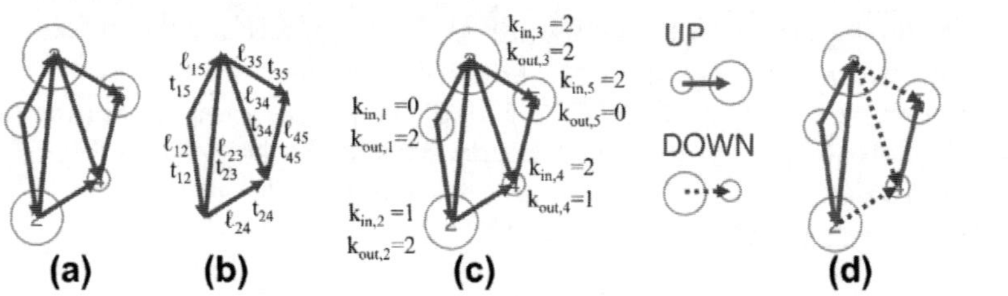

Figure 2. Metrics used in network construction. (a) Fully constructed network. (b) Separation distances ℓ_{ij} and temporal separations t_{ij} between connected events. (c) The number of links going into (k_{in}) and out of (k_{out}) an event. (d) Links were segregated into UP connections, when the connection was done from an event of lower magnitude to an event of higher magnitude, and DOWN connections, when the connection was done from a higher to a lower magnitude.

After the full construction of the network aver all events, as shown in figure 2 (a), the metrics obtained are the spatial separation ℓ_{ij} and the temporal separation t_{ij} of two connected events i and j as shown in figure 2 (b). The number of in going links $k_{in,i}$ and the number of out going links $k_{out,i}$ from an event i was obtained to characterize the structure of the network as shown in figure 2 (c). Lastly, the magnitude was incorporated in the network through a segregation of the connections to eitherlabelled as UP connections, when the connection was going from an event of lower magnitude to an event of higher magnitude, or being labelled as DOWN connections, when the link is from an event of higher magnitude to an event of lower magnitude. Statistical analyses were done from the metrics.

3. Results and Discussion

The Gutenberg Richter law states that the magnitude distribution of earthquakes follow a power law behaviour. As a first test, the magnitude distributions of the models should also follow the observed earthquake statistics. In figure 3, magnitude distributions of avalanche areas $p(A)$ obtained from the sandpile model and the Olami-Feder-Christensen (OFC) model was shown.

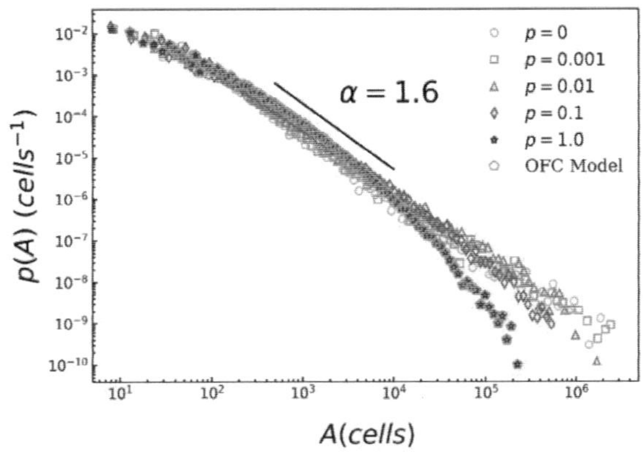

Figure 3. Magnitude distribution of avalanche events from the sandpile model for different targeted triggering probability $p = \{0, 10^{-3}, 10^{-2}, 10^{-1}, 1\}$ and of the avalanche events from the OFC model.

Due to the targeted triggering probability p, a bias in the spatial and temporal organization of events occurred compared to the original sandpile model. We investigate the cases for a broad range of values of the targeted triggering probability $p = \{0, 10^{-3}, 10^{-2}, 10^{-1}, 1\}$. The case where $p = 0$ is the original sandpile where the triggering is done on a random site and $p = 1$ is the completely targeted sandpile. From the magnitude distributions, the OFC model follows the distribution of the completely targeted sandpile model at $p = 1$ where the sites with highest states in the grid are driven. Compared to the power law behaviour observed in earthquakes, the completely targeted case shows a faster tail decay and an increase in the intermediate magnitude values A, this was observed since there is a slow accumulation of new susceptible sites, resulting to a small extent of avalanche magnitude produced.

Cases where targeted triggering was done only during a fraction of the intervals, such that $0 < p < 1$, the sandpile model captures the behaviour in between a simple sandpile and the OFC model. For the succeeding analyses, we focus on the case of $p = 0.004$, which shows characteristics similar to those obtained from the earthquake data.

3.1. Spatial and Temporal Distributions of Recurrences

The original sandpile recovers the power-law behaviour of the avalanche size distributions as observed in earthquakes since the stress in the system is released in a single cascade. However, the distribution of the spatial and temporal differences of correlated events results in random (Poisson) statistics. The effect of the targeted triggering is illustrated in the distributions of the separation distances and the waiting times between recurrences shown in figures 4(a) and 5(a).

The distribution of the separation distances from the OFC model and from the sandpile model with the nearest criterion of connection shows a similar behaviour. For the farthest criterion, the events with long separation in the OFC model closely follow the exponent of a power law observed in the behaviour of the original sandpile model. An abundance of events with small separation distances are observed which was not the case for earthquake catalogues. Temporal distributions from the two models behave in a similar trend as shown in figure 5.

3.2. K-degree Distributions of Recurrences

To complete the analysis and the comparison done, the structure of the constructed recurrence network was analysed from the distribution of the number of in and out going links from each event. Figure 6 presents the degree distributions $p(k_{in})$ and $p(k_{out})$ of the recurrence network for the nearest and farthest criteria.

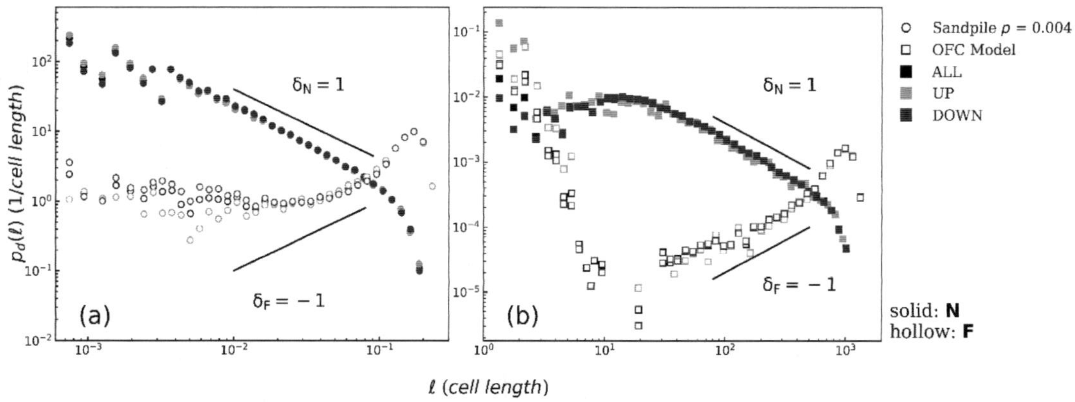

Figure 4. The distribution of separation distances of connected events in the constructed recurrence network for (a) the sandpile model with targeted triggering probability $p = 0.004$ and (b) the OFC model for the nearest and farthest criteria of connection. The connections were also segregated into the up and down magnitude criteria.

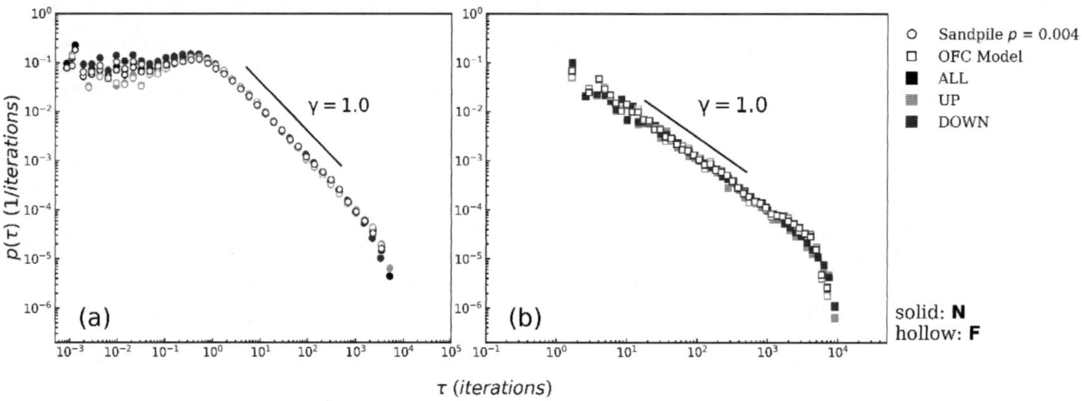

Figure 5. The distribution of temporal separation of the links in the constructed recurrence network with the nearest and farthest spatial criteria and the up and down magnitude criteria for the events from (a) the sandpile model with targeted triggering probability $p = 0.004$ and (b) the Olami-Feder-Christensen (OFC) model with dissipation parameter $\alpha = 0.15$. .

The distribution of the in going links (k_{in}) for the farthest criteria as shown in figure 6 (b) results to a scale-free distribution comparable to $p(k_{in}) \sim k_{in}^{-\beta}$, where $\beta = 2.1$. The distribution of the outgoing links (k_{out}) from the sandpile model closely follow a Poisson distribution regardless of the criteria of connection used, as shown in figure 4 (c) and (d). The distributions from the OFC model, however, show nontrivial properties. This could be due to the faster generation of events since all the sites in the pile is driven for every time step.

4. Summary and Conclusions

A sandpile model with a targeted triggering probability to target the most susceptible site in the grid and the Olami-Feder-Christensen (OFC) model for earthquake system was compared using a recurrence network analysis. The targeted triggering mechanism in the sandpile more closely captured the magnitude distribution of earthquake events, while the OFC model was observed to behave similar to a completely targeted sandpile model. The distribution of the temporal separation and the distribution of the spatial separation with the nearest criterion cannot be used as a basis for comparison of the two models due to the similar trend exhibited by both. The spatial separation of the OFC model using the farthest criterion of connection showed an abundance of short separation of connected events, which is not similar to the behaviour of earthquakes. The degree distributions of the out going links in the OFC model deviates a lot from a Poisson distribution showing non-trivial properties which is due to the faster generation of events.

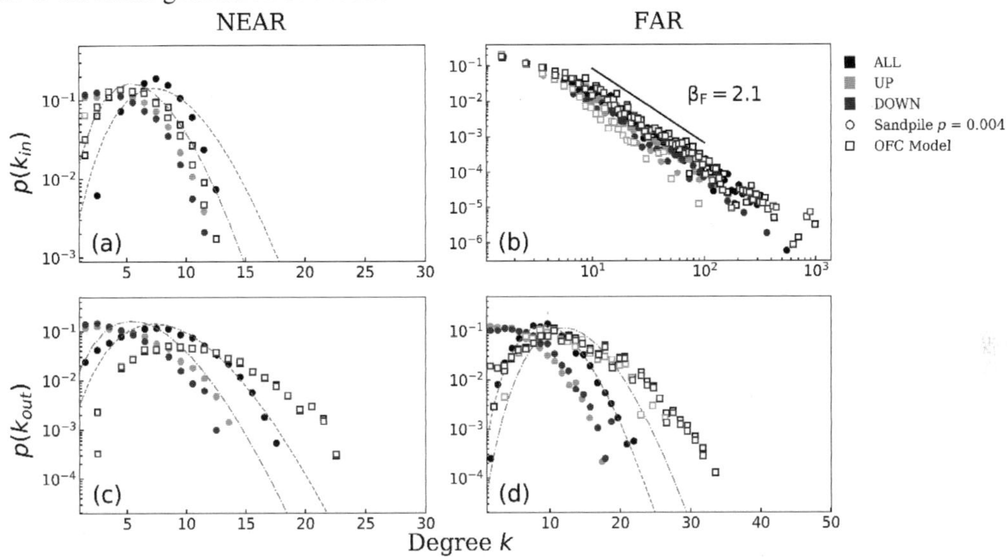

Figure 6. The distribution of the degree connections going into (k_{in}: a,b) and out (k_{out}: c,d) of the events from the sandpile model with targeted triggering probability $p = 0.004$ and the OFC model for the (a,c) nearest and (b,d) farthest criteria of connection.

5. References

[1] Bak P, Tang C and Wiesenfeld K 1988 *Phys. Rev.* A **38** 364
[2] Batac R C, Paguirigan Jr. A A, Tarun A B and Longas A G 2017 *Nonlin. Processes Geophys.* **24** 179
[3] Olami Z, Feder H J S and Christensen K 1992 *Phys. Rev. Lett.* **68** 1244
[4] Bak P, Tang C and Wiesenfeld K 1987 *Phys. Rev. Lett.* **59** 4
[5] Tarun A B, Paguirigan A A, and Batac R C 2015 *Physica A* **436** 293
[6] Janer C D, Biton D C and Batac R C 2017 *Acta Geophys.* **65** 1153
[7] Oteng-Ababio M 2012 *J Hous Built Environ.* **27** 2
[8] Allen R B, Bellingham P J and Wiser S K 1999 *Ecol.* **80** 708-14
[9] Nakagawa M, Saito M and Yamaga H 2009 *Jpn. Econ. Rev.* **60** 208-22
[10] Gutenberg B and Richter C F 1949 *Seismicity of the earth and associated phenomena* (London: Princeton University Press) p 16- 24
[11] Christensen K and Olami Z 1992 *Phys. Rev.* A **46** 4
[12] Zaliapin I, Gabrielov A, Keilis-Borok V and Wong H 2008 *Phys. Rev. Lett.* **101** 018501
[13] Lise S and Paczuski M 2001 *Phys. Rev.* E **63** 036111
[14] Davidsen J, Grassberger P and Paczuski M 2006 *Geophys. Res. Lett.* **33** L11304; Davidsen J, Grassberger P and Paczuski M 2008 *Phys. Rev.* E **77** 066104
[15] Baiesi M and Paczuski M 2004 *Phys. Rev.* E **69** 066106; Zaliapin I, Gabrielov A, Keilis-Borok V and Wong H 2008 *Phys. Rev. Lett.* **101** 018501

IOP Conf. Series: Journal of Physics: Conf. Series **1298** (2019) 012008 doi:10.1088/1742-6596/1298/1/012008

Computation of Cusp Bifurcation Point in a Two-Prey One Predator Model using Lagrange Multiplier Method

L Owen and J M Tuwankotta

Analysis and Geometry Group, Faculty of Mathematics and Natural Sciences, Institut Teknologi Bandung, Jl. Ganesha No 10 Bandung, Indonesia.

corresponding email : livia.owen@gmail.com

Abstract. In this paper we introduce a new method for computing cusp bifurcation point in a system of two prey one predator dynamical system. The method consists of rewriting the system of equations for the equilibria in a constrained optimization problem at where the classical Lagrange Multiplier Method can be applied to derive the fold bifurcation point. By having one extra parameter as free parameter we construct two fold curves which intersect each other non-transversally in a cusp bifurcation point. This result is verified by looking at three section in two-dimensional parameter space by taking one of the parameter constant and study the stability of the equilibria.

1. Introduction

The development of the predator-prey model has a very long history, originated from the work of A.J.Lotka and V.Volterra in 1920's (see [2] and the references there in). Variation of the model includes, among others: extending the dimension of the system (by considering more species, see for example: [7, 12, 14, 16, 18]) or by using various functional response (see for example [10, 11, 21]). Tripathi et al. [18] studied a dynamical system consisting of two species of prey and one predator. Here, the authors there assume that the prey species can help each other during predation and the predation rate for each of the population of prey is different. Elettreby [7] studied the global stability and persistence in a similar model to [18] without the assumption that different species of prey can help each other in preventing predation. Klebanoff and Hastings [14] investigated the chaotic dynamics of two prey one predator model.

In this paper, we follow the work initiated in [16] in where the classical Lagrange Multiplier Method is applied on finding fold bifurcation in two-prey one predator model. Note that fold bifurcation is a phenomena where two equilibria collide with each other and vanishes as the bifurcation parameter varies. Geometrically one can realize that an equilibrium can be seen as an intersection between the nullclines i.e. the zero level sets of the component of the vector field defining the dynamical system. Since the system depends on parameters, then computation of fold bifurcation is similar to looking for the maximum or minimum value such that the two nullclines intersect each other. Thus the problem of finding fold bifurcation is viewed as a constrained optimization problem, where Lagrange Multiplier Method can be applied.

This method is proposed as an alternative for computing fold bifurcation, which is usually done using numerical continuation software, for example: AUTO [6], XPPAUT [8], or MATCONT [3, 4]. Furthermore, the proposed method computes directly the fold bifurcation point, in contrast with using numerical continuation software where one start with general equilibirum. Another way of computing fold bifurcation can be found in [13, 17, 19, 20]. The latter depends highly on the existence of explicit formula for the equilibrium. In some case, this is not very easy to achieve.

The dynamical system which is considered in [16] (and also here) is three dimensional and it depends on ten parameters. Thus the system of equations for the equilibria is three dimensional. One can reduce it to two dimensional system depending on ten parameters by solving one of the equations for one of the variables. The solution is then substituted into the remaining two equations. As for the parameters, Marwan et al in [16] choose b, which is the group defence parameter, as the bifurcation parameter and fix the other nine parameter. As an extension to this work, we choose b and δ, which is the mortality rate of the predator, as free parameters and fix the other parameters.

It is shown in [16] that the system undergoes two fold bifurcations. By continuing the two loci of the fold bifurcations using another parameter, one can construct two fold curves. It is well-known in the literature that those two fold curves might coalesce non-transversally, and then disappear. This phenomena is known as the cusp bifurcation (see for example [15]), which is of co-dimension two. In this paper, our goal is to compute this cusp bifurcation point using the Lagrange Multiplier Method (see [5] for the detail and the proof of the method). This method we are proposing is an alternative for finding the cusp point in a system of two differential equations. However, the computation might still be too hard to do analytically. In that case, we will use a numerical solver for finding those solution.

The paper is organized as follows. In Section 2 we formulate the system of three differential equations which will be analyzed throughout the paper. This is the same system as in [16]. In Section 3 we briefly sketch the method we are using to compute the cusp bifurcation as an extension of the work in [16] which concentrates on finding fold bifurcations. Section 4 is devoted to study the stability of the equilibria in the neighborhood of the cusp bifurcation. This is to further confirm that the eigenvalues behavior near the cusp point match the theory. We end this paper with a few concluding remarks.

2. Model Formulation

Consider a three dimensional system of ordinary differential equations depending on ten parameters, i.e.

$$\begin{aligned}
\dot{x} &= x\left(1-\frac{1}{k_1}x-\frac{\mu z}{cx+1}\right) &&= x\,f(x,y,z) \\
\dot{y} &= \alpha x-\beta y-\frac{1}{k_2}y^2-\frac{\eta yz}{ay^2+by+1} &&= g(x,y,z) \\
\dot{z} &= z\left(-\delta+\frac{x}{cx+1}+\frac{y}{ay^2+by+1}\right) &&= z\,h(x,y)
\end{aligned} \tag{2.1}$$

where x, y and z are the population densities of the young prey, the adult prey, and the predator, respectively. Note that the growth rate of the young prey has been scaled to one, while the mortality rates of the adult prey and the predator are β and δ, respectively. Living in a large group, the adult prey can be able to defense from predator's attack. The group mechanism defense is done by the adult prey protect their population especially the young prey. As a consequence, the functional response of the young prey is of Holling type II while the the adult prey is of Holling type IV. The conversion rates are μ and η, respectively. The environment is assumed to react differently to the class of the prey, i.e. the carrying capacity for the young is k_1 while the adult. is k_2. We also includes three other parameters, namely c, a and b which determines the geometric shape of the functional response functions. All of these parameter take positive value but b which can be negative, as long as: ay^2+by+1 non-negative for $y \geq 0$. System (2.1) is so-called normalized two-prey one predator dynamical system.

In the absence of the adult prey, i.e. $y = 0$, system 2.1 equivalents with Bazykin's system in [1]. If $x = 0$, system 2.1 equivalents with a Predator-Prey systems with non-monotonic response function in [9, 21] but in our case, it doesn't have a biological meaning. For detail derivative of (2.1) see Marwan et al. [16].

The non-trivial equilibria of (2.1) (if exists) satisfies

$$\begin{aligned}(x-p)^2 - A(y-q)^2 &= r, \\ h(x,y) &= 0,\end{aligned} \tag{2.2}$$

where

$$\begin{aligned} A &= \frac{k_1\mu}{\eta\,(c\delta-1)\,k_2}, & p &= \frac{\eta k_1\,(c\delta-1) - k_1\mu\alpha - \eta\delta}{2\eta\,(c\delta-1)}, \\ q &= -\frac{\beta k_2}{2}, & r &= p^2 - Aq^2 + \frac{k_1\delta}{c\delta-1}. \end{aligned}$$

Following [16], we set a value of the parameters: $\alpha = 0.8$, $a = 7.29$, $\beta = 0.75$, $c = 1$, $\delta = 0.8$, $\eta = 0.25$, $k_1 = 6$, $k_2 = 4$, and $\mu = 0.75$. Using Lagrange Multiplier Method, two fold bifurcation points are found at $b = b_1 = -3.68591$ and $b = b_2 = -2.86798$.

3. Computation of the Cusp Bifurcation

In this paper, we choose δ as another free parameter apart from b. We will follow the procedure in [16] for computing fold bifurcation. Let us look at the second equation in (2.2), i.e. $h(x,y) = 0$. It is easy to see that

$$b = G(x,y,\delta) = -ay + \frac{cx+1}{x(c\delta-1)+\delta} - \frac{1}{y}. \tag{3.1}$$

solves $h(x,y) = 0$. Thus, the solution of the constrained optimization problem:

$$\begin{aligned}\text{Max/Min:}\quad & G(x,y,\delta) \\ \text{Given:}\quad & F(x,y,\delta) = (x-p)^2 - A(y-q)^2 - r = 0\end{aligned}$$

corresponds to a fold bifurcation. Using, Lagrange Multiplier Method, we solve:

$$\begin{aligned}\nabla G &= \lambda \nabla F, \\ F &= 0.\end{aligned} \tag{3.2}$$

Note that the solution (x, y, λ) of (3.2) depends in the free parameter δ. Thus, y-component of the solution (3.2) satisfies:

$$\sum_{i=0}^{10} A_i(\delta)\, y^i = 0, \tag{3.3}$$

see Appendix for the formula for the coefficient A_i.

Let $\{\delta_1, \delta_2, \ldots, \delta_N\} \in [0.7, 1]$ be chosen such that $0.7 = \delta_1 < \delta_2 < \ldots < \delta_N = 1$, for large enough natural number N. For each δ_k, $k = 1, 2, \ldots, N$ we solve (3.3) for y, $y_j(\delta_k)$, where m can be 1, or 2. Then we solve $F(x, y_m(\delta_k), \delta_k) = 0$ for x, $m = 1, 2$, $k = 1, 2, \ldots, N$. We derive two families of fold bifurcation points (or fold curves) (x_f, y_f). Lastly, we substitute x_f, y_f, and δ to (3.1), which defines b as a function of δ. These curves in (δ, b) is plotted in Figure (·1.·), the diagram on the left.

In Figure ·1.·, the diagram on the right shows the graph of the curve: $(b, y(b))$ for three different values of δ i.e.: $\delta_c = 0.734063212067$ (solid curve), $\delta = \delta_{f_2} = 0.75$ (dashed curve), and $\delta = \delta_{f_1} = 0.8$ (dotted curve). One can see that the fold points are getting closer to each other as δ approaches δ_c from above. This is a strong indication of the occurrence of cusp bifurcation.

4. Numerical simulation

In this section we will consider three cases, i.e. $\delta = \delta_{f_2} = 0.8$, $\delta = \delta_c = 0.734063212067$ and $\delta = 0.7$. For each of this cases we will follow the equilibria for the system (2.1) as the parameter b varies.

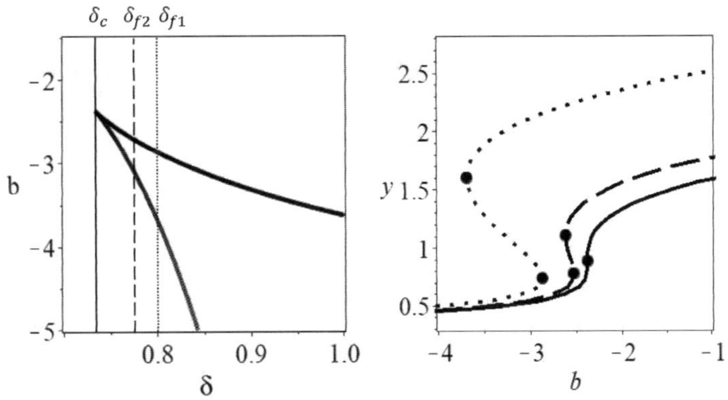

Figure ·1.· The diagram on the left shows two fold curves (solid curves) in (δ, b) coordinate of system (2.1) which intersect nontransversally at the cusp point at $(\delta_c, b_c) = (0.734063212067, -2.387660586)$. We have added three lines $\delta = \delta_c = 0.734063212067$ (solid line), $\delta = \delta_{f_1} = 0.75$ (dashed line), and $\delta = \delta_{f_1} = 0.8$ (dotted line). The diagram on right shows three curves $y(b)$ which solve (3.3), for $\delta = \delta_c$ (solid curve), δ_{f_1} (dashed curve) and δ_{f_2} (dotted curve).

4.1. *Two fold bifurcations at $\delta = \delta_{f_2}$*

For $\delta = \delta_{f_2} = 0.8$, we derive three branches of equilibria as a function of b, namely E_1, E_2 and E_3 in Figure ·2.·.

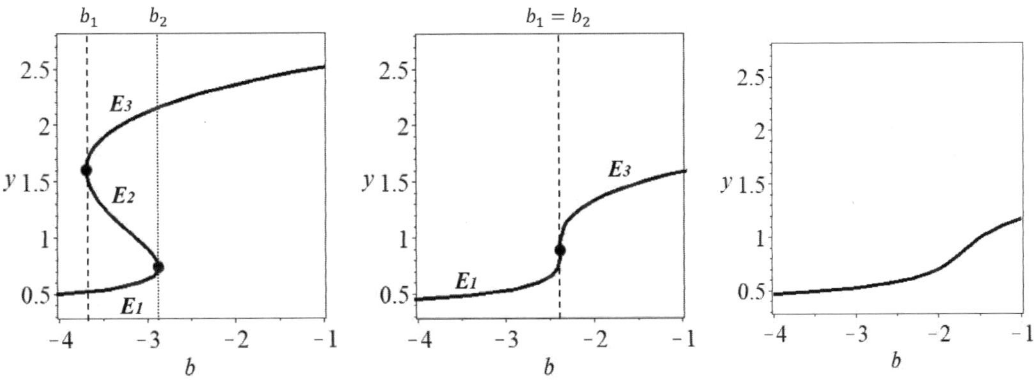

Figure ·2.· The branch of continuation equilibria as variation on parameter b for $\delta > \delta_c$. Two fold bifurcation points are $b = b_1$ and $b = b_2$. When $b < b_1$, the system (2.1) have one positive equilibrium E_1. Then at $b = b_1$, there appears a non-hyperbolic equilibria $E_2 = E_3$. Next when $b_1 < b < b_2$, this non-hyperbolic equilibria split into E_2 and E_3 while E_1 still exist. At $b = b_2$, E_1 and E_2 collapse into one non-hyperbolic equilibrium. And next, when $b > b_2$, the system (2.1) have one positive equilibrium E_3.

b	Equilibrium	Eigenvalues of the equilibrium
$b < b_1$	E_1	One negative and complex eigenvalues with negative real part
$b = b_1$	E_1	One negative and complex eigenvalues with positive real part
	$E_2 = E_3$	One zero, one positive, one negative
$b_1 < b < b_2$	E_1	One negative and complex eigenvalues with positive real part
	E_2	One negative and complex eigenvalues with positive real part
	E_3	One positive and complex eigenvalues with negative real part
$b = b_2$	$E_1 = E_2$	One zero, one positive, one negative
	E_3	One negative and complex eigenvalues with positive real part
$b > b_2$	E_3	One negative and complex eigenvalues with positive real part

4.2. *Cusp bifurcation at $\delta = \delta_c$*

For $\delta = \delta_c$, we get the branch of continuation equilibria as variation on parameter b.

b	Equilibrium	Eigenvalues of the equilibrium
$b < b_1$	E_1	One negative and complex eigenvalues with positive real part
$b = b_1$	$E_1 = E_3$	One zero, one positive, one negative
$b > b_1$	E_3	One negative and complex eigenvalues with positive real part

4.3. *No fold bifurcation point when $\delta < \delta_c$*

For illustration in this interval, we choose $\delta = 0.7$ and we get the branch of continuation equilibria as variation on parameter b.

5. Concluding remarks

In previous paper [16], Marwan et al. have developed the normalized two-prey one predator model. They found two fold bifurcation points that occur at $b = b_1$ and $b = b_2$ where $b_1 \leq b_2$. We continue their work by adding δ as a new parameter. Using Lagrange Multiplier Method, we compute a two parameter family of fold bifurcation curves and cusp bifurcation point in parameter space. Two bifurcation curves intersect each other non-transversally at a cusp point $(\delta, b) = (\delta_c, b_1)$ where $b_1 = b_2$. From the bifurcation diagram, we can analyze the number of equilibria and it's stability, for details, see section 4.

We still need to check another one and two parameters bifurcation by some continuation software for further stability analysis. This paper focus on the application of Lagrange Multiplier Method for computing fold bifurcation curves and the cusp bifurcation.

Acknowledgment Livia Owen acknowledges the financial support from The Indonesian Education Scholarship Program (LPDP), Ministry of Finance of the Republic of Indonesia. J.M. Tuwankotta's research is supported by research grant Riset KK B FMIPA ITB 2019, Institut Teknologi Bandung.

A. Appendix : Coefficients of equation 3.3

$$\begin{aligned}
A_{10}(\delta) &= 53808401250\,\delta^3 - 161425203750\,\delta^2 + 161425203750\,\delta - 53808401250 \\
A_9(\delta) &= 484275611250\,\delta^3 - 1452826833750\,\delta^2 + 1452826833750\,\delta - 484275611250 \\
A_8(\delta) &= 146478425625\,\delta^4 + 807908418000\,\delta^3 - 2249266459725\,\delta^2 \\
&\quad +1488893962950\,\delta - 194014346850 \\
A_7(\delta) &= 878870553750\,\delta^4 - 2460970410750\,\delta^3 + 8452627497900\,\delta^2 \\
&\quad -13081784678550\,\delta + 6211257037650 \\
A_6(\delta) &= 2215581629625\,\delta^4 - 10644623014950\,\delta^3 + 31086867794085\,\delta^2 \\
&\quad -34211693470860\,\delta + 11553642062100 \\
A_5(\delta) &= 2571269022000\,\delta^4 - 13535257269600\,\delta^3 + 32715458776080\,\delta^2 \\
&\quad -19417028388480\,\delta - 2914297434000 \\
A_4(\delta) &= 883829003400\,\delta^4 - 4011367429800\,\delta^3 + 3643682552856\,\delta^2 \\
&\quad +11312619948000\,\delta - 5438335626000 \\
A_3(\delta) &= -755036100000\,\delta^4 + 3855873980000\,\delta^3 - 9487870704000\,\delta^2 \\
&\quad +6182433324000\,\delta + 284048100000 \\
A_2(\delta) &= -369059670000\,\delta^4 + 1705830240000\,\delta^3 - 2826383122800\,\delta^2 \\
&\quad -911946600000\,\delta + 529043400000 \\
A_1(\delta) &= 52920000000\,\delta^4 - 268056000000\,\delta^3 + 664588800000\,\delta^2 - 449452800000\,\delta \\
A_0(\delta) &= 28224000000\,\delta^4 - 131328000000\,\delta^3 + 239708160000\,\delta^2
\end{aligned}$$

We show all fold bifurcation curves in Figure ·3.·.

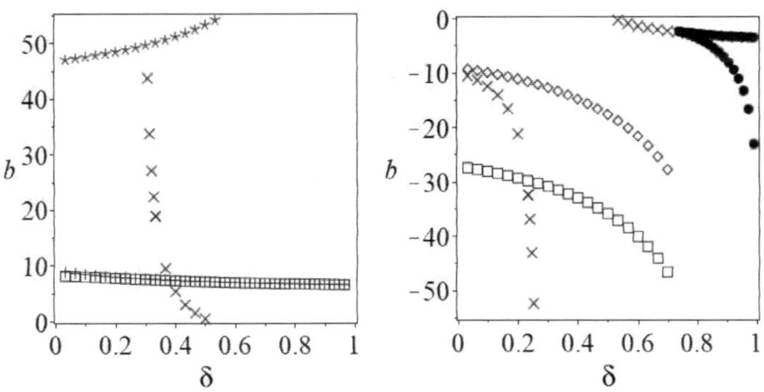

Figure ·3.· The bifurcation diagram for all fold bifurcation curves. The diagram on the left for $b \geq 0$ and the diagram on the right for $b \leq 0$. We use some symbols for points of fold bifurcation curves in the plot, i.e.: solid circle, asterisk, box, circle, cross, diagonal cross and diamond. The solid circle curves are the fold bifurcation curves in Figure ·1.·.

References

[1] Bazykin A D 1985 Mathematical biophysics of interacting populations *Nauka, Moscow.*

[2] Berryman A A 1992 *The origins and evolution of predator-prey theory* Ecology 73(5) pp 1530-1535.

[3] Dhooge A, Govaerts W and Kuznetsov Y A 2003 MATCONT: a MATLAB package for numerical bifurcation analysis of ODEs *ACM Transactions on Mathematical Software (TOMS)* 29(2) pp.141-164.

[4] Dhooge A, Govaerts W, Kuznetsov Y A, Mestrom W, Riet A M and Sautois B 2006 MATCONT and CL MATCONT: Continuation toolboxes in matlab *Universiteit Gent, Belgium and Utrecht University, The Netherlands.*

[5] Dineen S 2014 *Multivariate Calculus and Geometry* third edition (London: Springer Undergraduate Mathematics Series)

[6] Doedel E J and Oldeman B E 2012 *AUTO-07P: Continuation and Bifurcation Software for Ordinary Differential Equations* (Canada: Concordia University, Montreal)

[7] Elettreby M F 2009 Two-prey one-predator model *Chaos, Solitons & Fractals* **39** 5 pp 2018-2027

[8] Ermentrout B 2002 *Simulating, Analyzing, and Animating Dynamical Systems: A Guide to XPPAUT for Researchers and Students* (Philadelphia: SIAM).

[9] Harjanto E and Tuwankotta J M 2016 Bifurcation of periodic solution in a PredatorPrey type of systems with non-monotonic response function and periodic perturbation *International Journal of Non-Linear Mechanics* 85 pp 188-196.

[10] Holling C S 1959 The components of predation as revealed by a study of small mammal predation of the European Pine Sawfly *The Canadian Entomologist* **91** pp 293-320.

[11] Holling, C S 1959 Some characteristics of simple types of predation and parasitism *The Canadian Entomologist* **91** pp 385-398.

[12] Huang Y and Diekmann O 2001 *Predator migration in response to prey density: What are the consequences?* J. of Math. Bio. **11** 43 (6) pp 561-581.

[13] Jia Li and Zhien Ma 2009 *Dynamical Modeling and Analysis of Epidemics.* (Singapore : World Scientific Publishing Co. Pte. Ltd.).

[14] Klebanoff A and Hastings A 1994 Chaos in one-predator, two-prey models: cgeneral results from bifurcation theory *Mathematical biosciences* 122(2) pp 221-233.

[15] Kuznetsov Y A 1998 *Elements of Applied Bifurcation Theory* second edition Applied Math. Sciences 112 (New York : Springer).

[16] Marwan, Tuwankotta J M and Haryanto E 2018 Application of Lagrange Multiplier Method for Computing Fold Bifurcation Point In A Two-Prey One Predator Dynamical System *Journal Indonesian Mathematics Society* Vol. 24 No. 02 pp 07-19.

[17] Shigui Ruan, Wanbiao Ma, and Zhixing Hu 2012 Analysis of SIR Epidemic Models with Nonlinear Incidence Rate and Treatment *Mathematical Biosciences* Vol. 238 pp 12-20.

[18] Tripathi J P, Abbas S and Thakur M 2014 Local and global stability analysis of a two prey one predator model with help *Communications in Nonlinear Science and Numerical Simulation* 19(9) pp 3284-3297.

[19] Wang W and Ruan S 2004 Bifurcations in an epidemic model with constant removal rate of the infectives *Journal of Mathematical Analysis and Applications* 291(2) pp 775-793.

[20] Wang W 2006 Backward Bifurcation of an Epidemic Model with Treatment *Mathematical Biosciences* Vol. 201 pp 58.

[21] Zhu H, Campbell S A and Wolkowicz G S 2003 Bifurcation analysis of a predator-prey system with nonmonotonic functional response *SIAM Journal on Applied Mathematics* 63(2) pp 636-682.

IOP Conf. Series: Journal of Physics: Conf. Series **1298** (2019) 012009 doi:10.1088/1742-6596/1298/1/012009

Nonlinear random dynamical model for the stock market

Andrey Dmitriev and Vasily Kornilov

National Research University Higher School of Economics, Russia, Moscow

Email: a.dmitriev@hse.ru

Abstract. A new model of a stock market as a nonlinear random dynamical system with additive noise in three-dimensional phase space is offered. Implementations of this model in the adiabatic approximation possess all the key signs of fractal market, making the model a reasonable evolutionary model for a stock market. The use of adiabatic approximation allows us to model a stock market as a nonlinear random dynamical system with multiplicative noise with the power-law in one-dimensional phase space. This model shows fractality, long memory and 1/f noise in a stock market.

1. Introduction

An interdisciplinary research field, known as econophysics, was formed in the middle 1990s as an approach to solve various problems in economics, such as uncertainty or stochastic processes and nonlinear dynamics, by applying theories and methods originally developed by physicists. The term "econophysics" was coined by H E Stanley in order to describe the large number of papers written by physicists in the problems of (stock and other) markets (for econophysics reviews see refs. [1-4]).

Current state of theoretical economics allows one to effectively use advanced methods of physical and mathematical modelling for economical system. A remarkable example is applying nonlinear dynamics to analysis of financial time series [5, 6]. Moreover, in 1963 B B Mandelbrot [7] during his research of cotton prices found out that the prices follows a scaled distribution in time. These probability distributions are considered fractal. In 1994 was formed fractal market hypothesis (FMH) as an alternative investment theory to efficient market hypothesis (EMH). The FMH is a model of investor behaviour that unlike the EMH assumes investors have multiple time horizons and interpret information based upon their horizon.

According to the FMH, the financial time series (FTS) have the following key features: power law distribution, 1/f noise, and long memory.

The most comprehensive survey of mathematical models of financial markets can be found in the book of R J Elliott and P E Kopp [8]. Although the book and other relevant publications contain numerous conceptual models, we have not found any econophysical model of a stock market that can explain its fundamental functioning mechanisms and the key features. Thus, the purpose of this work is building of econophysical model of a stock market as a nonlinear random dynamical system which is explaining the key features of FTS.

2. Nonlinear random dynamical model for the stock market

2.1. Deterministic dynamical system

In detail, the construction of the nonlinear dynamical model of the stock is presented in works [9-11]. In this section, we only indicate the basic principles.

The evolution of a stock market can be described by the well-known Lorenz system of equations:

$$\begin{cases} \tau_\eta \dot{\eta}_t = -\eta_t + a_\eta h_t \\ \tau_h \dot{h}_t = -h_t + a_h \eta_t S_t \\ \tau_S \dot{S}_t = (S_0 - S_t) - a_S \eta_t h_t \end{cases}, \tag{1}$$

where $\eta_t \equiv \eta_t - \eta_{eq}$ is the variation of "ask" price (η_t) relative to equilibrium value (η_{eq}) is the "ask" price in equilibrium state); $h_t \equiv h_t - h_{eq}$ is the variation of "bid" price (h_t) relative to equilibrium value (h_{eq}) is the "bid" price in equilibrium state); $S_t \equiv N_{|a\rangle t} - N_{|p\rangle t}$ is an instantaneous difference between numbers of agents in $|a\rangle$-state and $|p\rangle$-state.

A particular market agent being in $|a\rangle$-state has maximum amount of valuable information about financial asset ($I_{|a\rangle}$) and has minimum information ($I_{|p\rangle}$) being in $|p\rangle$-state. The agent being in $|a\rangle$-state is able to generate local demand on deal with the asset and send an "ask-quantum" to other agents. If the agent is in $|p\rangle$-state (he/she does not have enough valuable information about the asset), then the agent's rational decision is do not generate demand on deal ("bid-quantum"). Moreover, for the agent in $|p\rangle$-state generating of a deal offer depends on the agent's reaction on received "ask-quantum" or can be his or her own decision. General pattern in stock markets is that local "ask" waves ("quanta") induce local "bid" waves ("quanta").

Indeed, stock market is an open system that continuously exchanges information and money flows with the external world. Sources of external information include corporate financial reports, financial news feeds, stock-ticker data and others. This information flow, in some sense, "pump up" the stock market, making inverse population of market agents: $N_{|a\rangle} >> N_{|p\rangle}$, where $N_{|a\rangle}$ is the number of agents being in $|a\rangle$-state, $N_{|p\rangle}$ is the number of agents being in $|p\rangle$-state.

Indeed, stock market is an open system that continuously exchanges information and money flows with the external world. Sources of external information include corporate financial reports, financial news feeds, stock-ticker data and others. This information flow, in some sense, "pump up" the stock market, making inverse population of market agents: $N_{|a\rangle} >> N_{|p\rangle}$, where $N_{|a\rangle}$ is the number of agents being in $|a\rangle$-state, $N_{|p\rangle}$ is the number of agents being in $|p\rangle$-state.

The system of self-consistent equations (1) is a well-known method to describe a self-organizing system. The Lorenz synergetic model was first developed as a simplification of hydrodynamic equations describing the Rayleigh-Bénard heat convection in the atmosphere; it is now a classical model of chaotic dynamics. Further research on the Lorenz system presented in a series of publications proved that the system provides an appropriate kinetic picture of the cooperative behavior of particles in any macroscopic dynamical system where the actualization of potential order is possible. Processes in such self-organizing complex systems in nonequilibrium state lead to the selection of a small number of parameters from the complete set of variables that describe the system; all other degrees of freedom adjust to correspond to these selected parameters. Following the terminology used in the synergy theory, these parameters are the order parameter (η_t), conjugated field (h_t), and control parameter (S_t). According to the Ruelle-Takens theorem, we can observe a nontrivial self-organization with strange attractors if the number of selected degrees of freedom is three or more.

In the system of equations (1) a_η is a coefficient, and positive constants a_h, a_S are measures of feedback in a stock market. Functions η_t/τ_η, h_t/τ_h, $(S_0 - S_t)/\tau_S$ describe the autonomous relaxation of the variation of "ask" price, variation of "bid" price and instantaneous difference between numbers of agents in $|a\rangle$-state and $|p\rangle$-state to the stationary values $\eta_t = 0$, $h_t = 0$, $S_t = S_0$ with relaxation times τ_η, τ_h, τ_S.

Financial interpretation of the results is as follows [9-11]:

- In case of relatively small intensity of external information pumping ($S_0 \to 0$), the stock market tends to stable equilibrium. However, practically this stable equilibrium state cannot be reached, since the market is an open system with permanent external information pumping.
- If $S_0 \to 28$, then the stock market functions as an open nonequilibrium system with deterministic chaos. Such behavior is typical for a financial market with considerably intense external information.
- 3-dimensional dynamical model (1) explains some properties of the stock market functioning such as fractality (fractal dimension equals 1.497) and chaotic nature (correlation dimension equals 1.896) of FTS [12].

2.2. Random dynamical system

The nonlinear dynamic model (1) explains the fractality and chaotic nature of empirical as well as the dissipative nature of the system. On the other hand, Eq. (1) cannot explain some other phenomena found in empirical data [13-17], and first of all, the key signs of fractal market signals: a power law of PDF, 1/f noise, and long memory. Let us consider different ways of improving (generalizing) Eq. (1) in order to adequately describe a stock market.

Taking into account stochastic terms and the fractionality of the order parameter, Eq. (1) takes the following form:

$$\begin{cases} \tau_\eta \dot{\eta}_t = -\eta_t^\alpha + a_\eta h_t + \sqrt{I_\eta}\xi_t \\ \tau_h \dot{h}_t = -h_t + a_h \eta_t^\alpha S_t + \sqrt{I_h}\xi_t \\ \tau_S \dot{S}_t = (S_0 - S_t) - a_S \eta_t^\alpha h_t + \sqrt{I_S}\xi_t \end{cases} . \quad (2)$$

In Eq. (2), I are noise intensities for each phase variable; ξ_t is white noise, where $\langle \xi_t \rangle = 0$, $\langle \xi_t, \xi'_t \rangle = \delta(t - t')$; $\alpha \in (0,1]$. The random dynamic system (RDS) (2) is a generalization of the deterministic dynamic system (1) where stochastic sources are added, the feedback is weakened, and the order parameter is relaxed. The replacement of the order parameter η_t by a smaller value (η_t^α) means that the process of ordering influences the self-consistent behavior of the system to a lesser extent than it does in the ideal case of $\alpha = 1$.

Let us analyze RDS (2) in adiabatic approximation when the characteristic relaxation time of the variation of "ask" price considerably exceeds the corresponding relaxation times of the variation of "bid" price and the instantaneous difference between numbers of agents in $|a\rangle$-state and $|p\rangle$-state: $\tau_\eta >> \tau_h, \tau_S$.

For a stock market functioning as an open nonequilibrium system, the adiabatic approximation means that when the external information feed tends to zero, the stream of "ask-quanta" slowly decreases and at the same time the "bid-quanta" and the number of agents being in $|a\rangle$-state decrease as well.

An adiabatic approximation is a necessary condition for the transformation of the three-dimensional RDS with additive noise (2) into a one-dimensional RDS with multiplicative noise of the following form:

$$\tau_\eta \dot{\eta} = f(\eta) + \sqrt{I(\eta)}\xi . \quad (3)$$

The terms of Eq. (3) corresponding to the drift and diffusion (intensity of the chaotic source) have the following form:

$$f(\eta) \equiv -\eta^{\alpha} + S_0 \eta^{\alpha} \left(1 + \eta^{2\alpha}\right), \tag{4}$$

$$I(\eta) \equiv I_{\eta} + \left(I_h + I_S \eta^{2\alpha}\right)\left(1 + \eta^{2\alpha}\right)^2. \tag{5}$$

The stationary probability distribution density of the deviation of the variation of "ask" price from the corresponding equilibrium value has the following form:

$$p(\eta) = Z^{-1}\left(1 + \eta^{2\alpha}\right)^{-2} \exp\left[\int \frac{\left(1 + \eta^{2\alpha}\right)^{-2}}{\eta^{\alpha}} d\eta\right], \tag{6}$$

where Z is a normalization constant.

Before we draw any conclusion about probability density function (PDF) (6), let's direct our attention to one significant fact that distinguishes theory from practice. Distributions of real systems and processes regardless of their nature cannot have an infinite expected value or variance. Therefore, power-law PDFs like $p(x) \propto x^{-2\alpha}$ (2α is chosen for the purposes of analysis of expression (6)) are approximate and not valid for large x. The exponential decrease of PDF corresponds to the intermediate asymptotics, and in practice instead of heavy tails we should have semi-heavy tails:

$$p(\eta) \propto \eta^{-2\alpha} \mathrm{P}\left(\frac{\eta}{\eta_c}\right), \tag{7}$$

where the scaling function $\mathrm{P}(\eta/\eta_c)$ is approximately constant at $\eta \cong \eta_c$ and quickly decreases when $\eta \to \infty$. Here the "heaviness of the tail" is shifted toward the intermediate range of η values. Thus, the dimensionless variation of "ask" price η scaled for η_c serves as a scaling variable η/η_c in (7). Since the integral in Fokker-Planck equation is regular at $\eta \to 0$, the PDF obtained has a power-law form.

The power law for PDF of the deviation of the variation of "ask" price, which is equivalent to for large times, was obtained and justified analytically. However, we could not obtain analytical expressions for power spectral density ($S(f)$), autocorrelation function (ACF), or the fractal dimensions (D_F). Therefore, we present below the results of numerical calculations for a family of realizations of RDS (3): $p(x) \propto x^{-1.18}$, $S(f) \propto f^{-1.36}$, $ACF(\tau) \propto \tau^{-0.04}$, $D_F = 1.235$.

3. Conclusions

The generalized Lorentz system (2) adequately models the evolution of a stock market as a complex system. The characteristics of η-realizations of RDS (3) are quantitatively close to the corresponding characteristics of empirical FTS.

For a description of the evolution of a stock market, the nonlinear dynamical system model (1) is a rough, not very accurate approximation. First, the model does not predict the occurrence of catastrophic values in a time series of the variation of "ask" price which would signify the complexity of a stock market, or the existence of long memory or the time series' tendency to follow trends. Despite this deficiency, Eq. (1) allows one to study stock market far from equilibrium, and it also explains the existence of dynamical chaos in a time series as well as their fractality.

The nonlinear random dynamical system (2) is a generalization of the model (1) that accounts for external stochastic sources and the fractionality of the order parameter (weakening of feedback and relaxation of the order parameter). This model adequately describes the evolution of a stock market.

IOP Conf. Series: Journal of Physics: Conf. Series **1298** (2019) 012009 doi:10.1088/1742-6596/1298/1/012009

Quantitative characteristics of the model (2) in adiabatic approximation are close to the corresponding characteristics of the empirical FTS. An adiabatic approximation allows us to reduce a three-dimensional random adiabatic system with additive noise (2) to a one-dimensional random dynamical system with exponential multiplicative noise (3).

4. References

[1] Chakraborti A, Toke I, Patriarca V and Abergel F 2011 *Quant. Fin.* **11** 991
[2] Chakraborti A, Toke I, Patriarca V and Abergel F 2011 *Quant. Fin.* **11** 1013
[3] Richmond P, Mimkes J and Hutzler S 2013 *Econophysics and Physical Economics* (Oxford University Press)
[4] Savoiu G 2013 *Econophysics. Background and Applications in Economics, Finance, and Sociophysics* (Elsevier)
[5] Hsieh D A 1991 *J. Finance* **46** 1839
[6] Small M and Tse C K 2003 *Stud. Nonlin. Dyn. Econom.* **7** 1
[7] Mandelbrot B B 1963 *J. Bus. Univ. Chicago* **36** 394
[8] Elliott R J and Kopp P E 2005 *Mathematics of the Financial Markets* (Berlin: Springer)
[9] Dmitriev A, Silchev V and Dmitriev V 2019 *Lecture Notes in Electrical Engineering* **489** 237
[10] Tverdislov V A, Dmitriev A V and Sidorova A E 2017 *Bulletin of the Russian Academy of Sciences: Physics* **81** 114
[11] Dmitriev A, Silchev V and Dmitriev V 2017 *WSEAS Transactions on Business and Economics* **14** 311
[12] Sparrow C 1982 *The Lorenz Equations: Bifurcations, Chaos and Strange Attractors* (Springer)
[13] Niu H and Wang J 2013 *Journal of Applied Statistics* **40** 2188
[14] Clementi F, Matteo T D and Gallegati M 2006 *Physica A* **370** 49
[15] Laherrere Jand Sornette D 1998 *Ero. Phys. J.* **2** 525
[16] Pisarenko V F and Sornette D 2006 *Physica A* **366** 387
[17] Altmann E G and Kantz H 2005 *Phys. Rev. E* **71** 056106

IOP Conf. Series: Journal of Physics: Conf. Series **1298** (2019) 012010 doi:10.1088/1742-6596/1298/1/012010

Revised Floyd-Warshall Algorithm to Find Optimal Path in Similarity-Weight Network and Its Application in the Analysis of Global Value Chain

Lizhi Xing[1,2], **Yujie Li**[2]

1 College of Economics and Management, Beijing University of Technology, Beijing, China
2 School of Online and Continuing Education, Beijing Foreign Studies University, Beijing, China

Email: itwasa@163.com; liyujie@bfsu.edu.cn

Abstract. This paper focuses on measuring the globally and nationally economic system's connectedness and industrial sector's function on the Global Value Chain (GVC), as reinforcements to the present studies on international trade. Firstly, we reconsidered the length-related and position-related measures in literatures about vertical specialization from the perspective of econophysics. Secondly, we redefined the inter-country and inter-sector propagating process of intermediate goods and proposed the concept of Strongest Relevance Path Length (SRPL) based on Revised Floyd-Warshall Algorithm (RFWA), which is the basis of new measurement. Thirdly, enlightened by betweenness centrality of node and edge, we introduced SRPL-based index to measure the Value-Added Pivotability of Industrial Sectors and Input-Output Relationships. Fourthly, we also proposed the concepts of Backward Closeness and Forward Closeness of Industrial Sectors based on closeness centrality.

1. Introduction

The rise of *Global Value Chain (GVC)* has naturally captured the attention of international trade economists eager to bridge the apparent gap between the new characteristics of the international organization of production and the standard methods used to collect, manipulate, and interpret international trade statistics. In particular, a remarkable body of work has devised ingenious empirical methods to disentangle the value-added and intermediate inputs of international trade flows.

According to the definition of value chain proposed by Porter, it represents value-adding process at every stage of production, from R&D, design, to the delivery of the final product and service to consumers. This process may just happen within a country or a region, but it is becoming increasingly international and sophisticated nowadays under the urge of globalization. Thus, trying to measure industrial sectors' position as well as consequent function on the GVC is not an easy job.

There is no doubt that measurement of length and upstreamness/downstreamness contributes a lot to the studies of vertical specialization. According to our study on the application of complex network theory [1,2,3], network-based algorithms and indices have great potential to shed new light on further understanding of industrial sector's position and function in consideration of the network-form architecture of GVC. However, only a few scholars adopted this kind of method to analyze countries positions in the international global value networks as far as now, such as Mesa-Arango et al. [4] and Cingolani et al. [5].

2. Construction of the Global Value Chain Network

To establish an industrial complex network, we consider a sector category within a region as a node and the inter-industry IO relationship as a tie, whose weight represents the sale and purchase relationships between producers and consumers. Thus, a graph $G=(V,E,W)$ containing n nodes is created, representing sectors within a nation or region, denoted as a node set V. Pairs of nodes are linked by ties reflecting their interdependencies, constituting an asymmetric tie set E. However, in valued graphs, a set E can be replaced by weight set W, which can be extracted from the region of inter-country inter-industry use and supply in ICIO table. We name the ICIO network model Global Industrial Value Chain Network (GIVCN), since its purpose is to reflect how economic shocks propagate and amplify along GVC, as well as to what extent the industrial impact generates at the inter-country level.

3. Optimal Path Searching Algorithm

We proposed a network-based framework to detect inter-sector relevance. In details, the information of sector's function and inter-sector relative position on the GVC is already embedded in the topological structure of GIVCN models, so the first thing is to redefine the optimal path to reflect the propagation process of intermediate goods on the premise of considering the properties of economic system.

3.1. The Issue of Path in Similarity-Weight Network

Both *Betweenness and Closeness centrality* rely on identifying optimal paths, which in the case of binary data can be unproblematically identified as shortest paths. But if the edges are weighted, there are a variety of possibilities in assessing the optimality of a path. In other words, there is no one generalization for weighted networks, and it all depends on what kinds of network process we are studying.

For many researchers studying ICIO networks, the most common style they used is to take the approach of binarizing the weighted edges and running the traditional *Breadth-First-Search (BFS)* algorithms, in consideration of the computational intension. In addition, someone defined the optimal path as the maximum shortest path length between any two pairs [6], as well as took the minimum shortest path length on the basis of taking the reciprocal of weights. They did so in part because of they wanted to obtain the metric of consecutive paths, however, it is inappropriate.

In our opinion, GVC's topological structure is very similar with supply chain, which sheds a light on the issue of path in GIVCN models. As we all know, the *Bullwhip Effect*, also known as the *Forrester Effect* [7], is a distribution channel phenomenon in which forecasts yield supply chain inefficiencies. It refers to increasing swings in inventory in response to shifts in customer demand as one moves further up the supply chain. For the sake of weakening it, the most direct and efficient way is to reduce the length of transaction process as well as improve the efficiency and effectiveness of each step. Therefore, in this paper, we designed a new sort of BSF algorithm to trace inter-sector crucial connection intermediate goods flowing on in respect of both propagation efficiency and effectiveness.

3.2. Revised Floyd-Warshall Algorithm

In network science, path plays a central role, and many network-based indices are developed around it. The *Shortest Path* between nodes i and j, also called *Distance* or *Geodesic Path*, is the path with the fewest number of edges linking them, denoted by d_{ij}. If there is no such path between them, it means $d_{ij}=\infty$. Besides, it is possible to find there are multiple shortest paths of the same length d_{ij}. In general, *Floyd-Warshall Algorithm (FWA)* as a classical BFS algorithms is used to find all inter-node shortest paths in a recursive way, which means it compares all possible paths through the network between each pair of nodes by incrementally improving an estimate on the shortest path until the estimate is optimal [8]. The core formula of FWA is:

$$d_{ij}^{(k)} = \min_{i,j,k\in\{1,2,\dots,N\}} \left\{ d_{ij}^{(k-1)}, d_{ik}^{(k-1)} + d_{kj}^{(k-1)} \right\} \quad (i \neq j) \tag{1}$$

Eq (1) works by first computing $d_{ij}^{(k)}$ for all pairs of source-to-sink nodes for $k=1$, then $k=2$, etc. This process continues until $k=N$, and we have found the shortest path using any intermediate nodes.

However, in weighted networks, particularly those with similarity weights such as IO system, we could use neither the objective function of minimization nor the iterative accumulation of paths. This is because it is meaningless to find the least efficient channel of intermediate goods, and we will get an infinite result inevitably if just taking it as a maximization problem. Hence, we must find one kind of fast and highly effective restraining algorithm to solve this operational research problem about network flow. In fact, it is possible to create a method to reconstruct the actual path between any two endpoint nodes with simple modifications to FWA, just like what scholars have done [9].

Therefore, we divided $w_{ik}w_{kj}$ by $w_{ik}+w_{kj}$ and named the result the *Relevance Path Length (RPL).* After transformation, we got $RPL_{ij}=1/\left(1/w_{ik}+1/w_{kj}\right)$. Different from edge weights that stand for the monetization of intermediate goods from sectors i to j, the RPLs in GIVCN model are not used to measure inter-sector relations in economic terms, but the gauge of spreading effectiveness of all paths. In sum, one RPL between nodes i and j is mathematically equivalent to the reciprocal of the summation of the reciprocals of relevant edge weights, namely:

$$RPL_{ij}=\left(\frac{1}{w_{ik}}+\frac{1}{w_{kl}}+\cdots+\frac{1}{w_{lj}}\right)^{-1}=\left(\sum_{i,j,k,l\in\{1,2,\ldots,N\}} w_{kl}^{-1}\right)^{-1} \tag{2}$$

where w_{kl} is the weight of an edge on the path connecting nodes i and j, and the exact number of RPL_{ij} is up to that of possible paths delivering the inter-sector intermediate goods. From the form, this formula is similar with parallel resistors. As for the self-loops existing in some networks, for instance the ICIO network, $RPL_{ii}=w_{ii}$ according to the Eq (2).

Then, we considered how to find the optimal one among RPLs, and named it the *Strongest Relevance Path Length (SRPL)*, which is the synthetic gauge of both spreading effectiveness and efficiency of a given path in the meanwhile as far as possible. For this purpose, we put forward the *Revised Floyd-Warshall Algorithm (RFWA)* as an iterative algorithm based on operations research:

$$SRPL_{ij}^{(k)}=\max_{i,j,k\in\{1,2,\ldots,N\}}\left\{w_{ij}^{(k-1)},\frac{w_{ik}^{(k-1)}\times w_{kj}^{(k-1)}}{w_{ik}^{(k-1)}+w_{kj}^{(k-1)}}\right\} \tag{3}$$

According to Eq (3), we will compare RPLs successively until picking out the largest one. If it is bigger than $w_{ij}^{(k-1)}$, we keep the record as it is, otherwise just make it equal to $w_{ij}^{(k-1)}$. When the $SRPL_{ij}^{(k)}$ happen to be equal to $w_{ij}^{(k-1)}$, it means the optimal path is just the most direct one between nodes i and j. In other words, there is no need to take even one more step via another node. Due to the nature of RPLs, the selected SRPL converges when finding the maximum. Besides, the self-loops contain non-negligible topological information, such as self-consumption of industrial sectors in ICIO network. Therefore, we incorporated this sort of edge into the comparison process, which means the source node and sink node of a RPL could be the same one.

3.3. Theoretical Basis of SRPL in GIVCN Model

By Porter's definition in 1985, an industry value-chain is a physical representation of the various processes involved in producing goods (and services), starting with raw materials and ending with the delivered product. It is based on the notion of value-added at the stage of production level. In fact, Wasilly Leontief's IO table, published in the 1950s, already provided estimates of the relative importance of each individual link in industry-level value chains. In recent decades, multinational enterprises locate a lot of activities, such as research, development, design, assembly, production of parts, marketing and branding, in different countries around the globe, to enhance efficiency and to optimize profits. It is the theory of comparative advantage behind the globalization, so here comes the concept of GVC. Furthermore, analytical tools stemming from complex network have been proven to be very effective in analysis of GVC, thus we proposed the SRPL in this paper to find the optimal path of intermediate goods propagation based on many theories and models.

The gravity model of international trade in international economics is a model that, in its traditional form, predicts bilateral trade flows based on the economic sizes and distance between two units. The model was first introduced in economics world by Walter Isard in 1954 [10]. The basic model for trade between two countries i and j takes the form of

$$T_{ij} = \frac{A \times Y_i \times Y_j}{D_{ij}} \tag{4}$$

where A is the constant, T_{ij} stands for trade flow, D_{ij} stands for the distance and Y_i, as well as Y_j, stands for the economic dimensions of the countries that are being measured, often using GDP measurements.

Before defining the RPL and searching SRPL, we noticed that many similar studies didn't take the length of value chain into account, which makes them overly qualitative or even meaningless. Then, the formula of gravity model shed light on the principle of our BFS algorithm, i.e., an intermediate goods transfer path is directly proportional to the relevant IO relations and inversely proportional to the length of value-added process. Thus, we replaced $Y_i \times Y_j$ with $w_{ik}^{(k-1)} \times w_{kj}^{(k-1)}$ to measure the influence from industrial structure, and D_{ij} with $w_{ik}^{(k-1)} + w_{kj}^{(k-1)}$ to provide distance information (the denominator of RPLs of two or more steps incorporates the number of relevant steps after employing fractionation, and increasing steps yield more transaction cost as well), explaining the dynamics of intermediate goods transferring from upstream sectors to downstream ones (sometimes they are the same one). As results, SRPL based on RFWA owns two aspects of economical meaning: one is to find all the optimal propagation paths of intermediate goods with both higher IO relationship and lower transaction cost between any pairs of sectors (including self-loops); another is to measure the inter-sector (maybe inner-sector) relationship strength from the standpoint of integral value chain.

4. Derivative Measures

As we all known, comparative advantage is the economic reality describing the work gains from trade for individuals, firms, or nations, which arise from differences in their factor endowments or technological progress. We believe that the GVC exactly stems from comparative advantage of nations all around the world, and SRPL can be used to identify these important value chains from ICIO network. Then, measuring with SRPL-based indicators and forecasting through simulation become feasible. As results, we can propose international trade policies from the angle of econophysics.

4.1. Betweenness Centrality

In similarity-weight network, however, the basis of information propagation differs from Boolean network, thus we proposed *Weighted Betweenness Centrality of Node based on RFWA* and denote it by C_B^{RFWA}. Then, we encountered an important issue here, that is, how to treat those nodes locating on both ends of every single SRPL. Can we just ignore them? The answer is no. As the framework has been expended to multi-layer network, in this case, all the industrial sectors processing value-added are between value-added level units and final demand level units. Correspondingly, they as nodes in GIVCN models are neither on the topest of value chain nor on the lowest. Therefore, there is no need to struggle with the function of either source nodes or sink nodes on the SPRLs, just focus on the frequency of appearance. We will assign different meanings or usages to source nodes and sink nodes when running more complex applications. The formula of C_B^{RFWA} is:

$$C_B^{RFWA}(i) = \sum_{i,j,k \in \{1,2,\ldots,N\}} SRPL_{jk}^{(N)}(i) \tag{5}$$

where $SRPL_{jk}^{(N)}(i)$ is the number of SRPLs within the scope of the whole network, which are connecting any pairs of others cross a given sector. Of special interest is how to measure the node with self-loop of its being a SRPL. The frequency seems to be 2 because this node appears as both the source node and sink node on this special path, which is, however, irrational. Thus, we set a rule that the number of appearances of the node on the same path is less than or equal to 1.

According to Eq (5), the most significant difference is we do not normalize it by the totality anymore, because we are more concerned with industrial sectors' propagation function on the GVC rather than their weakened heterogeneity reflected by ratio. Above all, we adopted it to measure the

Value-Added Pivotability of Industrial Sectors, with the purpose to evaluate the level of brokerage in the process of turnover of intermediate goods. For example, suppose that a given sector owns high pivotability, many sectors will go through it to reach the others via efficient and effective paths. In principle, this sector has power because it could threaten to stop transferring, making sectors use less efficient and effective paths to reach one another. But this superiority only works if the other sectors cannot easily enter into new trade relations around the world. In other words, not only can we use pivotability to evaluate the current global economic situation, but also offer favorable suggestions for the future.

To find the most important inter-sector economic relationships among numerous couples of sectors in GIVCN models, we introduced *Weighted Betweenness Centrality of Edge based on RFWA* to measure the *Value-Added Pivotability of Input-Output Relationships*:

$$C_E^{RFWA}(i,j) = \sum_{h,i,j,k \in \{1,2,\ldots,N\}} SRPL_{hk}^{(N)}(i,j) \tag{6}$$

where $SRPL_{hk}^{(N)}(i,j)$ is the number of SRPLs within the scope of the whole network, which are connecting any pairs of others cross a given couple of sectors i and j. Self-loop serving as a SRPL will plus 1 to this indicator of itself.

4.2. Closeness Centrality

GIVCN models took both weight and direction of nodes into account to make maximum use of ICIO data, thus we need to turn to new SRPL-based measures. Firstly, directed closeness centrality should be divided into two sorts: *Weighted In-Degree Closeness Centrality* and *Out-Degree Closeness Centrality based on RFWA*, denoted by $C_C^{RFWA-IN}$ and $C_C^{RFWA-OUT}$, according to nodes' position on the propagation path, serving as a sink node or source node respectively. Secondly, the shortest paths cannot effectively reflect the propagation efficiency of information in weighted network any longer as mentioned above, and it is therefore replaced with SRPLs while computation. Finally, yet importantly, the relative position of numerator and denominator needs to be changed, because SRPLs reflect the most efficient ways that information passes through in similarity-weight network, and the numerical value of closeness centrality should be proportional to the average value of it. Considering the above-mentioned, new closeness centralities in directed similarity-weight network based on RFWA are introduced:

$$C_C^{RFWA-IN}(i) = \frac{\sum_{i=1}^{N} SRPL_{ij}^{(N)}}{N} \tag{7}$$

$$C_C^{RFWA-OUT}(i) = \frac{\sum_{j=1}^{N} SRPL_{ij}^{(N)}}{N} \tag{8}$$

By the nature of Eq (7) and Eq (8), the summation of all the backward linkages is just equal to that of forward linkages, so there is a kind of conservation relationship between them in the closed economic system.

From the point of backward linkage, bigger $C_C^{RFWA-IN}$ means sectors rely much more on the intermediate goods from their upstream providers in the international division. And from the opposite side, the bigger $C_C^{RFWA-OUT}$, the more intermediate goods contributed to the downstream consumers. We defined the $C_C^{RFWA-IN}$ and $C_C^{RFWA-OUT}$ as *Backward Closeness and Forward Closeness of Industrial Sectors*, to quantify backward and forward inter-sector closeness degree starting from a given sector respectively. Note that, inner-sector self-consumptions denoted by self-loops in the network are included in both directions.

In sum, our hypothesis is that the relative position of industrial sector on the GVC could be reflected by backward closeness and forward closeness in this paper, and we may figure out what kind of comparative advantage different location will bring to it. Thus, we put forward an analytical framework and will adopt it in empirical studies.

5. Conclusions

The major difference between our framework and the others is that we argue there is no starting point (R&D and design) or ending point (the delivery of final products or services to consumer) in the economic system. Every single sector on the supply side is called the upstream sector only if it directly or indirectly provides intermediate products or services to one consumer at least, while each sector on the demand side is taken as the downstream sector only if it directly or indirectly consumes intermediate products or services from even sole provider.

6 Acknowledgment

The authors acknowledge support from Beijing Municipal Natural Science Foundation (Grant No. 9194024, Grant Recipient: Lizhi Xing), Beijing Municipal Social Science Foundation (Grant No. 17ZGC010, Grant Recipient: Lizhi Xing), Ri-Xin Talents Project of Beijing University of Technology (Grant Recipient: Lizhi Xing), and Technology Plan Key Program of Beijing Municipal Education Commission (Grant No. KZ201810005010, Grant Recipient: Jun Guan).

7. References

[1] Xing L, Ye Q, Guan J. Spreading Effect in Industrial Complex Network Based on Revised Structural Holes Theory [J]. PLoS ONE, 2016, 11(5): e0156270.
[2] Xing L, Dong X, Guan J. Global Industrial Impact Coefficient Based on Random Walk Process and Inter-Country Input-Output Table [J]. Physica A: Statistical Mechanics and its Applications, 2017, 471: 576-591.
[3] Xing L, Guan J, Wu S. Measuring the Impact of Final Demand on Global Production System Based on Markov Process [J]. Physica A: Statistical Mechanics and its Applications, 2018, 502:148-163.
[4] Mesa-Arango R, Narayanan B, Ukkusuri S V. The Impact of International Crises on Maritime Transportation Based Global Value Chains [J]. Networks & Spatial Economics, 2015:1-28.
[5] Cingolani I, Panzarasa P, Tajoli L. Countries' positions in the international global value networks: Centrality and economic performance [J]. Applied Network Science, 2017, 2(1):2-21.
[6] Tsekeris T. Global value chains: building blocks and network dynamics [J]. Physica A Statistical Mechanics & Its Applications, 2017, 488.
[7] Forrester J W. Industrial dynamics[M]. M.I.T. Press, 1961.
[8] Floyd R W. Algorithm 97: Shortest path [J]. Communications of the Acm, 1962, 5(6):345.
[9] Opsahl T, Agneessens F, Skvoretz J. Node centrality in weighted networks: Generalizing degree and shortest paths [J]. Social Networks, 2010, 32(3):245-251.
[10] Isard W. Location Theory and Trade Theory: Short-Run Analysis [J]. Quarterly Journal of Economics, 1954, 68(2):305-320.

Differences between English and Chinese in the Implementation of the RSA algorithm

Duoji Jiang[1], Kwame Yankson[2]
[1]Foreign Language School attach to Sichuan International Studies University, Chongqing,China
[2]School of Education, University of Michigan

Email: kyankson@gmail.com; 1013878088@qq.com

Abstract—In this paper, we explore the implementation of RSA Public Key Cryptography in both English and Chinese. We begin by examining the history of ancient and modern cryptography, public key cryptography, the key distribution problem and how the RSA algorithm solves the key distribution problem. We then show how the RSA algorithm works in practice, on both English and Chinese characters, words and sentences. We end with a comparison of differences in implementation between the two languages.

1. Introduction

Cryptography is a discipline that studies how to pass information secretly. It has been used by Royalty and by spies since ancient times. In modern times, it especially refers to mathematical research on information and its transmission. It is often regarded as a branch of mathematics and computer science. It is also closely related to information theory. Cryptography is the core of information security for related issues such as authentication and access control.

The primary purpose of cryptography is to hide the meaning of information but not to hide the existence of information, in other words, we do not hide the message but we make the message impossible to be understood without specific keys, so that a possible intruder can' t know the meaning of the information. Cryptography has been used in everyday life: chip cards including ATMs, and computer user access passwords in e-commerce as examples. The password is an important means of secrecy for the special transformation of information by both parties of communication according to the stipulated rules. In accordance with these rules, the transformed text into cipher text, known as encryption transformation; transform that back to the original text, is known as the decryption transformation.

Cryptography has gradually developed in the struggle between coding and deciphering. With the application of advanced science and technology, cryptography has become a comprehensive cutting-edge technological science. It has extensive and close relations with linguistics, mathematics, electronics, acoustics, information theory, computer science and so on.[1]

Since the 1970s, some scholars have proposed a public key system, that is based on, the mathematical principle of using one-way functions, so as to realize the separation of encryption and decryption, in which the encryption key is public and the decryption key is confidential in order to make cryptographic communication without exchanging keys possible. This new cryptosystem has aroused widespread attention and discussion in cryptology. [7]

This article focuses on one of the RSA public key cryptography algorithms based on one-way functions-RSA. The RSA public key encryption algorithm was introduced in 1977 by Ron Rivest, Adi Shamir and Leonard Adleman [2]. It was first published in the United States on July, 1987, when all three of them was working at MIT. RSA is the first letters of their surname spelled together. RSA is by far the most influential and popular public-key encryption algorithm that can resist most of the cryptographic attacks known so far and is recommended by the ISO as a public-key data encryption standard.

Today, only a short RSA key might be broken by a powerful way. In fact, using classical algorithms, there is no reliable way to attack the RSA algorithm even today [3]. The RSA algorithm is based on a very straightforward number theory: multiplying two large primes is easy, but it is extremely difficult to factorize their product, so the product can be exposed as an encryption key. Due to the way RSA algorithm was designed, in which the encryption and decryption do not share the same key, it becomes possible to distribute public keys across the internet. Before the invention of RSA algorithm, the Key distribution problem--how to transmit encrypted message without having to transmit the secret key-- had bothered people for a long time, the RSA algorithm is actually a solution to the problem at beginning, and the key distribution problem will be introduced in the next section.

As internet technology became wide spread across the world, the RSA algorithm is also applied by different peoples using different languages across the world, and our topic of this paper is to find the difference of the application of RSA in two specific languages-- Chinese and English.

2. The Key Distribution Problem

The key distribution problem was the most significant problem for postwar cryptographists. Before the 1970s, all of cryptography was based on symmetric functions. In a symmetric function $f(x) = y$, if we know the value of x or y, we can then compute the magnitude of the other one without too much effort. The way to get y if we have x is exactly reverse to the way to get x if we know y.

In practical terms, if Alice wants to send a message to Bob using symmetric-function-based cryptography, she firstly encrypts her message with her secret-key and sends the message to Bob. Bob then decrypts the encrypted message with the same key as Alice' s, however, Alice has to make sure that Bob has the same key as hers before she sends the encrypted message to him. Otherwise the message will be meaningless words. This would not be a serious issue if Alice and Bob are friends, live in the same city and can meet discretely to exchange the key. For governments and banks that have thousands of messages that need to be securely sent to all over the world, key distribution posed a complex problem.

The problem lies in the fact that in general, if two parties have to communicate secretly with a third-party to deliver the secret key, a weak link is created in the exchange process. This is because the key can be compromised during exchange between the two parties. The risk of the secret-key being stolen during delivery cannot be ignored. This is the weak link. Otherwise it is nearly impossible for intruders to break governments and banks' ingenious encryption algorithms even if they were symmetric-function-based. Therefore, the banks and governments before and during the 1970s had to make sure that they had the most trustworthy people to deliver keys face-to-face for them, in case of any potential intruder. The cost of doing that could be enormous.

For example, in the 1970s, banks hired trustworthy employees as full-time key carriers, these carriers traveled around the world with a locked box, delivering the keys to the clients face to face, and the clients will therefore receive encrypted documents next week. The cost of daily key-delivering increases and with the expanding of the business network, more messages had to be sent and more keys to be delivered. Finally, the cost of key distribution became a nightmare for the banks.

To some extent, militaries and governments of many countries can also spend enormous amounts of money on key distribution, simply because their messages are too significant that they need to protect the safety of these messages at any possible cost. During the Second World War, the German high command had to deliver monthly magazine with daily keys to all the operators of enigma machine-a cipher machine of Nazi German army. The secret keys of United States government were administered and delivered by COMSEC (Communications Security Establishment). In the 1970s, COMSEC delivered tons of secret keys every day.

Hellman and Diffie from Stanford University first proposed the solution to the key distribution problem. They assumed a scenario involving three people： Alice, Bob and Eve. In their scenario, Alice wants to communicate secretly with Bob, but Eve always tries to steal their keys. How can Alice interact with Bob without passing the key to Bob in person for security? Hellman and Diffie thought of a scenario where Alice loads her message in a box, then puts on her own lock and sends it to Bob. Bob puts his lock on and sends it back to Alice. Alice can remove her lock and send it back to Bob again, then Bob only needs to unlock his own padlock to get the message. Through this means of interaction, Eve has no chance at all to steal their key because the message is always encrypted while delivering. A problem that plagued people for two millennia seems to have a proposed solution.

However, in practice, key distribution is not as simple as this theory might suggest. In the hypothetical situation as mentioned, you can choose to open the lock of Alice or Bob firstly, and you can finally open the box successfully. However, for most cryptography algorithms using symmetric functions, you have to unlock in the exact reverse order as you locked them up. Take a simple example: Alice' s encryption method is to double the original number. Her decryption method is to divide the ciphertext by 2. Bob' s encryption method is plus 2 on the text, his decryption method is to subtract 2. Assuming that the initial information is 2, Alice encrypted first: $2x2 = 4$. Bob then encrypted by adding 2 to get the ciphertext: $4 +2 = 6$. Now, in order to decrypt a symmetric-key based ciphertext, you must follow exactly the opposite order of encryption. First reverse Bob' s encryption: 6 - 2 = 4. Then reverse Alice' s: 4 ÷ 2 = 2, in order to get the original information. If you do not follow the order, say you decrypt Alice' s encryption firstly: 6 ÷ 2 = 3, then solve Bob' s: 3-2 = 1, the result you get,1, will be completely different from the original text, 2). Diffie and Hellman had to find one way that is different from symmetric key algorithm, which needs an exact reverse order of encryption to decipher a cyphertext.

To solve this problem, Hellman and Diffie used a mathematical concept: one-way functions, that are asymmetric rather than symmetric. For a one-way function, f(x)=y, it is relatively easy to calculate y when x is known, but it is extremely difficult to calculate the corresponding x if y is known. In other words, it is much more difficult to calculate the inverse. In particular, Hellman and Diffie used the modulo arithmetic idea that A (mod B) equals to the reminder when A is divided by B (See Appendix for explanation of modular arithmetic). For example 16(mod5)=1 because 3 times 5 plus 1 equals to 16, which means 1 is the remainder after 16 is divided by 3 times 5.When ask you the value of 16(mod 5),it is pretty easy to get the only result-1.However,it becomes pretty complicated if you want to track back to know the original number, x with knowing x(mod 5)=1,because x could be many numbers in this case, it only has to be positive integer times 5 plus 1,it could be 6,11,16...the only mean to find the original cipher is to try them all one by one. Because of the special nature of the one-way function: it simple to execute in one direction but to attack or break through it from another direction is extremely difficult, ···. It becomes the basic idea of public-key cryptography, and the detailed procedure will be introduced in the following article.

It is noteworthy that the RSA algorithm is not the first successful case of a public-key cryptography algorithm. In fact, Hellman and Diffie first proposed an algorithm of their own[5], the basic procedure is as follows:

1.Alice and Bob appoints one random number g, for example, g=5, one prime number p, say p=23, together and both g and p are public.

2.Alice and Bob generates their own random number, say a and b, individually and privately, for example a=6 and b=15.

3.Alice calculates ga(mod p),which in here equals to 56(mod23)=8,and send it to Bob. Bob calculates gb (mod P), in here is 515(mod 23) =19, and send it to Alice. (Here Alice and Bob had to send each other some information before they could have the same key, which requires them to be online at the same time.)

4.Alice calculates {[gb(mod p)]a}(mod p) with her private key a. Bob calculates {[ga(mod p)b]}(mod p)with his own private key b. They will then successfully have the same key without tell each other the key in fact,196(mod23) = 815(mod23) = 2.

This article will not spend space here explaining its proof since our main focus is the RSA algorithm. However, one thing worth noting is that there was a flaw with the algorithm: it requires both sides of communication to be online at the same time in order to share the same secret key, as mentioned in step 3, Alice and Bob had to send each other the result of ga(mod p) and gb(mod p)before they could share the same key . Diffie and Hellman couldn' t ignore this flaw.

Diffie then proposed a cutting-edge conception-asymmetric key. Simply speaking, it means the encryption key and decryption key of one algorithm don' t share the same value. The point is, Alice could then make public the encryption key so that her friends all over the world could send e-mail to her using this public key. She could keep the decryption key secret to ensure that any potential intruder wouldn' t be able to decrypt the ciphertext. Though it was a great proposal, Diffie never made his proposal come true. It was three young men from MIT-Rivest, Shamir, and Adleman who implemented Diffie' s proposal. These three young men created one algorithm that is still being widely-used even today - the RSA algorithm.

3. Prepare Your Paper Before Styling

Based on one-way function, Rivest, Shamir and Adleman implemented Diffie' s asymmetric key conception.[6]

The most significant part about RSA is it used the concept of one-way functions to achieve Diffie' s asymmetric key proposal. They used some properties of prime numbers and modular arithmetic in their one-way function. In order to understand the algorithm, we have to introduce these specific properties before going into the algorithm.

3.1 Prime factorization as a one-way function

A prime number is a kind of integer that has only two factors. For example,12 is not a prime number because it could be divided by 1, 2, 3, 4, 6 and 12. However, 17 is a prime number because it can only be divided by 1 and 17. Numbers that are not prime are called composite.

To understand the specific property of prime numbers that RSA uses, we need to understand factorization. Factorization is finding all the numbers that divide a composite number. Factorizing a big composite number could be unimaginably time consuming because the only way of factorization till today is still to try to divide the composite number by all the known prime numbers one by one. For example, dividing the composite number 428,207,868, which equals to 22 x 32x 112 x 197 x 499, takes a lot of time because you have to try with all the known prime numbers one by one, and it will take a lot time trying with smaller factors until you find the bigger factors 197 and 499.

Theoretically, when the composite number is big enough, it will take forever to find all the factors of it. However, compare to the tedious process of factorization, it is much easier to construct the number 428,207,868---just randomly choose a few big numbers on the prime number table and multiply them, you will have a big composite number that is difficult and time consuming to factorize. In other words, it is relatively easy to construct a composite number that could be difficult to factorize---an ideal one-way function.

As we can see in the RSA algorithm below, N is the product of two big prime numbers, p and q, which means N itself is considerably big too. One of the principles for RSA is, you can encrypt the message if you know N, but you can decipher it only if you know p and q. Getting N from p and q is easy, you just multiply one with another, but knowing N, finding p and q becomes much more complicated. As discussed in the previous paragraph, it is hard to divide big composite numbers, especially when their factors are relatively big like p and q here. Therefore, N could be the public key. Alice could publicize them in order to let Bob send messages to her without having to worry about a potential intruder because only she has the two numbers needed to decipher the ciphertext-p and q.

3.2 Modular arithmetic

Modular arithmetic is a process to get the reminder of a division of integers, the divisor of division is called modulus. For example,80(mod 50) = 30 because 30 is the reminder of the division 80÷50=1 reminder 30. In modular arithmetic, calculation from one side is relatively easy for it is still essentially a division of integers. In contrast, it is more difficult to get the original number even knowing the reminder and modulus because there is an infinite amount of numbers that lead to the same reminder. For example, 80 (mod 50) = 30 because 30 is the reminder of the division 80÷50, 130(mod 50) also equals to 30 because 130÷50=2 remainder 30, and so on, and in general, 30+50×N (mod 50) where N is a random integer will always equal to 30. In other words, it is an algorithm that is easy to calculate from one side but nearly impossible to calculate from the other side-another ideal one-way function.

In the RSA algorithm, M stands for the original text and C stands for the ciphertext, in the encrypting step. C is calculated by C = Me (mod n), where n and e make up the public key. To calculate Me (mod n) = C is relatively easy, it is just a deformation of division as discussed before. However, based on the characteristic of modular arithmetic discussed in the former paragraph, there could be infinite Ms even if we know e, n and C. The steps that are required to execute the RSA algorithm are outlined below:

1. STEP 1: CREATE PRIVATE AND PUBLIC KEYS
   First, select two (typically VERY large) and distinct primes, p and q Then compute n = pq.
   Then compute z = ϕ(n) = (p – 1) (q – 1)
   Then choose a number "e", where e > 1 and gcd (e, z) = 1
   The number e is called the "public encryption exponent"
   Then find the unique number d, such that de = 1 (mod z)
   The number "d" is called the "private decryption exponent"
   Make n and e public, and keep p, q, and d secret

2. STEP 2: ENCRYPT A MESSAGE
   Pick a number M, where gcd (M, n) = 1 and 0 < M < n. "M" is called the message text
   Calculate C, where C = Me (mod n) using the public key (i.e. n and e). "C" is called the Cipher text. Send C to Bob.

3. STEP 3: DECRYPT A MESSAGE
   Decrypt C with the private key (i.e. d), by calculating (C)d (mod n) = M (mod n) to get the message text M.

In order to explain how RSA works, we can work through a simple example using small prime numbers. The example is encrypting and decrypting only one letter:

1.We firstly select two random prime numbers, here we choose two small prime numbers p=17 and q=23 in order to explain how RSA algorithm works in a simple way, but p and q are typically very big in practical use. Here n = pq =17 x 23 = 391.

2.We then calculate z = ϕ(n), which means integers smaller and relatively prime to it, and according to Euler's phi function, equals to (p – 1)(q – 1)=352. This step is important for the one-way function.

3.We then choose another number e that is larger than 1 and relatively prime to z, which is 352. We can choose e=503 in this case.

4.Next we find d where de=1(mod z), 503d=1(mod352), in this case we have d=7 [important for one-way function as well, for encryption and decryption]

5.Public key part: n, e as public key, keep p, q, d as private key.

6. Encryption part: Pick a message M where $0<M<n$ and gcd (M, n) = 1, in this case pick 111 as our M. Keep in mind here that RSA only deals with numbers so messages to be encrypted first have to be converted into words. For our example, we encrypt by calculating C = Me (modn), C = 111503 (mod 391) = 359. N and e are public keys. Alice then sends C to Bob.

7.Decryption: by calculating (C)d (mod n) = M (mod n), 359 7 (mod 391) =M (mod 391), the result is M = 111.

As we can see, the asymmetric conception was perfectly achieved in the RSA algorithm, therefore Bob could send a secure message to Alice no matter whether or not Alice is online. Based on the principle of the one-way function-modular algorithm and factorization of big numbers, the security of communication was also ensured. Just as the Diffie-Hellman algorithm, with RSA, the key distribution problem was also no longer bothersome since Alice and Bob only send one-way-function-encrypted ciphertext to each other during the communication.

4. Implementing Rsa – English Messages Using Ascii And Hexadecimal Code

The RSA example from the last section is for only one letter. In that example, we encrypted one letter at a time. However, in the single letter case, we still have to assign a number to the letter we want to encrypt. In the example, that number, M, is 111. Because RSA is a purely mathematical algorithm, we have to use numbers as representatives of the letters in order to encrypt the messages.

In fact, RSA can actually encrypt more than one letter at a time. RSA can encrypt an entire word by combing letters. To encrypt a sentence, we can encrypt words and punctuation. To accomplish that, we have to assign numbers to the letters and punctuation. In English, we use ASCII as numerical medium to transfer the letters and punctuation to numbers, and ASCII uses hexadecimal (see appendix for hexadecimal), range from hexadecimal 1 to ff, which is decimal 1 to 256.

For example, if we want to transmit the English message “To be or not to be?”, we firstly search for each letter or symbol’s ASCII numerical representative. [4]

T:54 o:6f blank:20 b:62 e:65 blank:20 o:6f r:72 blank:20 n:6e o:6f t:74 blank:20 t:62 e:65 ?:3f

Forming the words with letters, we got

To:546f blank:20 be:6265 blank:20 or:6f72 blank:20 not:6e6f74 blank:20 to:546f blank:20 be:6265 ?:3f

Then we combine the words and punctuation together as a sentence, which is 546F206265206f72206e6f742062653f, use this number as M and repeat the RSA procedure, we can successfully encrypt the sentence.

5. Implementing Rsa – Chinese Messages Using Hexadecimal

In order to transmit messages with languages other than Latin-letters-based languages, we use Unicode as their numerical representatives. In order to include numerous languages around the world, the Unicode ranges from 1 to ffff in hexadecimal format, which is 1 to 65535 in decimal format.

In Chinese, there is no letters forming the words. The characters, as the basic units of Chinese written language, directly form sentences with punctuation.

For example, if we want to transmit the Chinese message “生存还是毁灭?”(means “To be or not to be?”),the first step is to find each character and symbol’s numerical representative in Unicode.

生:751f 存:5b58 还:8fd8 是:662f 毁:6bc1 灭:706d ?:003f

Combining together they form Hexadecimal 751f5b588fd8662f6bc1706d003f, and then we just replace M with this number and repeat the process of RSA, we can implement RSA with this Chinese sentence.

6. Discussion And Conclusion

This paper aimed to provide a comparison of the application of the RSA algorithm in Chinese and English. Indeed, RSA algorithm is mathematics-based. Messages in English and in Chinese languages have to be translated into numbers firstly and only then after can we apply the RSA algorithm. However, because of the different characteristics of the two languages themselves, the application of RSA in the two languages are different. There are mainly two differences.

The first difference has to do with the process of constructing sentences. In English, the translation happens letter by letter, the letters make up words and the words then build up the sentences while in Chinese, the translation happens character by character and the characters themselves directly compose sentences, so the application of RSA in Chinese is slightly more troublesome because it takes one more step than in English to form letters to words.

The second difference involves the length of representatives of basic units of written Chinese and English. This second difference derives from the difference of phonography and ideography between the two languages. Latin letters which English and many other western European languages use are based on phonography. In phonography, the alphabets were designed to represent sound therefore the number of basic units of phonography is comparatively limited because of the limitation of variety of pronunciation of human language. There are only 107 letters, 52 diacritics and 4 prosodic marks in the International Phonetic Alphabet 2005 edition and only 26 letters in English.

Chinese characters on the other hand are represented as ideographs. The characters were designed to represent ideas and have little connection with the pronunciation. Compared to the simplicity of pronunciation, there are too many complicated ideas people want to express and the complicity of human thinking generated the impressive amount of Chinese characters. There are 106,230 different characters in The Dictionary of Chinese Variant Form compiled by the Taiwan(ROC) Ministry of Education in 2004.

So apparently the amount of numbers to represent letters and characters are different for English and for Chinese. The ASCII has only 128 different symbols in total, two-digits long in hexadecimal. However, the part of Unicode to represent Chinese characters are four-digits long in hexadecimal which will make the calculation more complicate.

However, the RSA algorithm is a pure-mathematical algorithm. The differences in implementation in English and Chinese only lie in the differences of languages themselves. To computers, the processes of application of RSA in English and Chinese are virtually the same. At the basic level, it is logic and mathematics. Maybe this last assertion can provide us with some clues to the beauty of mathematics. It is also one of the reasons why so many talented people spend their life time study mathematics: its precision and logic go beyond all divergence and differences in human society. It is blind to all different intentions, ideas of good and evil, and racial differences among people who study and apply mathematics.

7. Appendix

Mathematics in RSA encryption and decryption

1.Prime numbers

Prime number is one kind of integer that only has two factors. A factor of a number divides that number exactly. For example 17 is a prime number because only 1 and 17 could exact divide it. However 15÷5=3, so 3 and 5 are factors of 15, and this also proves that 15 is not a prime number., Integers that are not prime numbers are called composite numbers.

2.GCD

GCD simply means Greatest Common Divisor. Divisor and factor are synonyms. They have the same meaning. If A divides B exactly, A is B’ s divisor or factor. For example, 20 has divisors 1, 2, 4, 5, 10, 20 and 24 has divisors 1, 3, 4, 6, 8, 12, 24. 1 and 4 are common divisors of both 24 and 20, therefore 4 is their greatest common divisor. Some countries say greatest common factor instead of greatest common divisor. When two integers share 1 as their GCD, those two numbers are relatively prime to each other.

3.Factorizing

Factorizing means to find one number’ s factors. Unfortunately, although there appears to be certain characters with small factors, we still have to try all the prime numbers one by one to find all of a number’ s factors. That is the reason we could view factoring multiple two big prime numbers as a one-way function. Multiplying is quite easy but finding the factors of their product is much more difficult.

4.Modular arithmetic

Modular arithmetic is actually in our daily life., Time is the most apparent example. For example, it is 1:30pm now, it should be 2:20pm after 50 minutes instead of 1:80. Because apparently minutes goes from 0 to 59 only, when it comes to 60 it will jump into another hour, so when 30 minutes have passed from 1:30 it has already jump into 2:00. Modular arithmetic means you divide X by Y, you get the remainder after it is divided. For example, in this case,30 plus 50 is 80,and you divide 80 with 60,you get the result 1 and the remainder 20,80(mod 60)=20 because 80=1 x 60+20.Modular arithmetic is a cycle, for example, 20 here can appear infinite times,140(mod60) also equals to 20 because 140=2 x 60 +20, in other words, if we know the value of a and b and the equation x=a(mod b) and we want to find x, it becomes quite difficult, because x could equal to any only if a + bc only if c is a integer, there could be infinite amount of values for x, that makes modular arithmetic a one-way function.

5.Euler’ s phi function

Φ(n)=the number of all the integers smaller than a positive integer N and relatively prime to it (means they only share 1 as their only common divisor). For example, ϕ(8)=4,because 1,3,5,7 are smaller than 8 and relatively prime to it.

Euler’ s phi function states that when a, n are positive integers and GCD (a, n) = 1, a ϕ(n)=1(mod n)

Let n and a be positive, relatively prime integers. Then a ϕ(n) = 1 (mod n).

6.Fermat

It is actually a special case of Euler’ s phi function. When n is prime, ϕ(n)=n-1 because prime numbers are relatively prime to all integers besides 1.

So when GCD (n, a) = 1 a(n-1) = 1(mod n).

7.Chinese Remainder Theorem

Let m, n be integers satisfying gcd (m, n) = 1. Then the simultaneous congruence x = b (mod m) and x = c (mod n) have exactly one solution for $0 < x < mn$.

8.Number Systems.

For any X-digits number system, when the digit comes to x, plus 1 to the higher digit place and the number on the original number position goes back to 0. For example, in decimal number system, when 9 plus 1, adding 1 to the tens digit and the units digit goes back to 1, which is 10.

This becomes abstract to understand when it comes to number systems other than decimal. For example, in the hexadecimal format, we use a to f to represent 10 to 15, when 9 pluses 6,it becomes f (Decimal 15)instead of 15 because there could be 16 digits in the units digit! ff in hexadecimal format equals 15 x 15+15=255 in decimal format.

8. Acknowledgment

We would like to thank Illume.

9. References

[1] Luciano & Prichett (1987) Cryptology - from Caesar ciphers to public key cryptography
[2] Singh, S. (2000). The code book: the science of secrecy from ancient Egypt to quantum cryptography. London: Anchor
[3] Gardiner, M (1977). A new cipher that would take millions of years to break. Scientific American, 237(8), 120 - 124.
[4] Unicode Consortium. C0 Controls and Basic Latin.(2018)http://unicode.org/charts/
[5] Diffie & Hellman (1976) New directions in cryptography
[6] Rivest et al. (1978) A method for obtaining digital signatures and public key cryptosystems
[7] Hellman (1979) The mathematics of public key cryptography

IOP Conf. Series: Journal of Physics: Conf. Series **1298** (2019) 012012 doi:10.1088/1742-6596/1298/1/012012

A New Method Of Human Brain Segmentation Utilizing A Class Of Power Series Solutions Of Fractional Differential

SAMIR B. HADID[1], RABHA W. IBRAHIM[2] & NORSHALIZA KAMARUDDIN[2]

[1] Ajman University, College of Education and Basic Sciences, UAE
[2] Cloud Computing Center, University of Malaya, Kuala Lumpur, 50603, Malaysia

Email: s.hadid@ajman.ac.ae

Abstract: A biological dynamic system is very useful in engineering possessions such as control systems, signal processing, and bio-molecular communication networks. On the other hand, fractional differential equations are one of the main subjects that are used to solve problems in biological dynamic systems. In the current paper, we investigate a new class of fractional dynamical systems of brain segmentation. The application shows the improvement of segmentation of a class of brain images.
MSC (2010): 26A33; 34K37

1. Introduction

Fractional operators [1] induced the possessions of memory and heredity of deliveries. Applied difficulties include descriptions of fractional concept allowing the procedure of initial conditions. Fractional time results are related with sub-diffusion of fractional idea, where elements feast more gradually than a traditional diffusion. The fractional operators are operated to model fractional diffusion or distribution where elements buffet at a rate not in arrangement with the traditional Brownian motion model [2]. This operator showed its soundness in all sciences including the medical sciences. The area of medical images has been established and adapted by all fractional operators.

In image processing, the image segmentation is a technique of separating a digital image into multiple pixels. The purpose of this idea is to reduce the illustration of a feature into roughly that is more important and informal to study [3]. Image segmentation is naturally utilized to find bits, pieces and boundaries such as curves, lines, area,... etc in features. Specially, the technique of transmission in a feature such that all pixels remain with fixed segment. Fractional image segmentation is planned by applying different methods from fractional calculus [4-12]. In the current study, we consider the existence of solution of the nonlinear fractional problem with special type by using power series solution. The application is suggested to improve the segmentation of a class of brain images. Comparisons are illustrated with recent studies in fractional segmentation.

2. Fractional processing

Consider the differential equation

$$y^{(\alpha)}(x) = f(x,y), \qquad \alpha \in \Re^{+}, \tag{1}$$

with the initial condition

$$x^{(\alpha-1)}(0) = C\,, \tag{2}$$

where $\Re^+$ is the set of positive real numbers and f is a function from $I \times E$ into E, E being an Euclidean space, and C is a constant. In this section, we shall obtain a solution for the differential equation:

$$y^{(\alpha)}(x) + Cx^{k-\alpha}y(x) = 0\,, \tag{3}$$

where k is a real constant and $\alpha \in \Re^+$.

First, we required the following definition [1]:

2.1 Definitions

Let f be a function which is defined almost everywhere on [a,b]. For $\alpha > 0$, define the following integral operator

$$\overset{b}{\underset{a}{I}}{}^{\alpha} f = \frac{1}{\Gamma(\alpha)}\int_a^b (b-t)^{\alpha-1} f(t)dt \tag{4}$$

provided that this integral (Lebesgue) exists, where Γ is the Gamma function.

Consequently, the derivative is defined as follows: If $\alpha \in R$, and f is defined on $[a,b]$, we denote $f^{(\alpha)}(x) = \overset{x}{\underset{a}{I}}{}^{-\alpha} f$ for all $x \in (a,b]$, provided that $\overset{x}{\underset{a}{I}}{}^{-\alpha} f$ exists. Fix $\alpha > 0$, and let $n \in N$ be such that $n-1 \le \alpha < n$. From the previous definitions, we have $f^{(\alpha)}(x) = \overset{x}{\underset{a}{I}}{}^{-\alpha} f = D_x^n \overset{x}{\underset{a}{I}}{}^{\alpha-n} f = \frac{d^n}{dx^n}\frac{1}{\Gamma(n-\alpha)}\int_0^x (x-t)^{n-\alpha-1} f(t)dt$ (4). Let $b > -1$. Note that for a power function we have:

$$(x^b)^{\alpha} = \begin{cases} 0 & \text{if } b \in \{\alpha-1, \alpha-2, \dots, \alpha-n\} \\ \Gamma(b+1)/\Gamma(b-\alpha+1)\}x^{b-\alpha} & \text{otherwise} \end{cases}$$

Hence, we shall prove the following proposition:

2.2 Proposition

Let $k \ge n$, and let C be a real constant. Then the equation:

$$y^{(\alpha)}(x) + Cx^{k-\alpha}y(x) = 0\,, \tag{5}$$

has a solution of the form

$$y(x) = x^{\alpha-n}\sum_{i=0}^{\infty} a_i x^i\,, \tag{6}$$

where

$$a_i = \begin{cases} C & \text{if } 0 \le i \le n-1 \\ 0 & \text{if } n \le i \le k-1 \end{cases}, \tag{7}$$

where C is a constant and

$$a_{i+k} = -Ca_i(i+k-n)!/\Gamma(i+k-n+\alpha+1), \text{ if } i \geq k. \tag{8}$$

Proof: Let $y(x) = x^{\alpha-n}\sum_{i=0}^{\infty} a_i x^i$, where $a_i (i = 0,1,2,...,)$ are define in (7). We consider independently k different subsequences of the sequence $\{a_i\}_{i=0}^{\infty}$. In general, we obtain the sequence $\{a_{j+ki}\}_{i=0}^{\infty}$, $j = 0,1,2,...,k-1$. If $n \leq j < k$, then the subsequence $\{a_{j+ki}\}_{i=0}^{\infty}$ consists of zeros, i.e. $a_{j+ki} = 0$ for all i. Suppose now that $0 \leq j < n$. We have

$$\lim_{i\to\infty} \frac{a_{i+k}}{a_i} = 0, \qquad \lim_{x\to\infty} \frac{\Gamma(x)}{\Gamma(x+\varepsilon)} = 0,$$

for all $\varepsilon > 0$. Therefore, by using the d'Alamberts criteria of convergence of the series, the series is convergent for all $x \in R$. Moreover, it is uniformly convergent in every finite interval $(-r,r), r \in R.$ Thus, we have:

$$y^{(\alpha)}(x) = \frac{d^n}{dx^n}\frac{1}{\Gamma(n-\alpha)}\int_0^x (x-t)^{n-\alpha-1} t^{\alpha-n}\sum_{i=0}^{\infty} a_i t^i dt$$

use the change of variables $t = xs$, we get $y^{(\alpha)}(x) = \frac{d^n}{dx^n}\frac{1}{\Gamma(n-\alpha)}\sum_{i=0}^{\infty} a_i x^i \int_0^1 (1-s)^{n-\alpha-1} s^{\alpha-n+i} ds$

$$= \frac{d^n}{dx^n}\sum_{i=0}^{\infty} a_i x^i \frac{\Gamma(\alpha-n+1+i)}{i!}$$

$$= \sum_{i=n-k}^{\infty} \frac{\Gamma(\alpha+i+k-n+1)}{(i+k-n)!} a_{k+i} x^{i+k-n}$$

$$= \sum_{i=0}^{\infty} \frac{\Gamma(\alpha+i+k-n+1)}{(i+k-n)!} a_{k+i} x^{i+k-n}$$

$$= -Cx^{k-\alpha}\left(x^{\alpha-n}\sum_{i=0}^{\infty} a_i x^i\right) = -Cx^{k-\alpha} y(x)$$

We could do the above calculation (in particular, the change $(\int \sum = \sum \int)$ because all the series considered above are uniformly convergent on a finite subintervals in the set of real number. This completes the proof.

3. Results and Discussion

In this section, we apply the numerical solutions (6) in a domain. The aim of the technique is to reduce the energy at the outline position in the boundary location where the energy is reduced. The fractional Gaussian kernel (see [4-6]) is utilized in the suggested process to improve the feature texture and retain its construction. Normally, as the usual Gaussian filter is employed, reduce of information sits down as the parameter rises. The fractional Gaussian kernel also increases the arrangement of the inhomogeneous substances in a region. Moreover, it inserts the local feature data within the region to remove the body border throughout the segmentation procedure. In our analysis, we utilize the fractional Gaussian kernel of order $0<\alpha<1$ for two variables (x, y):

$$G_\alpha(x,y)= (1/ (2\pi)^{n/2}\alpha^n) \exp (-(x^2+y^2)/2\alpha^2)$$

The four-sided window (rectangle) changes during the feature to develop the image texture. The window's tool in the planned method does not have specific sizes as it depends on the topology movements in the image. The window effort is basically measured inside the definite domain of the segmented segment, which covers the segment margin angle. The steps of the technique are as follows:

1. Initialization of contour : $I_0(x,y)$, where I(x,y) is the image;
2. Computing the fractional Gaussian kernel : $G_\alpha(x,y)$ for different values of the fractional power $0<\alpha<1$;
3. Illustrate the window by using the coefficient of the solution in (6);
4. Calculate $a_{i+k} = -Ca_i(i+k-n)!/\Gamma(i+k-n+\alpha+1)$, if $i \geq k$. We shall use C=1.
5. Evaluate the fractional gradient of the image Δ^α I(x,y) from equation (5);
6. Minimize the energy, by applying the fractional integral equation

$$E_\alpha = \int [G_\alpha(x,y) * \Delta^\alpha I(x,y)]\, dx\, dy$$

7. Measure the contour, by minimizing E_α ; otherwise go to step 2.

The experiment here is considered utilizing Matlab, R (2008b) on a 2.5 GHz Intel Processor i5 giving to the application outline.

For $\alpha= 1$, we get the exact result of the problem (5). Fig. 1 displays that whenever α growth the means absolute error (MAE) reductions and therefore, we obtain the exact outcome at $\alpha= 1$. From above applications, we deduct that the outcomes are approximated to the extended polynomial (6). There are three categories of assessments which are sensitivity, accuracy and specificity [6]. Nevertheless, we focus on the accuracy measure. To manage the assessment, two features of identical medical image are utilized. The first feature is a segmented feature formed by the certain entire active contour model involving ours. The second feature is the medical image with physically tired border. The correctness of the segmentation result is estimated by calculating the space of the physically tired border to the automatic boundary shaped by the active contour model.

4. Conclusions

An investigation has been made of the effects of fractional calculus in segmentation. We considered a new method based of polynomial outcomes of fractional power. The best value of this fractional power is $\alpha=0.88<1$.

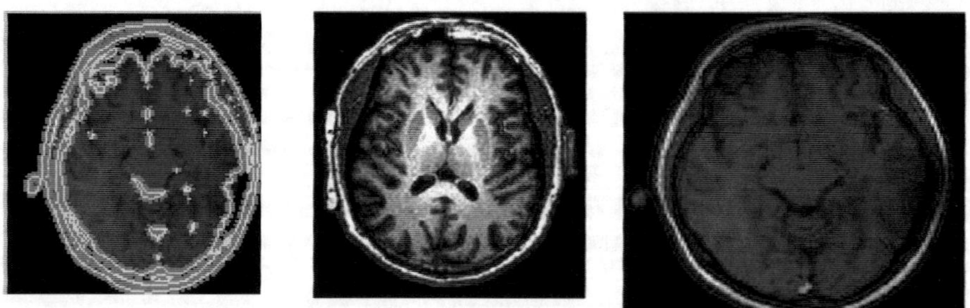

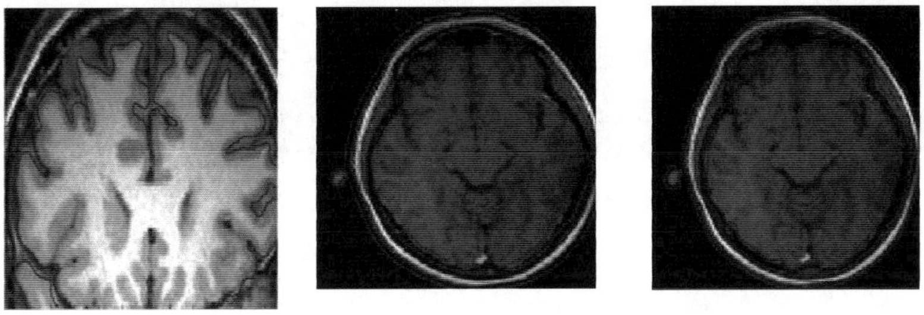

Figure 1. The first row is the experiments and results on images of brain α=0.88, C=1, 0.1< a_i <1 by using the proposed method. The second row is the results given in [4], [5] and [6] respectively.

5. References

[1] Kilbas A.A., Srivastava H. M. .; and Trujillo J.J.(2006), *Theory and Applications of Fractional Differential Equations*. North-Holland, Mathematics Studies, Elsevier.

[2] Ibrahim R.W., (2017) *Fractional Calculus of Multi-objective Functions & Multi-agent systems.* LAMBERT Academic Publishing, Saarbrcken, Germany.

[3] Jalab, H. A. .; and Ibrahim, R. W.(2015), Fractional Alexander polynomials for image denoising, *Signal Processing*, 107, 340-354.

[4] Ibrahim, R. W.; Moghaddasi, Zahra; Jalab, Hamid A J. (2015) Fractional differential texture descriptors based on the Machado Entropy for image splicing detection, *Entropy,* 17(7), 4775-4785.

[5] Kamaruddin N., Abdullah A., .; and Ibrahim R.W.(2016.), Local region-based ACM with fractional calculus for boundary segmentation in images with intensity inhomogeneity, *Malaysian* Journal *of Computer Science*,29(2) 124-144,

[6] Ibrahim, R. W. Nashine, H. K.; and Kamaruddin, N.(2017), Hybrid time-space dynamical systems of growth bacteria with applications in segmentation, *Mathematical Biosciences*, 291, 10-17.

[7] Ghamisi, P., et al. (2014): Multilevel image segmentation based on fractional-order Darwinian particle swarm optimization. *IEEE Transactions on Geoscience and Remote sensing* 52.5 , 2382-2394.

[8] Acharya, A., Uddipan M., and Charless F. (2016), On Image segmentation using Fractional Gradients-Learning Model Parameters using Approximate Marginal Inference." *arXiv preprint arXiv:1605.02240.*

[9] Clark, H. D., Reinsberg, S. A., Moiseenko, V., Wu, J. and Thomas, S. D. (2017). Prefer Nested Segmentation to Compound Segmentation. *arXiv preprint arXiv:1705.01643.*

[10] Hemalatha, S., and Anouncia, S. M. (2018). Partial Fractional Derivative (PFD) based Texture Analysis Model for Medical Image Segmentation. In *Knowledge Computing and Its Applications* (pp. 213-225). Springer, Singapore.

[11] Hadid, S.B. and Merad A., Variational Iteration Method for Solving Problems with Integral Boundary Conditions", (2018) Springer Proceeding in Mathematics and Statistics, Differential and Difference Equations with Applications, Vol. 230, and pp: 133-140.

[12] Ibrahim, R. W and Hadid S. B (2018)., The EOQ Model: A differential cyclic system for calculating Economic Order Quantity, International Journal of Analysis and Applications (ISI), 16(3) (2018), 437-444. Canada.

Image Processing of Case Study of Solar Radio Burst Type III and Type IV Burst Using Sunpy

Z. S. Hamidi[1,3] , N.Ramli[2] ,N. N. M. Shariff [1,4]

[1]Institute of Science, Universiti Teknologi MARA, 40450, Shah Alam, Selangor, Malaysia
[2]Department of Physics, Faculty of Science, University of Malaya, 50603, Kuala Lumpur, Malaysia
[3]School of Physics and Material Sciences, Faculty of Applied Sciences, Universiti Teknologi MARA, 40450, Shah Alam, Selangor, Malaysia
[4]Academy of Contemporary Islamic Studies (ACIS), Universiti Teknologi MARA, 40450, Shah Alam, Selangor, Malaysia

Email:zetysh@salam.uitm.edu.my

Abstract. Solar radio burst at low-frequency lies in the fact that they originate in the same layers of the solar atmosphere in which geo-effective disturbance probably originate Type III bursts have consistently the fastest drift rates of bursts at metric wavelengths and was created by an electron beam through plasma emission. Meanwhile, type IV burst is broadband quasi-continuum features associated with the decay phase of solar flares. A geomagnetic storm is a major disturbance of Earth's magnetosphere that occurs when there is a very efficient exchange of energy from the solar wind into the space environment surrounding Earth. The aim of this study is to find out whether type III bursts preserved energy to solar bursts type IV thus led to the formation of geomagnetic storm. The solar radio burst type III and IV were observed by CALLISTO Spectrometer. CALLISTO is an acronym from Compound Astronomical Low cost Low frequency Instrument for Spectroscopy and Transportable Observatory. The Sunpy by Python was used to analyze the spectrogram image by eliminate the unwanted signal or noise in the image which is in the presence of noise, the spectrum gives inaccurate reading of frequency and time. Although the study did not show much correlation between bursts type IV and type III in producing these storms, there are several possible explanations from this result. It may be the case that solar events such as solar flare, CMEs and active region ejected from the solar disc is not heading towards Earth and the energy has eventually disperses into the interplanetary medium.

1. Introduction

One of the main reason detecting solar radio burst because it is the important tool for specifying magnetic, thermal and density structures at the time. Solar radio burst at low-frequency lies in the fact that they originate in the same layers of the solar atmosphere in which geo-effective disturbance probably originate [1]. This is where the solar flares and coronal mass ejections (CME) are believed to be launched. Solar flare is sudden large energy releases in the magnetic active region are the most active manifestation of solar activity [2, 3]. It might define as a rapid brightening in H alpha, but simultaneously can have manifestations right across the electromagnetic spectrum and might eject the high energy particles and blobs plasma into the solar wind. Coronal mass ejections (CMEs) are huge explosions of magnetic field and plasma from the Sun's corona [4]. A CME is a large scale coronal structure ejected from the Sun and propagate into the interplanetary space [5]. Certain CMEs may intercept the Earth orbit if they move toward the direction of the Earth, causing space weather. Fast CME eruptions are often accompanied with flares and filament eruptions [6].

This study was focused on solar burst type III and type IV in relation to solar activities [7]. Solar burst type III is the radio signatures of the solar flare accelerated electron beams ejected into the corona and interplanetary medium which are characterized by very fast frequency drifts in the dynamic spectra [3]. Another potential use of type III burst is that they always have a significant relationship with solar flare events [8]. Type III bursts have consistently the fastest drift rates of bursts at metric wavelengths and was created by an electron beam through plasma emission [9]. The radio emission tracks the electron beam as it travels through the decreasing plasma density of the solar corona and solar wind.

Meanwhile, type IV burst is broadband quasi-continuum features associated with the decay phase of solar flares [10]. Solar burst type IV always related to the development of sunspot groups [11]. It has been well established that type IV bursts have a high probability of being followed by geomagnetic disturbance [12]. It generally believed to be synchrotron radiation from relativistic electrons spiraling in magnetic fields and results from trapped electron populations [13]. There are several subclasses of type IV bursts, the type most tightly coupled to flares is called type IV continuum. The centimeter-wavelength bursts, particularly the intense ones, are closely associated with solar flares [14].

The Earth's surrounding known as a Space Weather is the conditions of the solar wind speed and density, magnetic field strength and orientation and energetic particle levels [15]. The high energies and magnetic field strengths from the Sun travelling at up to thousands of km s^{-1} impact the Earth's magnetosphere that might be harmful to the Earth technology [16]. The correlation of solar radio burst with solar activity play an important role in space weather study. The information gathered from the relation of solar burst type III and type IV with solar activity may be useful.

The relationship between solar radio burst type III and IV being followed by geomagnetic disturbance cannot be denied anymore [17]. The authors mention the significant relationship between type IV burst and geomagnetic storm. But, does the presence of burst type III preceded type IV burst affect the production of geomagnetic storm? The valuable information may be obtained from this study regarding the occurrence of geomagnetic storm with solar radio burst type III. In addition, such information may also be very important and useful for solar monitoring.

2. Geomagnetic Storm

A geomagnetic storm is a major disturbance of Earth's magnetosphere that occurs when there is a very efficient exchange of energy from the solar wind into the space environment surrounding Earth [18]. These storms result from variations in the solar wind that produces major changes in the currents, plasmas and fields in Earth's magnetosphere. The solar wind conditions that are effective for creating geomagnetic storms are sustained (for several to many hours) periods of high-speed solar wind.

Geomagnetic storm is defined by changes in disturbance storm time (Dst) index, has been used to characterize the size of a geomagnetic storm. However, there are currents produced in the magnetosphere that follow the magnetic field and connect to intense currents in the auroral ionosphere. These auroral current produce large magnetic disturbances on the ground used to generate a planetary geomagnetic disturbance index called Kp. This index is used globally to describe space weather that can disrupt systems on Earth. The mean planetary Kp index for geomagnetic storm is nT with interval for 3-hours. The Kp-index ranges from 0 to 9, where a value of 0 means very little geomagnetic activity and a value of 9 means an extreme geomagnetic storm. For geomagnetic storm, it has 5 indexes of Kp started from level G1 (minor storm), G2 (moderate storm), G3 (strong storm), G4 (severe storm) and G5 (extreme storm).

3. Methodology

The radio spectrometer of the CALLISTO (Compound Astronomical Low-cost Low-frequency Instrument for Spectroscopy and Transportable Observatory) was used in this research for monitoring the solar radio burst [19]. The instruments observe automatically, their data is collected every day via internet and stored in a central data base. The 'burst mode' can be triggered from the spectrograph of CALLISTO automatically [20]. The e-CALLISTO system has already proven to be a valuable new tool for monitoring solar activity and for space weather research [21].

The solar radio burst type III and IV were observed by CALLISTO Spectrometer. CALLISTO is an acronym from Compound Astronomical Low cost Low frequency Instrument for Spectroscopy and Transportable Observatory. The CALLISTO spectrometer was designed by an electronics engineer, Christian Monstein from the Institute for Astronomy of the Swiss Federal Institute of Technology Zurich (ETH-Zurich). The radiometric bandwidth is about 300 kHz and the integration time for this spectrometer is 1 msec with 0.25 sec time resolution of 200 channels per spectrum (800 pixels per second). The instruments were placed at a low-level noise location to reduce system noise and the preamplifier was connected to the spectrometer to amplify a gain. In term of the sensitivity, it is depending on the instrument whether it connected to the small or the large antenna.

The Sunpy by Python was used to analyze the spectrogram image by eliminate the unwanted signal or noise in the image which is in the presence of noise, the spectrum gives inaccurate reading of frequency and time. In order to remove the unwanted signal, the important thing needs to consider is to ensure the valuable signal not removed. Thus, give a quality data that has not affected the value of the data. Reduced the noise while maintaining its spectral characteristics. This problem can be overcome by adjusting the V_{min} to normalize the colour.

```
%%%%EXAMPLE FROM LOCAL FOLDER

from matplotlib import pyplot as plt
from matplotlib import cm
import sunpy
from sunpy.spectra.sources.callisto import CallistoSpectrogram
image = CallistoSpectrogram.read("M:\MyPython\BLEN7M_20120813_123000_24.fit.gz ")
image = image.linearize_freqs(1)
nobg = image.subtract_bg()
nobg.plot(cmap=cm.gist_ncar,vmin=5)# cm.hot, cm.spectral, cm.gist_ncar, cm.jet
plt.ylabel("Frequency [MHz]")
plt.xlabel("Time [UT]")
plt.title("Bleien Switzerland")
plt.show()
```

Figure 1. Coding of sunpy for data analysis

4. Results and Discussions

The 3rd August 2011 solar burst event showed a type IV burst was preceded and followed by type III burst. A group of solar radio burst type III at 0817-0818 UT from the frequency range of 60-70 MHz that was detected by BIR station. Two hours later, a smooth continuum moving type IV with strong structure appear at 1030-1349 UT at the frequency range of 160-900 MHz that were detected from BLEN7M, HUMAIN, INPE and MRO stations. This type IV burst then was followed by single burst type III which is occur above 30-700 MHz at 1200-1904 UT. A single type III burst were recorded at BIR, BLEN7M, and OOTY stations.

There was a flare M6-class at 1348 UT blast from sunspot 1261. The magnetic storm associated with this complex radio activity had level G4 from 2000 to 2400 UT are about 48 hours later. A CME produced by this event is heading directly toward Earth as recorded by SOHO. It was travelling slowly with speed 356 km/s with 122 sfu radio flux. The number of sunspot is 98 and the interplanetary

IOP Conf. Series: Journal of Physics: Conf. Series **1298** (2019) 012013 doi:10.1088/1742-6596/1298/1/012013

magnetic field is 3.0 nT appear at 00:30:05 and onset at 21:28:51 UT with mass 7.20e+13 g. There are 3 sunspot during this event which are AR1263, AR1261 and AR1260.

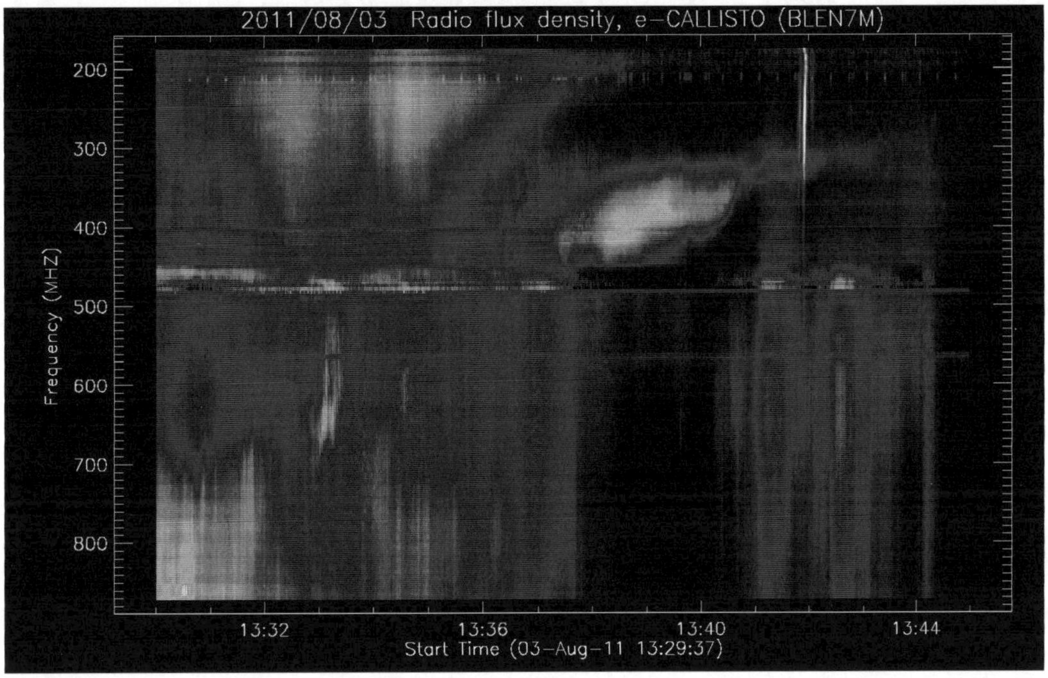

Figure 2. Solar Radio Burst Type IV on 3th August 2011 and a group of solar radio burst type III from 13:32 UT to 13:44 UT.

The event was associated with M6 class flare erupted from AR1261 active region. The type IV burst stop almost exactly at the time of the soft X-ray peak at 1348 UT, regarded as the time at which energy release in the flare ends. As CME travelling towards Earth with speed 296 km/s, it reached Earth 48 hours later. In addition, severe geomagnetic storm level G4 and strong storm level G3 were recorded from 1800 UT to 0000 UT on 5 August 2011. There are 5 Active Region recorded during this event with 98 sunspot number. It was reported an ionospheric waves and VHF radio noise from the listening station above the Arctic Circle in Norway. The M1 class eruption from the active region AR1261 was observed by three spacecraft: SOHO, STEREO A and STEREO b. It was also recorded that the CME elft the Sun with the speed of 900 km/s and reach the Earth on 5 August 2011. A mild moderate geomagnetic storm are detected on 5th and 6th of August 2011.

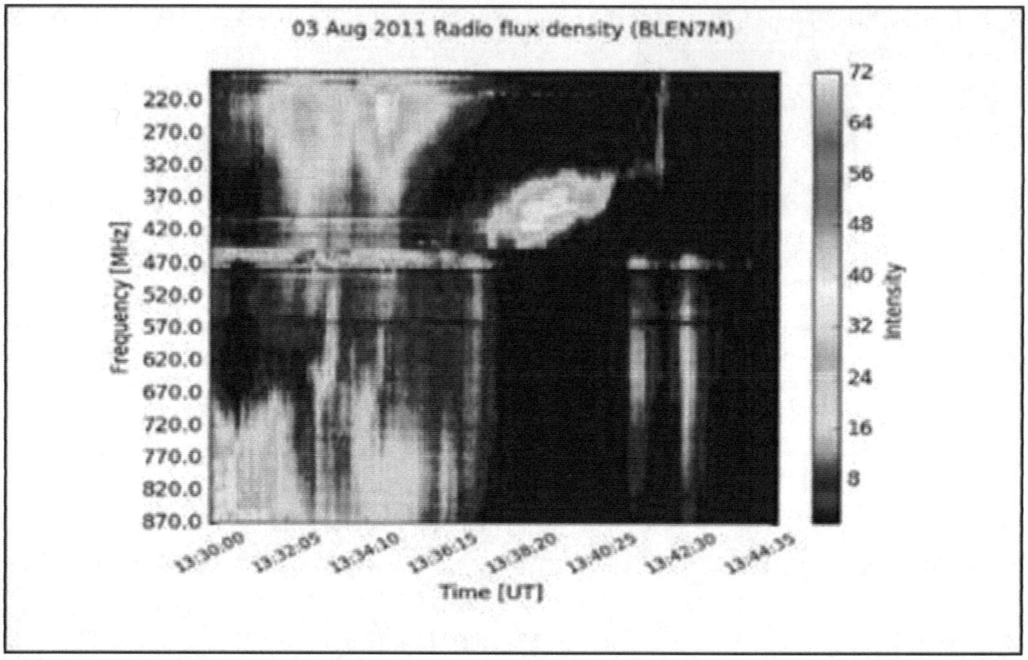

Figure 3. Solar Radio Burst Type IV on 3th August 2011 and a group of solar radio burst type III from 13:32 UT to 13:44 UT after using Sunpy

The data of geomagnetic storms was obtained from website Laboratory of X-Ray Astronomy of the Sun, LPI (link: http://www.tesis.lebedev.ru/en/). The TESIS is a set of solar imaging instrument developed by the Lebedev Physical Institute of the Russian Academy of Science and launched aboard the Russian spacecraft CORONAS-PHOTON in January 30, 2009. It showed the data from the last 3 days of magnetic storms with three indicators such as geomagnetic calm, geomagnetic disturbances, and geomagnetic storm. Other than, the data of geomagnetic storm from WDC for Geomagnetism, Kyoto also collected for comparison.

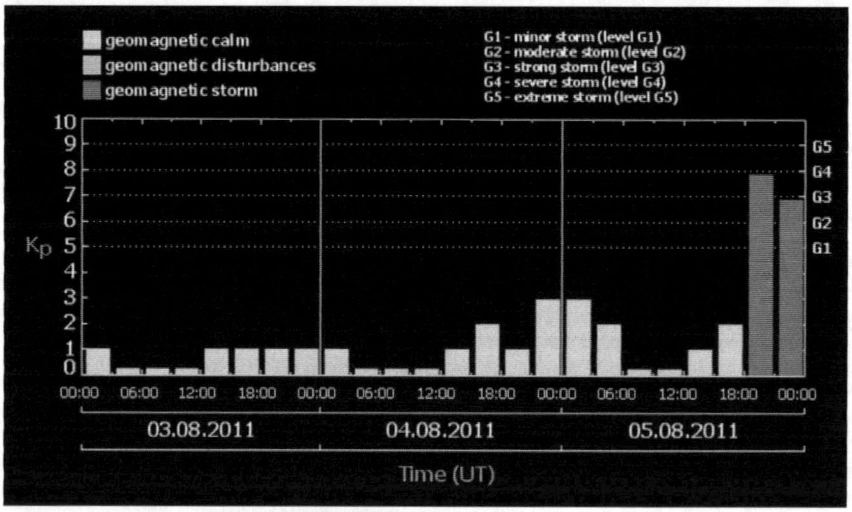

Figure 4. The geomagnetic storm on 5th August 2011

5. Conclusions

In this study several solar and geomagnetic activity indices were analyzed along with the events of solar, solar wind and magnetospheric origin. It has been showed that type IV bursts have a high probability of being followed by geomagnetic disturbance. The occurrence of type IV alone (without type III) is not enough, since the corresponding solar plasma cloud behind the shock front has too week a particle density to cause an observable geomagnetic effect. Similarly, the occurrence of a type III without type IV burst may not be sufficient to influence the geomagnetic activity, since it has no shock front associated with it. This association between type III and type IV solar radio burst can be explained by the fact that both of this solar radio burst are generated by fast electrons. The disturbance which gives rise to type III bursts have high velocities and may be either streams of electrons or fast, transverse shock. While, the electrons which give rise to type IV are principally have a similar velocity range.

6. Acknowledgments

We are grateful to CALLISTO network, STEREO, LASCO, SDO/AIA, NOAA and SWPC make their data available online. This work was partially supported by the grant, 600-IRMI/DANA 5/3/BESTARI (067/2017), UiTM grants and Kementerian Pengajian Tinggi Malaysia. Special thanks to the National Space Agency and the National Space Centre for giving us a site to set up this project and support this project. Solar burst monitoring is a project of cooperation between the Institute of Astronomy, ETH Zurich, and FHNW Windisch, Switzerland, Universiti Teknologi MARA and University of Malaya.

7. References

[1] Hamidi, Z., et al. *Dynamical structure of solar radio burst type III as evidence of energy of solar flares*. in *AIP Conference Proceedings*. 2013. AIP.

[2] Carmichael, H.J.N.S.P., *A process for flares*. 1964. **50**: p. 451.

[3] Hamidi, Z. and N. Shariff. *The Non-thermal of a Large Solar Flares Associated with High Energy of Solar Burst Type III on 4th September 2017*. in *Journal of Physics: Conference Series*. 2018. IOP Publishing.

[4] Shariff, N., Z. Hamidi, and N. Zainol. *Analysis of Slow Partial Halo CME Events with Velocity of 100-200 km/s: An Observation through CALLISTO System*. in *Industrial Engineering, Management Science and Application (ICIMSA), 2017 International Conference on*. 2017. IEEE.

[5] Gosling, J.T.J.P.o.m.f.r., *Coronal mass ejections and magnetic flux ropes in interplanetary space*. 1990. **58**: p. 343-364.

[6] Kahler, S.J.A.r.o.a. and astrophysics, *Solar flares and coronal mass ejections*. 1992. **30**(1): p. 113-141.

[7] Ramli, N., et al. *The relation between solar radio burst types II, III and IV due to solar activities*. in *Space Science and Communication (IconSpace), 2015 International Conference on*. 2015. IEEE.

[8] Hamidi, Z., et al. *Signal detection of the solar radio burst type III based on the CALLISTO system project management*. in *Industrial Engineering, Management Science and Application (ICIMSA), 2016 International Conference on*. 2016. IEEE.

[9] Hamidi, Z., et al. *Magnetic Reconnection of Solar Flare Detected by Solar Radio Burst Type III*. in *Journal of Physics: Conference Series*. 2014. IOP Publishing.

[10] White, S.M.J.A.J.o.P., *Solar radio bursts and space weather*. 2007. **16**: p. 189-207.

[11] Sabri, S., et al. *The dependence of log periodic dipole antenna (LPDA) and e-CALLISTO software to determine the type of solar radio burst (I-V)*. in *Industrial Engineering, Management Science and Application (ICIMSA), 2016 International Conference on*. 2016. IEEE.

[12] Melrose, D.J.S.P., *Plasma emission due to isotropic fast electrons, and types I, II, and V solar radio bursts*. 1975. **43**(1): p. 211-236.

[13] Dulk, G.A.J.A.r.o.a. and astrophysics, *Radio emission from the sun and stars*. 1985. **23**(1): p. 169-224.

[14] Donnelly, R.F.J.S.p., *Extreme ultraviolet flashes of solar flares observed via sudden frequency deviations: experimental results.* 1971. **20**(1): p. 188-203.

[15] Mead, G.D.J.J.o.G.R., *Deformation of the geomagnetic field by the solar wind.* 1964. **69**(7): p. 1181-1195.

[16] Echer, E., et al., *Introduction to space weather.* 2005. **35**(5): p. 855-865.

[17] Veselovsky, I., et al., *Solar and heliospheric phenomena in October–November 2003: causes and effects.* 2004. **42**(5): p. 435-488.

[18] Kappernman, J.G. and V.D.J.I.s. Albertson, *Bracing for the geomagnetic storms.* 1990. **27**(3): p. 27-33.

[19] Abidin, Z.Z., et al., *Radio frequency interference in solar monitoring using CALLISTO.* 2015. **67**: p. 18-33.

[20] Zainol, N., et al. *Effective data collection and analysis of solar radio burst type II event using automated CALLISTO network system.* in *Industrial Engineering, Management Science and Application (ICIMSA), 2016 International Conference on*. 2016. IEEE.

[21] Husien, N., et al. *Solar Radio Bursts Detected by CALLISTO System and Their Related Events.* in *Industrial Engineering, Management Science and Application (ICIMSA), 2016 International Conference on*. 2016. IEEE.

IOP Conf. Series: Journal of Physics: Conf. Series **1298** (2019) 012014 doi:10.1088/1742-6596/1298/1/012014

Image Feature Extraction Method to Analyze Soft-Mode Turbulence Fluctuation in Nematic Liquid Crystal

Risti Suryantari[1], Yusril Yusuf[2], and Flaviana[3]

[1,3] Department of Physics, Faculty of Information Technology and Science, Parahyangan Catholic University, Jl Ciumbuleuit 94, Bandung , Indonesia 40141
[2] Department of Physics, Faculty of Mathematics and Sciences, Gadjah Mada University, Jl Bulak Sumur, Yogyakarta, Indonesia

Email: ristisuryantari@unpar.ac.id

Abstract. An electroconvective pattern called soft-mode turbulence (SMT) of homeotropic alignment in MBBA (4'-Methoxy-benzylidene-4-butyl-aniline) nematic liquid crystal appears when an external electric field with a certain frequency applied on the nematic liquid crystal sample at a threshold voltage (V_{SMT}). The threshold voltage will increase with increasing of frequency. SMT fluctuation occurs when V_{SMT} continues to increase at the same frequency. This phenomenon can be analized by applying an image processing tecnique. Based on image feature extraction method, inhomogeneity of SMT patterns can be observed through contrast, correlation, energy, homogeneity, and entropy parameters. Those parameters represent a result in accordance with the image visualization. Significant changes in the value of contrast, energy, homogeneity, and entropy parameters occur at $V > V_{SMT}$ which fluctuation begin. In this experiment the fluctuation begin at $V = 27.3$ V.

1. Introduction

Liquid crystal has a good response to the external field such as temperature, electric fields, and magnetic fields. In equilibrium and without an external field influences, nematic liquid crystal molecules are oriented in one direction called the (n) director. The responses of nematic liquid crystal to electric field can be studied by placing liquid crystal material between an ITO (Indium Tin Oxide) parallel glass plates as an electrode. There are two typicals of nematic liquid crystal alignments namely planar and homeotropic. Director n has an initial director in the x-axis (n = 1, 0, 0) in a planar alignment, and z-axis alignment (n = 0, 0, 1) in a homeotropic alignment [1,2]. The external force can cause distortion along the uniaxial axis of the director in nematic liquid crystal. There are three typicals of deformations namely splay, bend and twist, as shown in figure 1 [2].

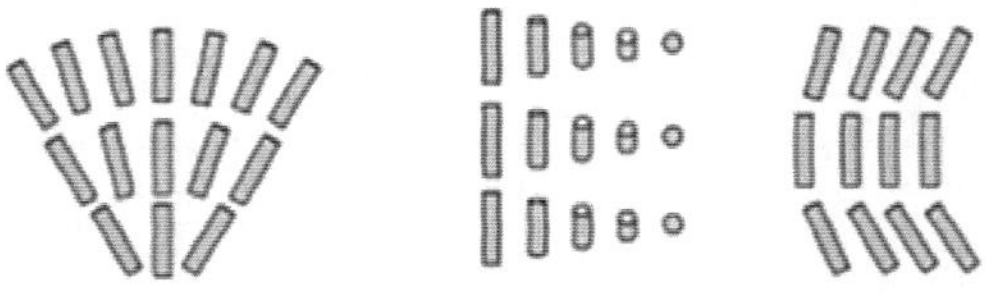

(a) (b) (c)

Figure 1. (a) splay (b) twist and (c) bend deformation of nematic liquid crystal

When a DC or AC electric field is applied between two electrodes, the ionic particles in nematic liquid crystal will move periodically due to the influence of the field. This phenomenon is known as electro-hydrodynamic instability (EHD). EHD appears when the voltage greater than threshold voltage (V_{EHD}) applied to nematic liquid crystal layer [2]. The effect of EHD is a response that given by a nematic liquid crystal because of its anisotropy properties. The effect of EHD is indicated by the formation of various convective patterns, in the form of dark-light lines. The effect of EHD occurs because of the reorientation of the direction of the liquid crystal molecules. The physical mechanism in EHD is described by Carr (1969) and Helfrich (1969) in the Carr-Helfrich (CH) mechanism as shown in figure 2.

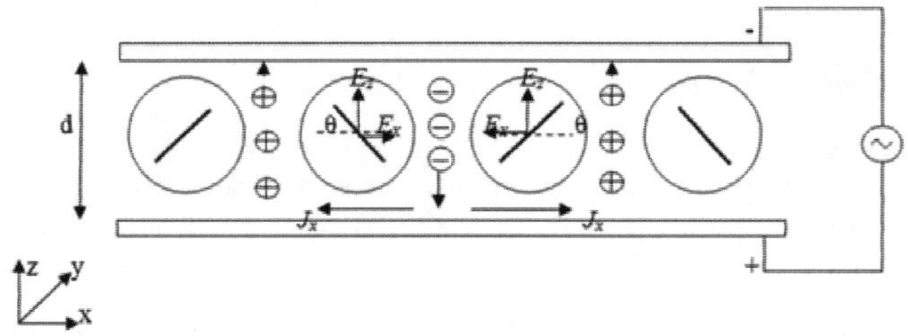

Figure 2. Carr-Helfrich mecanism in planar case

When the voltage greater than the threshold voltage V_{EHD}, the periodic modulation of the director causes convection in nematic liquid crystal associated with the charge motion. E_z is the AC external field that given to the sample, E_x is the internal electric field that produced by the distribution of the periodic charge and J_x is the current density along the x-axis. The director n is represented by a thick bar in the circle. In the conductive region, the director cannot follow the direction of the given external field, the space charge oscillates with the outer field and stationary convection occurs. A regular line pattern is formed at the threshold voltage V_{EHD} [2].

$$V_{EHD}^2 = V_0^2 \frac{\left(1 + f^2\tau_c^2\right)}{\left(\xi^2 - 1 - f^2\tau_c^2\right)} \tag{1}$$

ξ is Helfrich constant and τ_c relaxation time of charge.

Equation 1 shows that the threshold voltage increases with increasing of frequency and tends to infinite when the frequency approaches the critical frequency value f_c.

$$f_c = \frac{1}{\tau_c}\sqrt{\xi^2 - 1} \tag{2}$$

Zaman conducted a study on the effects of EHD on homeotropic samples of MBBA nematic liquid crystals [3].

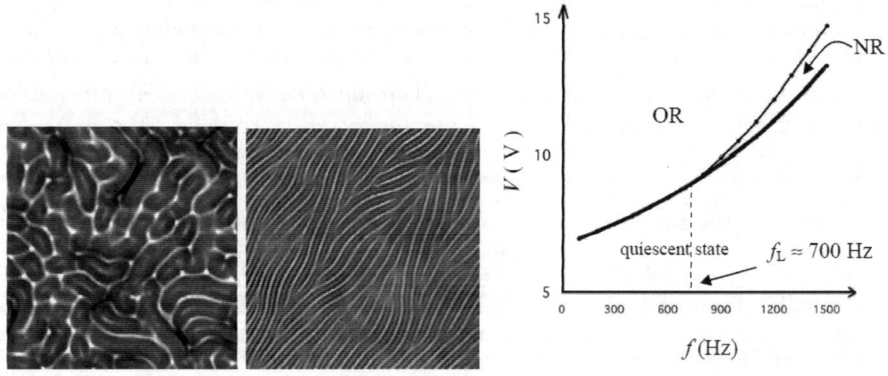

Figure 3. Types of SMT (a) *oblique rolls* (OR) and (b) *normal rolls* (NR), and (c) the graph of threshold voltages as a function of frequency modulation [3]

Zaman show the convection phenomenon when the voltage exceeds the threshold voltage at a certain frequency, which is called the soft-mode turbulence (SMT). There are two types of SMT namely oblique rolls (OR) and normal rolls (NR) which shown in Figure 3. The graph of threshold voltages as a function of frequency modulation also shown in Figure 3 [3].

There are many ways to analyze the image of SMT patterns captured using polarizing microscope. Narumi, et all, observed SMT relaxation by measuring a temporal autocorrelation function of patterns dynamic in SMT to observe the spatial-temporal chaos in homeotropically aligned nematic liquid crystal. They reported that the relaxation well fitted by a compressed exponential function [4]. In our study, we will analyze the patterns of SMT at various voltages. We introduce a method in image processing namely image feature extraction to distinguish the texture of patterns based on statistical data extraction. There are two typicals of image feature extraction, first order feature extraction (for example mean, variance, skewness, kurtosis and entropy parameters) and second order feature extraction (for example contrast, correlation, angular second moment, and energy parameters) [5].

2. Method

Nematic liquid crystal sample with 50 μm of thickness and homeotropic alignment placed in a hot stage at a constant temperature. An electric field in the *z* axis direction applied to the sample. The EHD effect that appears on the sample is observed using a polarizing microscope by adjusting the frequency modulation. At the certain frequency, the voltage is adjusted until a soft mode turbulence (SMT) pattern appears. At the same frequency, the voltage continues to be increased for fluctuations observation.

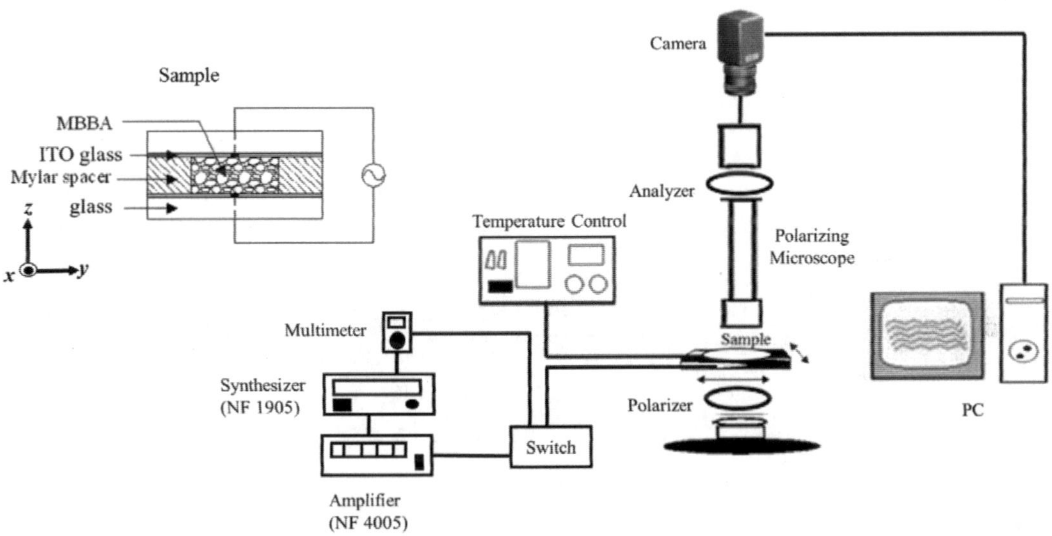

Figure 4. Experimental set-up

We use Matlab software to analyze the images. The RGB images is converted to HSV images to simplify the values. HSV contains of the hue, saturation, and value components for the image. The value parameter is selected to show the level of darkness or brightness of images (value-image). The value components express the brightness of the color. The value parameter is in the range 0-1 from black to white. Then the gray-level co-occurrence matrix (GLCM) process applied to value-image. According to the gray-level co-occurrence matrix (GLCM) process, quantitative data can be extracted, namely image feature extraction, with statistical parameters like contrast, correlation, energy, and homogeneity. We also compare with another statistical data parameter like entropy value [6].

3. Result and Discussion

Figure 5 shows the graph of SMT voltage (V_{SMT}) as a function of frequency. Approach a critical frequency, the threshold voltage value will be greater than before. The graph as in figure 5 accordance with equation 1 which the increasing of threshold voltage is not linearly proportional to the critical frequency. Since the voltage is applied to the sample, then it is raised slowly, the liquid crystal molecules continues to move because its physical properties as an anisotropic substances. The emergence of the SMT pattern is characterized by the formation of a darker color on the sample than the previous one. After a contrast different color appear, the voltage is raised more slowly so that the threshold voltage which SMT pattern appearance can be determined precisely. Electroconvection pattern appears in the homeotropic alignment in the same manner as the tilted director appears in the planar alignment.

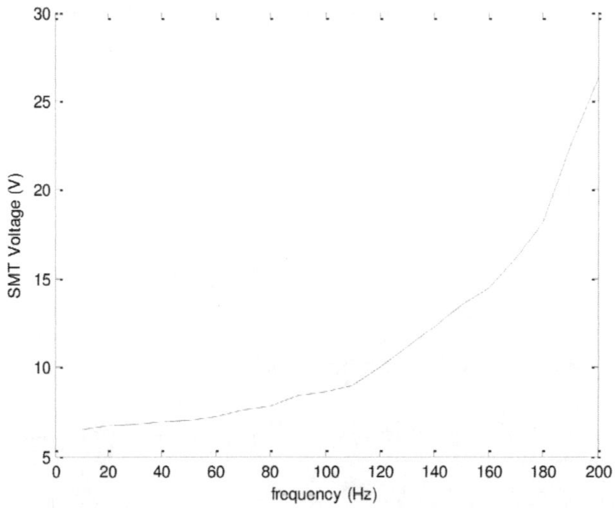

Figure 5. SMT voltage (V_{SMT}) as a function of frequency

If the voltage continues to be increased, there will be fluctuating, as shown in Figure 7. In the homeotropic sample, if the voltage is increased greater than the critical threshold voltage, the elastic free energy density also tend to increase and the orientation of the director becomes more fluctuating.

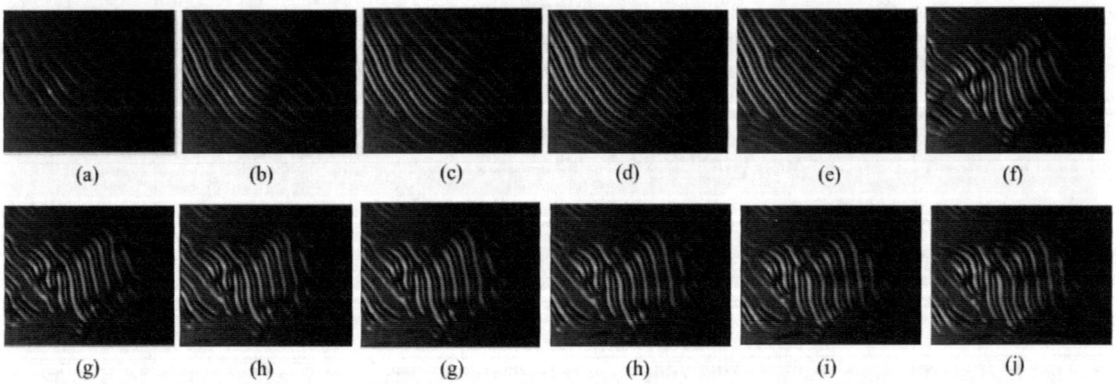

Figure 6. SMT fluctuations at $f = 200$ Hz. (a) SMT pattern at the threshold voltage $V_{SMT} = 26.6$ V. The voltage continues to increase with $\Delta V = 0.2$ V. (j) SMT pattern at Vj= 28.5 V

Fluctuations in the SMT pattern as shown in figure 6 can be quantitatively shown by applying feature extraction method to the image using Matlab software.

The result of the first order feature extraction, according to entropy parameter, can be shown in the graph of entropy as a voltage function in figure 7. Entropy is a measurement of the image intensity randomness. The greater of the entropy the more irregular intensity of each pixel, as shown in figure 7. The decreasing of entropy parameter occurs when fluctuations begin at $V = 27.3$ volts.

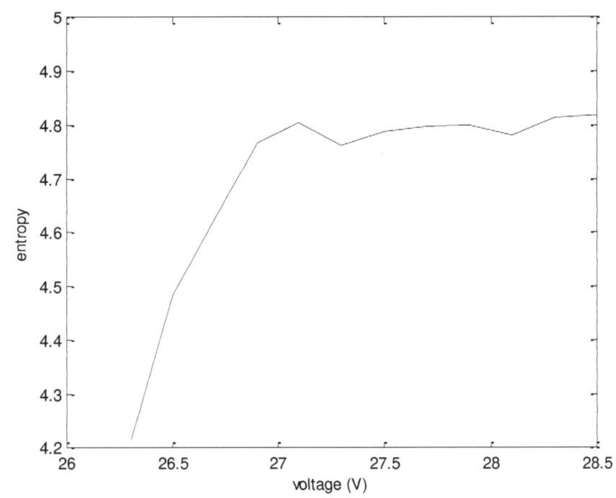

Figure 7. Texture entropy as a function of voltage when 200 Hz frequency applied (V > V_{SMT})

Second-order feature extraction can be used to distinguish the texture of an image from another images through parameters such as contrast, correlation, energy, and homogeneity. The results of the second order feature extraction based on statistical data of these parameters are shown in figure 8 and figure 8 and 9.

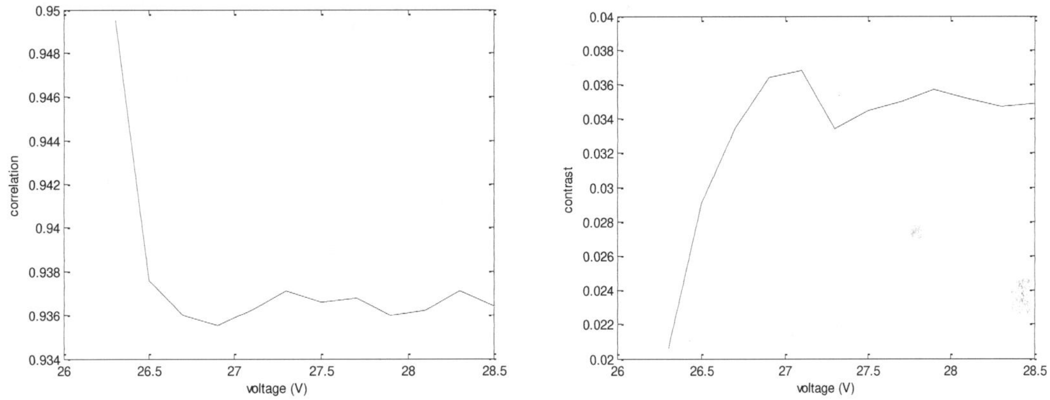

(a) (b)

Figure 8. (a) Texture contrast parameter as a function of voltage (V > V_{SMT}) , and (b) Texture correlation parameter as a function of voltage, for 200 Hz in frequency (V > V_{SMT})

Contrast parameter is a measurement of the local variations in the gray-level co-occurrence matrix. It measures of the intensity contrast between a pixel and its neighbor over the whole image. Contrast parameter is also known as variance and inertia. The graph in figure 8a shows that the value of the contrast increases with increasing of voltage, it indicates the greater of the difference in intensity between pixels, in accordance with Figure 7. At $V = 27.3$ volt, the contrast value tend to decrease (figure 7f). Based on the image in figure 7, it appears a change of significant pattern in figure 7e to 7f where the fluctuations occur.

Based on the graph of correlation, energy and homogeneity parameter, in figure 8a, 9a and 9b, it can be shown that the values of those parameters decrease as a function of the voltage. Correlation parameter is a measurement of the joint probability of occurrence in specified pixel pairs. Decreasing of correlation values shows that the relationship between pixels has a low recurrence probability.

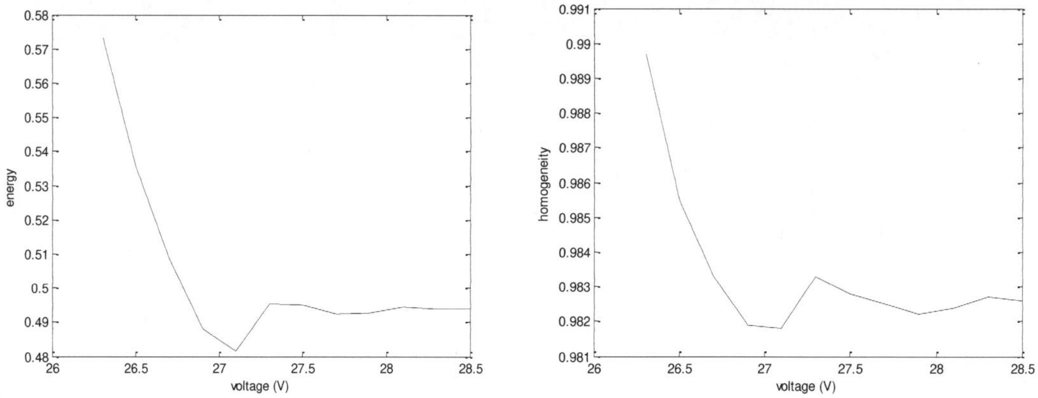

(a) (b)

Figure 9. (a) Texture energy as a function of voltage, and (b) Texture homogeneity parameter as a function of voltage, for 200 Hz in frequency ($V > V_{SMT}$)

Energy parameter is also known as angular second moment (asm) which it measures the uniformity of pixels. Decreasing of energy parameter values show an increasing of inhomogeneous pattern. Homogeneity parameter is a measurement of the closeness of the elements distribution overall to the diagonals so that the lower homogeneity values show the non-homogeneity of the distribution of pixels.

The graph of energy and homogeneity parameters in figure 9 also shows that at $V = 27.3$ V the intensity value increases again. This behavior appropriate with the contrast parameter where the increasing of intensity occur significantly when fluctuation begin, as shown in figure 7e to 7f.

4. Conclusion

Fluctuations in soft-mode turbulence patterns at the same frequency occur when the voltage greater than the threshold voltage. Soft-mode turbulence fluctuation can be observed by extracting image characteristics, both first order and second order feature extraction. Contrast, energy, homogeneity and entropy parameters represent a result in accordance with image visualization. There is a significant changes of the contrast, energy, homogeneity, and entropy parameters when the voltages greater than the threshold voltage which fluctuations begin to occur. In this experiment the fluctuation begin at $V =$ 27.3 V. Based on the values of contrast, energy, homogeneity and entropy parameters it can be shown that when the voltage that applied to the sample greater than the threshold voltage, the irregularity in intensity of images will be increase and also indicates the fluctuation of the molecules movement on sample because of an anisotropic properties in nematic liquid crystal.

5. References

[1] Yang, Deng-Ke & Wu, Shin-Tson 2006 Fundamentals of Liquid Crystal Devices (John Wiley &Son Ltd)
[2] Chandrasekhar, S 1992 Liquid Crystal Cambridge. University Press
[3] Zaman, R 2008 New Type of Order-Disorder Transition in Soft-Mode Turbulance of Electroconvection in Nematics (Master Thesis, Kyushu University, Japan).
[4] Narumi T, Yoshitani J, Suzuki, Hidaka Y, Nugroho, F, Nagaya, T, and Kai, S 2013 Memory function of turbulent fluctuations in soft-mode turbulence *Phys. Rev.* D 87 012505
[5] Gonzales, RC, & Woods, RE 2002 Digital Image Processing (Prentice Hall)
[6] Information on https://www.mathworks.com/help/images/
[7] Rinto Anugraha, Koyo Tamura, Yoshiki Hidaka, Noriko Oikawa, and Shoichi Kai 2008 Order-Disorder Phase Transition in a Chaotic System, *Phys. Rev.* D. 100, 164503

Optimization of pathology diagnosis by applying chaos theory and fractal analysis to EEG-signal processing

Vasily Kornilov, Maxim Ostroukhov and Andrey Dmitriev

National Research University Higher School of Economics, 33 Kirpichnaya St., 105187 Moscow, Russia

E-mail: a.dmitriev@hse.ru, vkornilov@hse.ru

Abstract. By 2018, there are more than 70 million people suffering from various forms of one of the most common neurological diseases – epilepsy. In fact, epipelsy is a central nervous system (neurological) disorder, manifesting itself in anomalous brain activity. Nowadays, most of the patients do not have the ability to foresee the onset of an attack in advance, which is due to the complex symptoms that are difficult to predict. One of the best way to analyze such kind of behaviour is electroencephalography (EEG). This research paper considers the problem of EEG epileptic seizures prediction from the point of the theory of nonlinear dynamical systems. According to the study, signals of cerebral cortex neural networks corresponds to multifractal nature which gives an opportunity to analyze changes between states in phase space during the abnormal electrical activity. Therefore, monitoring the preictal stage and early indication of an attack may help patients to avoid problems related to sudden seizures. This research will provide valuable information regarding the mechanism of epilepsy onset and introduce a prediction model using machine learning algorithm based on fractal analysis of RQA.

1. Introduction

Epilepsy is understood as a sudden excessive excitation of neurons in the cerebral cortex, the external manifestation of which consists in repeated convulsive attacks, manifested in violations of the nervous and cardiovascular systems of the patient, loss of consciousness, partial or complete loss of control over their own movements and, as a result, the inability to control their condition during the attack [1]. According to the survey published by the World Health Organization (WHO), the incidence or the number of new cases of epilepsy is increasing every year, ranging from 30 to 50 per 100,000 in high-income countries and 80-100 in low and middle-income countries, which form about 80% of all epilepsy cases.

By its complex nature, the disease is almost untreatable. Antispasmodic drugs which are aimed at reducing the probability of the seizure onset, often lead to severe side effects such as disruption of the pancreatic system, the cardiovascular system, etc. Moreover, after brain surgery, which can be helpful in some cases, 30% of people continue falling into periodic seizures. As a result, people suffering from this disease are constantly in a state of anxiety because of the possibility of being in a state of seizure at the wrong time from the medical and social points of view.

One of the most common ways to monitor current state of disease is to analyse brain activity within electroencephalography (EEG) due to its sensitivity and neurological nature of pathology. Considering EEG as time series research scientists distinguish four stages of disease: 1) interictal stage - stable brain activity within normal limits, 2) preictal stage – slight increase in brain activity, also called the aura, 3) ictal stage - epileptic seizure, and 4) postictal stage - recovery from an attack, weakness, reduced brain and physical activity [2].

An important approach to support the patient's life cycle is to detect the presence of an epileptic attack. A positive aspect of this method is the timely notification of others in order to create the necessary conditions at the time of epileptic seizures. In other words, the main goal of this article is an early identification of the preictal stage.

2. Practical implementation

2.1. Dataset description
CHB-MIT Scalp EEG Database was chosen as a source data set [3, 4]. The data collected at Boston children's hospital consists of EEG records of patients with resistant epilepsy. Each patient was followed up for 1-3 days, some of which were given anticonvulsant drugs to determine the type of epilepsy and evaluate the candidate for surgery to remove the part of the brain that provoked the attack. All signals were recorded at the standard frequency of 256 Hz with 16-bit resolution. A distinctive feature of this set is the presence of epileptic seizures' periods labelling, which allows to track the dynamics of the brain activity on time series up to a second.

2.2. Initial fractal analysis
To begin with, it's needed to test the hypothesis that EEG data is an example of a nonstationary data set with long memory which is a simple indicator of correspondence to nonlinear dynamics.
To test the assumption, autocorrelation function (ACF) was constructed (figure 1):

$$R(\tau) = \frac{E[(X_t - \mu)(X_{t+\tau} - \mu)]}{\sigma^2} \#(1)$$

where τ is the lag at which the correlation is considered, μ-expected value of the quantity under consideration, σ^2 - the variance.

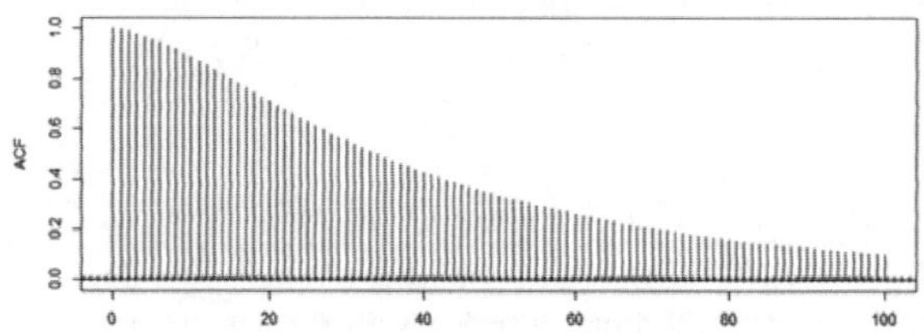

Figure 1. Autocorrelation function.

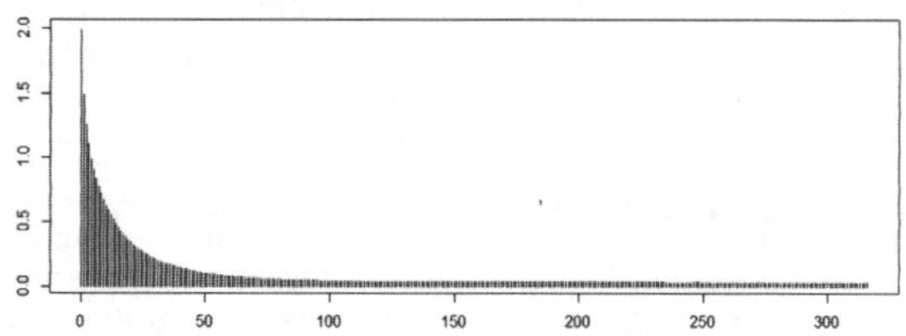

Figure 2. Average Mutual Information.

The value of the autocorrelation function has a rather weak downtrend and slowly converges to zero, which indicates the presence of characteristic features of the model with long memory and the possibility of applying the theory of nonlinear dynamic systems to the set.

An alternative metric for ACF is Average Mutual Information (AMI) - the average mutual information calculated through the entropy of two values:

$$I(X;Y) = H(X) - H(X|Y) = H(X) + H(Y) - H(X,Y)\#(2)$$

This metric (figure 2) gives an ACF-like result and demonstrates a slow decline to higher benefits, which gives additional reason to believe that the series has a long memory.

Moving to the stochastic characteristics of the time series to understand the structure of the available data correlation dimension is calculated by the formula:

$$C(r) = \frac{1}{\frac{m(m-1)}{2}} \sum_{i=0}^{m-2} \sum_{j=i+1}^{m-1} \theta\left(r - p(x_i, x_j)\right) \#(3)$$

where Θ is the Heaviside function: $\Theta(\alpha) = \begin{cases} 1, \alpha \geq 0, \\ 0, \alpha < 0, \end{cases}$

p – distance in n-dimensional phase space, m is the number of points on the attractor.

Due to the fact that there is no precise information about the phase space, the Takens' method [5] was used for the attractor's reconstruction. Eventually, correlation dimension equal to 8.5 was received.

To check the system for the presence of persistence, we calculate the Hurst index (H):

$$E\left[\frac{R(n)}{S(n)}\right] = Cn^{H}, n \to \infty \ \#(4)$$

where *R(n)* the scope of n-values of the time series, *S (n)* – the standard deviation, n-the value of the time interval, *E[...]* –expected value. The numerical estimation of the indicator in this case can be calculated as the slope coefficient of the line describing the change in relation to n, constructed in logarithmic coordinates. When the parameter H<0.5, the system is considered to be antipersistent, i.e. no trend tends to be saved, but on the contrary – it changes to the opposite. The resulting estimate-0.2427±0.1838 gives the opportunity to confidently state the absence of any persistence in the data and the presence of a long-term trend to vary between high and low values of the series, constantly changing the relative direction.

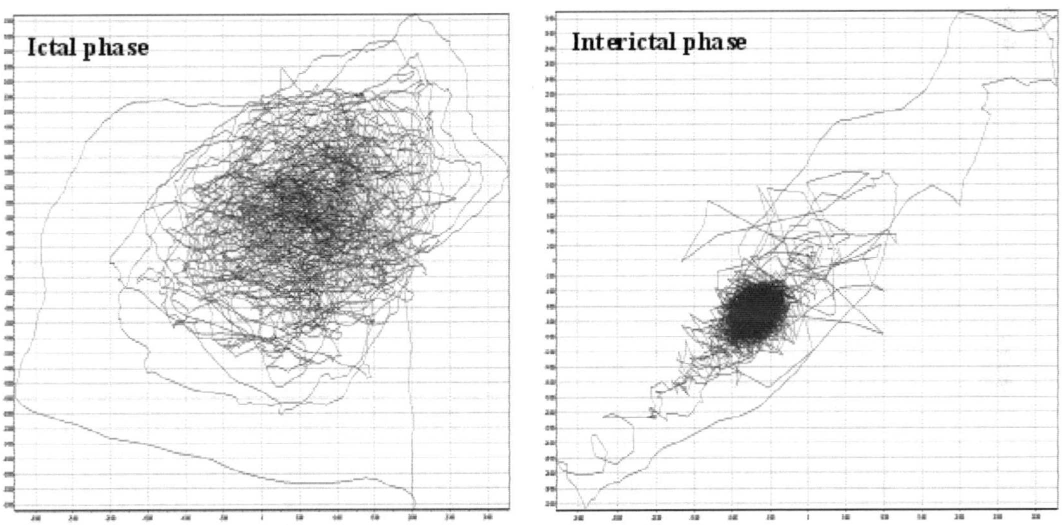

Figure 3. Attractors for the ictal and interictal stages.

Figure 3 shows the attractors built separately for the ictal phase and the phase of normal activity. To build such an attractor, windows were chosen, covering ~1 minute of EEG recording. The window size was chosen based on the average duration of the attack in the patient. Estimates of the resulting set of trajectories are presented in table 1.

Table 1. Comparison of fractal measures between ictal and interictal stages.

	Seizure	Dormant state
Correlation dimension	4.856	8.775
Phase space dimension	6	12

The obtained data shows the confirmation of formulated hypothesis [6] about significant decrease of correlation dimension during turbulence and allow us to conclude that the metrics of nonlinear dynamics for different phases of attack can vary.

2.3. *Recurrence Quantification Analysis (RQA)*

The main objective of this work was exposed to the application of the RQA approach to the prediction of seizures. It is important to note that RQA is a parametric method and depends on several factors: embedding dimension, time delay, ε radius of neighborhood, distance metric and the minimum length of the linear structures. The selection of each of these parameters can be devoted to the whole article, so for the purposes of this article the classical approaches to find optimal parameters were chosen.

- **Embedding dimension**: according to Takens' theorem [5], embedding dimension is $2m+1$, where m is the correlation dimension, will be sufficient to preserve the properties of the original phase space.
- **Delay time:** common heuristic for calculating the delay time is to select the first lag of the autocorrelation function whose value is zero or reduced to 1/e from the original value.
- **Radius ε of the neighbourhood:** the key parameter is the allowable radius, beyond which it can be assumed that the system has moved to another state.
- **Distance metric:** Euclidean distance is commonly used, but there are lots of different options for different tasks.
- **Minimum length of linear structures:** selected exclusively by manual selection for the purpose of high-quality visualization of statistics.

It is necessary to pay attention to the fact that the above parameters are selected on the basis of the interictal phase to find non-standard patterns of behavior during the transition of EEG into the phase of attack.

Using the approach:

- **Embedding dimension**: 12. It is calculated as the average of several calculated investments at different stages of rest.
- **Delay time**: 70.
- **The radius of the neighborhood ε**: 350.
- **Metric**: Euclidean distance.
- **Minimum length of linear structures**: 8.

The results of the calculated statistics are shown in figure 4.

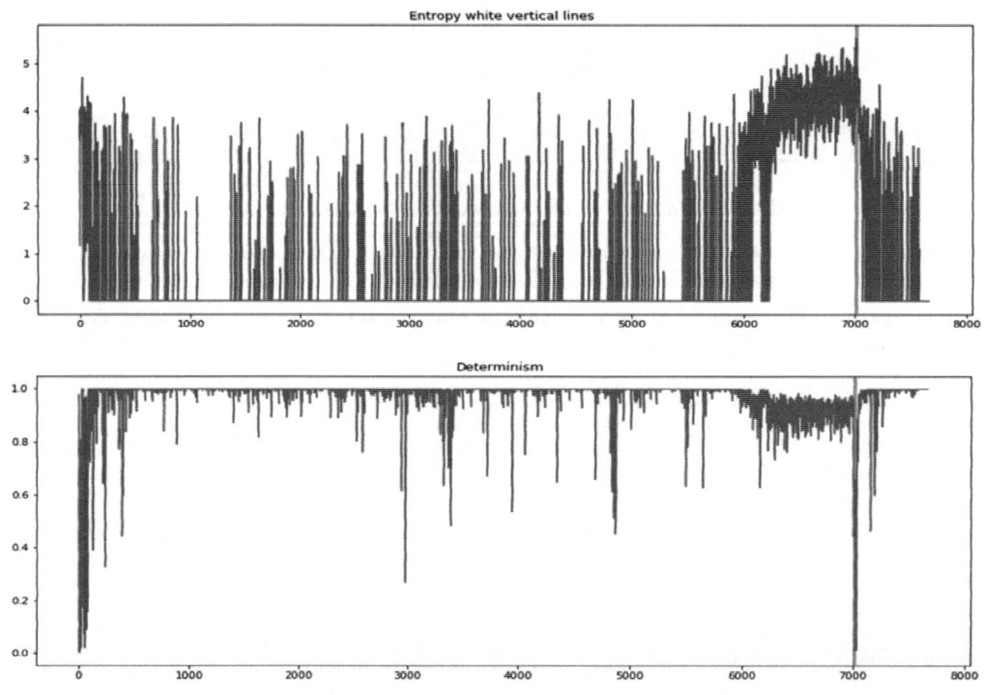

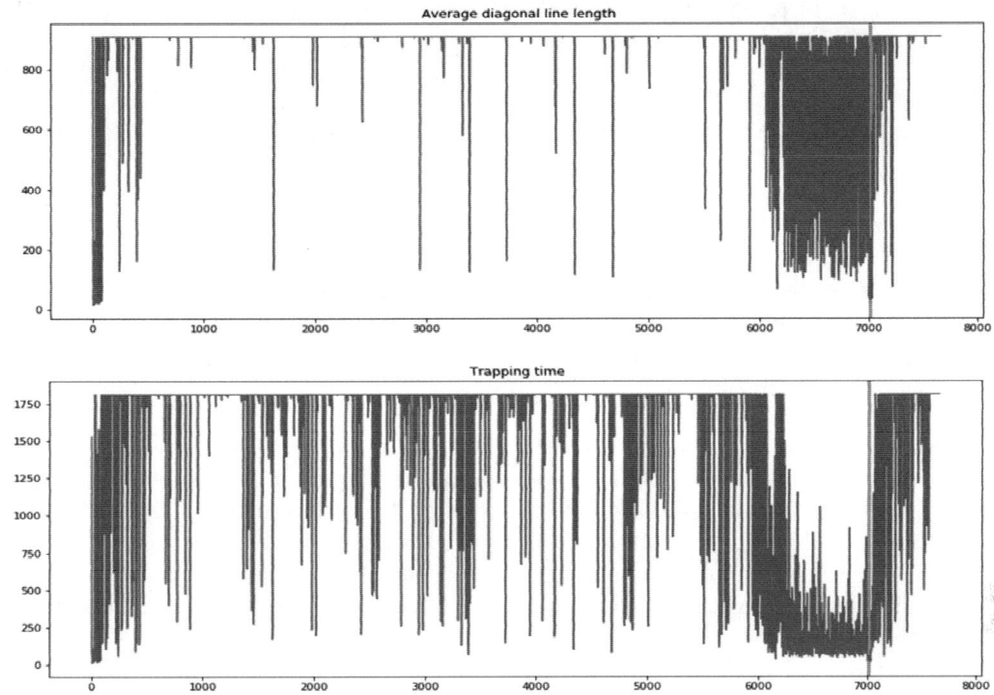

Figure 4. RQA metrics at 4 hour period.

The vertical line on the charts shows the seizure period. It is clearly visible that RQA reflects the changes in brain activity during preictal stage, which allows to confirm the hypothesis of the presence of structural changes in the state of the system obtained on the basis of RQA for a certain period of time before the attack. The practical value of the results is that, using such statistics, basic algorithm that detects the onset of an attack in advance can be implemented. Thus, it can be concluded that an approach based on the study of RQA metrics utilizing rolling window for some types of seizures can provide information about the presence of preictal phase. However, it should be noted that this approach cannot guarantee 100% accuracy, due to sufficient number of artifacts on in data which potentially leads to false positives.

In [7], published using the same data set, the authors propose an algorithm for classifying outliers based on the application of the support vector machine (SVM), whose purpose is to detect the preictal phase using spectral, spatial and temporal features for estimation. The model was evaluated using a leave-one-out cross-validation based on a data set consisting of a set of two-hour pairs so that one hour refers to the data in which the seizure is not recorded, and the other necessarily contains the seizure. Three metrics were chosen for measuring algorithm's quality:

- Sensitivity – percentage of predicted seizures, recall.
- Specificity–true negative rate.
- Delay- the delay between the expert evaluation of the onset of the attack and the evaluation of the algorithm.

The peculiarity of this implementation is that predictive model were trained on each person individually. In the context of this study, it is interested to make a comparison with an algorithm that would not depend on a particular patient. For the experiment integrity, the same SVM model was trained, formed according to the description of the article, supplemented with RQA metrics and tested on three patients. Since it is impossible to restore the implemented models exactly, a comparison with the local model was made, which gives the results not much worse than those given in the article. In table 2 you can see the comparative statistics of the applied model with and without new features.

Table 2. Evaluation of RQA features using SVM

	Basic SVM	SVM with RQA
Sensitivity	73%	84%
Specificity	88%	86%
Delay	6,3 s.	5,1 s.

The main achievement is a significant increase in the recall of the detection preictal stages and a slight decrease in the average delay in the determination of emissions. Analyzing the obtained results, it can be assumed that the model became more “confident” in classification which has been achieved at the expense of new quantitative characters. This "confidence" is also expressed in a slight decrease in the specificity of the model, since the increase in the “sensitivity” of the model always leads to an increase in the number of false positives.

3. Conclusions

Summing up the results of the research, it can be said that the approach based on the application of chaos theory to EEG is a good addition to existing studies, the quality of which is justified by the multifractal nature of the signal. At the same time, it is worth noting that the task of building a predictive model, expanding the existing one was achieved mainly through the approach based on the analysis of the quantitative characteristics of the Recurrence Plot with manually selected hyperparameters.

From the point of practical application, an important achievement of the study is a significant increase in the model recall, which gives rise to rethink the current methods of detecting epileptic seizures used on real people in devices implanted surgically. The main advantage of this work is not only the growth of different metrics, but also the proof that the electrical signals inside the human brain can be analyzed and predicted by fractal analysis, which is a positive outlook for further research in this area.

Obviously, this article is just an initial step towards the development of an adaptive predictive algorithm utilizing chaos theory. In this regard, there are several ways which the future improvements can be devoted to such as: analysis of dependencies between different parts of the brain, additional RQA metrics and at the same time using advanced machine learning models which can be a great basis improvement for the final results and a significant push for the implementation of such models to the patients’ medical devices.

4. References

[1] Singh A and Trevick S 2016 *J. Neurol. Clin.* **34** 837–47

[2] Świderski B, Osowski S, Cichocki A and Rysz A 2007 *Adaptive and Natural Computing Algorithms* ed B Beliczynski, A Dzielinski, M Iwanowski and B Ribeiro (Warsaw: Springer) vol 4432 pp 373-81

[3] Goldberger A L, Amaral L A, Glass L, Hausdorff J M, Ivanov P C, Mark R G, Mietus J E, Moody G B, Peng C K and Stanley H E 2000 Physiobank, physiotoolkit, and physionet: components of a new research resource for complex physiologic signals *J. Circulation.* **101** 215-20

[4] Shoeb A H 2009 Application of Machine Learning to Epileptic Seizure Onset Detection and Treatment (Boston: Massachusetts Institute of Technology)

[5] Takens F 1981 Detecting strange attractors in turbulence *Lecture Notes in Mathematics* vol. 898, ed D A Rand and L-S Young (New York: Springer-Verlag) pp 366-381

[6] Babloyantz A and Destexhe A 1986 *J. Proc. Natl. Acad. Sci.* **83** 3513-7

[7] Hamadene W, Peyrodie L and Seidiri H 2006 *8th international Conference on Signal Processing* (Beijing: IEEE)

IOP Conf. Series: Journal of Physics: Conf. Series **1298** (2019) 012016 doi:10.1088/1742-6596/1298/1/012016

Impact of vehicular lateral placement on speed under non-lane discipline traffic stream

Dibyendu Pal[1], S. Sreenivasulu[2]

[1]Assistant Professor, Department of Civil Engineering, NERIST, Nirjuli, Arunachal Pradesh, India.
[2]Executive Director, Eclat Engineering Consultants, Hyderabad, India

[1]Corresponding Author's Email: pal_here00@yahoo.co.in

Abstract. The vehicles plying on the roads of developing nations such as India are mixed in nature and do not follow any lane markings. The vehicles can easily take place laterally on any free space across the road section anticipating better maneuverability on the next time step. The vehicles usually maintain different speed for faster and safer forward movement on the next time step. Therefore, this study aims to find the relationship between vehicular speed and lateral placement. For this purpose, two hrs traffic video data were collected from VIP Road, Kolkata, India. From the traffic video film, vehicle trajectories were extracted using image processing software TRAZER after applying appropriate corrections. From the lateral placement distribution of each category of vehicles, it is observed that LMV prefers median lane, whereas MTW and MThW prefer shoulder lane. Speed-placement models were developed for all vehicles clustered and each category of the vehicle using the first hour data. It was observed that the relationship between average speed and lateral placement is linear up to a certain distance from the median edge and quadratic for the remaining portion of the road width. The speed-placement models' validation using second-hour data indicate that the developed models are highly accurate in terms of MAPE. The results obtained in this study may be used as an input parameter while assigning vehicular speed based on lateral placement for analyzing non-lane discipline traffic stream using the simulation model.

1. Introduction

The traffic stream plying on the roads of developing nations such as India are mixed in nature. The vehicles plying on the road sections do not follow any lane markings. This type of stream is generally termed as non-lane discipline traffic stream. Since on Indian highways no dedicated lane is available for a specific type of vehicle, hence the usage of road space is generally mixed in nature. The same road space is shared by different kinds of vehicles such as Motorized two-wheelers (MTWs), Light motor vehicles (LMVs) such as Car, Motorized three-wheelers (MThWs), and Heavy motor vehicles (HMVs) such as Bus, Truck. The static (dimensions) and dynamic (speed, acceleration, deceleration, lateral maneuverability) characteristics of these vehicles are not unique. The mixed vehicle interacts with each other and produces a complicated situation which affects the stream speed, capacity, safety, and level of service. The dissimilarity between the characteristics of various types of vehicles and ability of few vehicles to maneuver at live situation enforce the driver to look for the opportunities to overtake or pass the leading vehicle. A driver of the subject vehicle may change its current lateral position if he/she is not happy with the current speed in the present lateral position because of the presence of a leading vehicle [1]. Smaller size vehicles, such as MTW and MThW can easily pass through the lateral gap between other vehicles or other vehicle and road edges. However, sometimes the available lateral gap in the forward direction of the subject vehicle is not adequate to pass through the current corridor of the subject vehicle. In such cases, the driver shifts the vehicles in the lateral direction to adjust its position for a comfortable movement on the next step. Vehicles occupy any place across the road section depending on its suitability and accessibility for faster progress on the next step. In real traffic stream movement, the driver simultaneously drives the vehicle in the lateral direction as well as in the longitudinal direction by maintaining a certain speed. Analyzing these traffic stream to mitigate the various operational glitches is a complex task. One cannot study these stream behavior just by standing beside the road section or implementing a solution without understanding its effect on traffic stream.

Traffic simulation model plays a vital role to analyze the complex interactions of non-lane discipline traffic streams. Simulation models are widely used to recommend alternative solutions. These models are used to understand the impact of substitute measures on traffic stream behavior and finally recommend to implement the most feasible and acceptable alternatives. Nevertheless, only a realistic and appropriate traffic simulation model would produce the best solution to the existing problems. However, the development of a non-lane discipline traffic simulation model requires enormous information on existing traffic stream. Among several parameters, vehicular lateral placement distribution is one of the crucial parameters for developing the non-lane discipline traffic simulation model. Hence, this study is concentrated on finding the lateral placement distribution for different types of vehicles using vehicular trajectory data collected from non-lane disciplined traffic streams and its impact on vehicular speed.

The lateral placement of a vehicle on the road section is the lateral position of the center line of the vehicle measured from the median edge. [2] estimated the lateral distribution of HMVs. They also tried to find the impact of lateral placement on the design of road pavements. They used “shift factor (SF)” to explain the relationship between speed and rutting. [3] studied the lateral placement data of Delhi-Gurgaon expressway collected during the daytime. They observed that vehicles speed reduces from median-side lane to shoulder-side lane. [4] studied the lateral placement and speed of vehicles on the two-lane bi-directional road section. They found that the speed of MThW, HMV and slow-moving vehicle increases from road edge to the center line. However, due to the effect of the vehicles coming from the opposite direction, the speed of car and MTW initially increase but later decreases toward the center line. The speed of MThWs, HMVs, and slow-moving vehicles have a linear relation with lateral placement, whereas, cars and two-wheelers have a non-linear relationship with lateral placement. [5] studied lateral positions of cars and HMVs at a two-lane, two-way road in the Port of Brisbane. They observed that opposing traffic and length of opposing vehicle effect the lateral positions of cars, utility vehicles, and semitrailer.

[6] analyzed the lateral distribution of mixed traffic on two-lane roads. They concluded that the lateral distributions of mixed traffic of two-lane roads consist of a unimodal or bimodal distribution. From their study with 19 sections of Indian highways, they found that lateral placement follows a unimodal curve if the “placement factor (PF)” and “skewness range (SR)” are less than 1.30 and 0.54, respectively. [7] conducted the study for placement distribution of U turning vehicles collected from 13 different median openings of six lanes divided roads. The placement data of five sections followed the normal distribution, i.e., unimodal when “placement factor (PF)” was less than 0.86.

The reviews of the above literature suggest that lateral placement has a significant effect on traffic flow analysis. Mostly the impact of this parameter was analyzed for the two-lane road section, eight lanes expressways and six-lanes U-turn on median openings. However, for urban arterial roads, the effect of lateral placement on the speed of different types of vehicles is not being explored in the literature specifically for the case of three-lane roads. Hence, this study has attempted to analyze the vehicular lateral placement distribution on non-lane discipline traffic stream plying on urban arterials. It is also planned to check the impact of lateral placement of vehicles on the speed of the traffic stream.

The remainder of this article is organized as follows. The next section described the brief description of data collected for this study and its relevant information. Section three illustrates the analysis of collected data for finding the relationship between speed and lateral placement. Validation of the developed speed-placement models is given in section four. The final section summarizes the important conclusions drawn from the present study, its application and future works.

2. Data collection

In this study, traffic video films were captured from VIP Road, Kolkata, India. The video camera was mounted on foot over bridge located across the road section. The video camera was installed in such a way that it can focus toward the front of the traffic stream. The video camera was adjusted to focus on a stretch of about 50 m in the longitudinal direction of the arterials while filming the traffic flow. The road stretch was marked using cello tape for a length of 30 m in the longitudinal direction. The road width of the section was 10.80 m. The vehicles plying on this section were mixed and not following lane markings, i.e., lateral movements were not restricted. The mixed traffic contains various types of vehicles such as cars, buses, trucks, two-wheelers, three-wheelers, and light commercial vehicles. The traffic video films were collected between 9:30 hrs and 11:30 hrs on a working day with clear weather.

The collected video films were processed to extract the vehicle trajectories automatically using image processing software, TRAZER. Manual intervention is not required in TRAZER to track the vehicle trajectories. However, it also allows manual involvement and, this permits the user to augment unidentified vehicle trajectories or wrongly detected vehicle trajectories. TRAZER can collect about 20 to 40 m long trajectories. The frequency of the vehicle trajectory tracking is 25 Hz, i.e., vehicle's positions were detected by TRAZER at every 0.04 sec interval. Longitudinal and lateral positions of the vehicles were tracked separately over 0.04 sec interval. The accuracy of vehicle detection by TRAZER is about 85-90% [8]. TRAZER classifies the vehicles in four different categories such as MTW(two-wheelers), LMV (cars, light commercial vehicles), MThW (three-wheelers, Auto) and HMV (buses, trucks). Accordingly, the trajectories of four categories of vehicles such as MTW, LMV, MThW, and HMV were extracted from the traffic videos.

Trajectory data collection using image processing methods, specifically under non-lane discipline traffic streams consists of several errors. The reason may be due to the occlusions, illumination changes, and perspective angle changes of the camera [9]. The extracted trajectories were corrected to overcome the anticipated errors in position data of the vehicle trajectories. The method used to correct the trajectories is called complete ensemble empirical mode decomposition with adaptive noise (CEEMDAN). The corrected trajectories were used to find the hourly flows and compositions of the vehicles. These data were also used to obtain the lateral placements of the vehicles across the road width. The hourly flows for both hours traffic data were found to be 5170 and 4727 vehicles, respectively. The compositions of the traffic stream and descriptive statistics (mean and standard deviation (S.D.)) of lateral positions of the vehicles are shown in Table 1. The corrected trajectories were used to estimate the speed of the vehicles using the method of Wavelet Transforms (WT). The details of both the techniques are available in [9]. The descriptive statistics of the vehicular speeds are given in Table 1.

Table 1. Traffic compositions and descriptive statistics of lateral positions and speeds

SL. No	Data collection time	Vehicle type	Flow (Veh/hr)	Vehicle Composition (%)	Lateral Placement (m)		Speed (km/hr)	
					Mean	S.D.	Mean	S.D.
1	9:30 hrs to 10:30 hrs	ALL	5170	100	4.57	2.27	40.13	7.65
		LMV	2955	57.16	3.65	1.75	41.59	7.66
		HMV	290	5.61	4.68	1.86	37.81	7.15
		MTW	1478	28.59	5.73	2.47	39.04	7.39
		MThW	447	8.65	6.79	1.46	35.55	5.77
2	10:30 hrs to 11:30 hrs	ALL	4727	100	4.45	2.26	42.53	12.81
		LMV	2820	59.66	3.51	1.61	45.04	14.33
		HMV	278	5.88	4.74	1.87	37.87	9.59
		MTW	1332	28.18	5.88	2.52	39.11	8.65
		MThW	296	6.26	6.72	1.49	38.39	9.41

IOP Conf. Series: Journal of Physics: Conf. Series **1298** (2019) 012016 doi:10.1088/1742-6596/1298/1/012016

3. Analysis of vehicular lateral placement and speed

After obtaining the corrected lateral position data, the frequency of all vehicles and each vehicle category for lateral placement is plotted over each meter of lateral positions of vehicles. Figure 1 shows the frequency plot of lateral placement for both the hours. It can be seen from the figure that around 25% to 28% LMVs are plying through the lane close to the median. This may be due to the reason that because of left hand side driving in the Indian traffic stream, overtaking/passing through right side is more convenient. Hence, to overtake/pass the vehicles, LMVs are trying to pursue through the median lane. The MTW and MThW are plying close to the shoulder lane. Since MTW and MThW are smaller in size; hence, they have the flexibility to pass through any small spaces ahead of them. Even these two vehicles may utilize the shoulder partially to overtake/pass the other vehicles. This may be due to the fact that majority of the MTWs and MThWs are utilizing shoulder lanes. In the case of HMV, due to its size, the maneuverability is very less. Hence, they mostly allow the other vehicles to overtake/pass them. After obtaining the lateral placement, and the average speed of all vehicles the analysis have been carried out to obtain the relationship between speed and lateral placement.

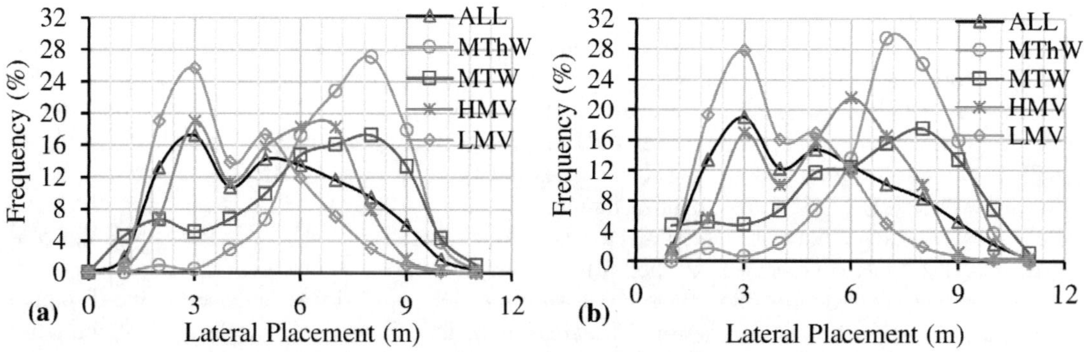

Figure 1. Lateral placement distribution of vehicles during (a) 9:30 to 10:30; and (b) 10:30 to 11: 30

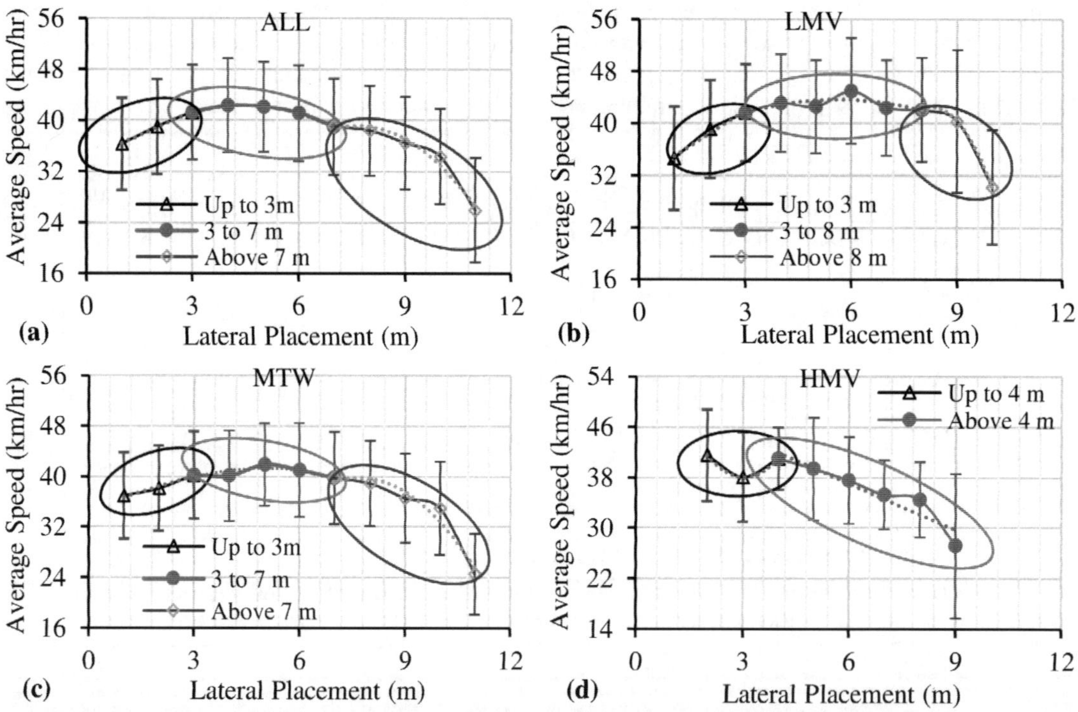

Figure 2. Variation of speed over each meter interval of lateral placement for (a) ALL vehicles; (b) LMV; (c) MTW; and (d) HMV for the traffic stream data collected between 9:30 and 10:30 hrs.

Figure 2 shows the variation of average speed over each meter interval of lateral placement for all vehicles clustered and each category of vehicles for the traffic data collected between 9:30 hrs and 10: 30 hrs. The figure also shows the error bars of standard deviation. It can be seen from figure 2(a) that there are three distinct zones of trends of average speed. Up to 3 m, the average speed is increasing as the vehicle is placing away from the median edge. The reason may be due to the fact that vehicles feel less safety while situated close to the edge. Hence, they maintain less speed as close to the median. Similar is the case for shoulder lanes (above 7 m). At the middle of the road section (3 m to 7 m), the speed is more or less flat. Keeping this in mind and from the visual observation, the lateral placement is divided into three groups. Up to 3 m from the median edge (0 to 3 m) the average speed is linearly related with lateral placement. Between 3 to 7 m, the average speed has quadratic relation with lateral placement. The last distinct zone, i.e., above 7 m, the average speed also has a quadratic relationship with lateral placement. The speed-placement model developed for three distinct zones are given in Table 3. The R-square values obtained for these three relationships are 0.9975, 0.9986 and 0.9705, respectively. Similar analysis has been carried out for LMV, MTW, HMV, and MThW. The average speed and lateral placement relationship for LMV, MTW and HMV are shown in Figure 2(b), 2(c) and 2(d), respectively. The speed-placement models for the different distinct zone for each category of vehicles and their R-square values are presented in Table 2.

Table 2. Models (speed-placement) developed for speed over each meter interval of lateral placement

Vehicle type	Models [$y = average\ speed (km/hr)$, $x = lateral\ placement\ (m)$]	The range of lateral placement (m)	R^2
ALL	$y = 2.4855x + 33.911$;	$0 < x \leq 3$	0.9975
	$y = -0.5094x^2 + 4.528x + 32.32$;	$3 \leq x \leq 7$	0.9986
	$y = -1.1197x^2 + 17.133x - 26.406$;	$10.8 \geq x \geq 7$	0.9705
LMV	$y = 3.501x + 31.476$;	$0 < x \leq 3$	0.9748
	$y = -0.309x^2 + 3.474x + 33.974$;	$3 \leq x \leq 8$	0.5151
	$y = -4.197x^2 + 69.604x - 246.1$;	$10.8 \geq x \geq 8$	1
MTW	$y = 1.6332x + 35.174$;	$0 < x \leq 3$	0.9719
	$y = -0.347x^2 + 3.475x + 32.608$;	$0 \leq x \leq 7$	0.5888
	$y = -1.312x^2 + 20.181x - 37.768$;	$10.8 \geq x \geq 7$	0.9501
HMV	$y = 3.188x^2 - 19.379x + 67.527$;	$1 < x \leq 4$	1
	$y = -2.461x + 51.809$;	$9 \geq x \geq 4$	0.8865
MThW	$y = 2.037x + 29.993$;	$1 < x \leq 5$	0.9677
	$y = 0.251x^2 - 4.876x + 57.386$;	$10 \geq x \geq 5$	0.7246

4. Validation

Traffic stream data collected between 10:30 hrs and 11:30 hrs from VIP Road, Kolkata, India is used to validate the models developed for average speed and lateral placement of all category of vehicles. Figure 3 shows the average speed over each meter lateral placement of all vehicles clustered for the observed data and the estimated average speed using the developed models (as shown in Figure 2(a) and presented in Table 3). It can be seen from the figure 3 that both observed and estimated average speed are reasonably matching. Similar exercises have been carried out for each category of vehicles and found that the observed and estimated average speeds are fairly matching.

IOP Conf. Series: Journal of Physics: Conf. Series **1298** (2019) 012016 doi:10.1088/1742-6596/1298/1/012016

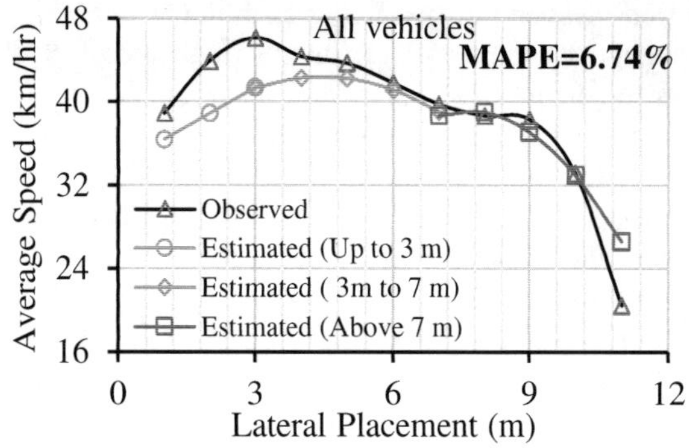

Figure 3. Observed and estimated average speed (km/hr) of all vehicles

The performance of the estimated models was carried out using the mean absolute percentage error (MAPE) value as given in equation (1)

$$MAPE = \frac{1}{N}\sum_{i=1}^{N}\left|\frac{O_i - E_i}{O_i}\right| * 100\% \quad (1)$$

Where N = number of observations, O_i = observed average speed (km/hr), E_i = estimated average speed (km/hr) using the developed model. The MAPE values for all vehicles clustered, LMV, MTW, HMV and MThW were found to be 6.74%, 9.69%, 4.04%, 5.47% and 9.16%, respectively. From the Lewis scale of interpretation of estimation accuracy [10] a prediction with a MAPE value less than 10% is considered to be highly accurate, 11%–20% as good, 21%–50% as reasonable, and 51% or more as inaccurate. The MAPE values for average speed and lateral placement were found in the range of 0%–10%. Hence, it can be inferred that the developed models are found to be in highly accurate range and can be used as a good estimator for average speed (km/hr) based on lateral placement (m).

5. Summary and conclusions

Non-lane discipline traffic stream data collected from VIP Road, Kolkata, India were used to analyze the impact of vehicular lateral placement on average speed. For this purpose, the vehicle trajectories were extracted from traffic video film using image processing software TRAZER. The raw trajectories were corrected using CEEMDAN techniques. The corrected speed from smoothed trajectories was obtained using WT. Lateral placement distribution of all vehicles and each vehicle category of first hour data indicates that at wider road (10.80 m) vehicles are placed across the road width. LMVs prefer to ply close to the median lane, whereas, MThW and MTW prefer shoulder lane. The impact of lateral placement on average speed shows that vehicles don't maintain constant speed across the road width. It depends on the lateral placement of vehicles. In case of all vehicles clustered, LMV and MTW three different distinct zones across the road width were observed. Close to the median (up to 3 m from the median edge), the average vehicular speed is linearly related with lateral placement. For the other two zones (3m to 7m, and above 7m), the relationship is quadratic. For HMV and MThW two distinct zones were observed. The second-hour data collected from the same road stretch was used to validate the developed speed-placement models using the MAPE between observed and estimated average speed over each meter interval of lateral placement. The MAPE were found to be 6.74%, 9.69%, 4.04%, 5.47% and 9.16% for all vehicles clustered, LMV, MTW, HMV, and MThW, respectively. According to the Lewis scale of interpretation of estimation accuracy [10] a prediction with a MAPE value less than 10% is considered to be highly accurate. Hence, it can be said that the developed models are found to be in highly accurate range and can be used as a good estimator of average speed (km/hr) based on lateral placement (m). The results of this study may be used as an input parameter for assigning vehicular speed based on lateral placement for analyzing non-lane discipline traffic stream using the simulation model of three lanes road. To further verify the results, more data covering different road width from various arterials consisting non-lane discipline traffic stream may be analyzed.

6. References

[1] Asaithambi G, and Joseph J 2018 *Journal of Transportation Engineering Part A: Systems* **DOI: 10.1061/JTEPBS.0000170.**

[2] Blab R, and Litzka J 1995 *Proc. of the Int. Symp. on Heavy Vehicle Weights and Dimensions* (Road Transport Technology University of Michigan) **Ann Arbor MI** 389–395.

[3] Bharadwaj N, Kumar P, Arkatkar S, Maurya A, and Joshi G 2016 *Current Science* **110** 808-822.

[4] Balaji K, Bharadwaj M R K, and Dey P P 2013 *Indian Highways* **41** 49-53.

[5] Bunker J M, and Parajuli A 2006 *Transport Engineering in Australia* **10** 129–139.

[6] Dey P P, Chandra S, and Gangopadhyay S 2006 *Journal of Transportation Engineering* **132** 597-600.

[7] Mohapatra S S, and Dey P P 2015 *Transportation letters* **7** 252-263.

[8] Mallikarjuna C, Phanindra A, and Ramachandra R K 2009 *Journal of Transportation Engineering* **135** 174-182.

[9] Pal D, and Chunchu M 2018 *Canadian Journal of Civil Engineering* **45** 435-445.

[10] Kenneth D L, and Ronald K K 1982 *Advanced in Business and Management Forecasting* Emerald Book Bingley UK.

An Analysis of Active Region as a Trigger of Solar Flares

ZS Hamidi[1,2],M.F.M.Noh[2], W.N.Toni[2], N.N.M.Shariff[3]
[1,2]Institute of Science, Universiti Teknologi MARA, 40450, Shah Alam, Selangor, Malaysia
[2] Faculty of Applied Sciences, Universiti Teknologi MARA, 40450, Shah Alam, Selangor, Malaysia
[3]Academy of Contemporary Islamic Studies, Universiti Teknologi MARA, 40450, Shah Alam, Selangor, Malaysia

Email: zetysh@salam.uitm.edu.my

Abstract. Sunspot number, solar radio flux and solar wind involve in interpreting the movements of CMEs and solar flare. Rotating sunspots are an extremely efficient way to inject energy into the magnetic field of the sun's atmosphere. A solar flare is essentially a blast on the surface of the sun going from minutes to hours long. Therefore, the objective of this study is to state that the active region sunspot waves as a trigger solar flare. Equally important, the energy efficiency associated with solar flares may take several hours or even days to build up, but most flares take only a matter of minutes to release their energy. Detailed analyzed of solar flare event on 2016 and 2017 based on e-CALLISTO, the higher the number of active region sunspots, the higher the class of flares produced and released. Four classes of solar are observed in categories of frequency, interplanetary magnetic field and sunspot number emitted by the solar. By using the interplanetary magnetic field data, the magnetic energy contained in the active region sunspots has been calculated which this magnetic energy triggered the emission of solar flare. This study exposed us to analyze the active region waves as a trigger of solar flares and also the factor influenced by it. The physical element that triggers the solar flare by measuring the magnetic energy in the flaring site.

Key words. Sun; Active region; Sunspot; Solar Flare; Data analysis

1. Introduction

In the past, researcher had more focus on the mechanism of the solar flare and nuclear fusion, but rejected due to the low densities in upper chromosphere or lower corona. The positive sunspots, composed of proton particles, and the negative sunspots composed of electron particles, are produced and ejected from tachocline zones, then appearing on the solar surface and moved towards the polar region. However, it processes takes a longer period of time, the nuclear fusion occurred along the first or multiple inner orbits of gyrating protons, depends on the magnitude of producing external magnetic field [1]. Therefore, the objective of this study is to state that the active region sunspot waves as a trigger solar flare. Sunspot number, solar radio flux and solar wind involves in interpreting the movements of CMEs and solar flare [2]. Solar flares are greatly related to the sunspot appearance, it is also found that flares usually occur near and along the dividing line (neutral line) between areas of oppositely directed magnetic fields of sunspots and the development of flares [3].

2. Literature Review

IOP Conf. Series: Journal of Physics: Conf. Series **1298** (2019) 012017 doi:10.1088/1742-6596/1298/1/012017

The disturbed area around and above the bipolar sunspot pairs or groups which are highly magnetized is called an active region A solar flare is essentially a blast on the surface of the sun going from minutes to hours long. Flares occur when the magnetic fields in and around the sun reconnect. There are 5 classes of solar flare (A, B, C, M and X) [4]. They are typically connected with active regions, frequently observed as sun spots, where the magnetic fields are strongest. Solar flare and Coronal Mass Ejections (CMEs) are well known as one of the most massive eruptions which potentially create major disturbances in theinterplanetary medium and initiate severe magnetic storms when they collide with the Earth's magnetosphere [5]. Flares and CMEs do not occur when potential magnetic field is in stable equilibrium (a) state because there is no free energy that can be released [6]. Sheared magnetic fields or flux ropes that store free energy when in metastable state (b) and loss-of-equilibrium or loss-of-stability can occur owing to some disturbances [7]. By storing more free energy, the system moves to an unstable equilibrium state (c). Finally, the system reaches the non-equilibrium state (d), and the system cannot stay in a static state any longer and moves to a dynamic state. In this case, sheared magnetic arcades reconnect owing to emerging flux or horizontal motion in the photosphere, and flux ropes suddenly rise upward if the system reaches the loss-of-equilibrium state; then, flares and CMEs occur. Interference with convection magnetic field darkening the sunspots. The spikes clustered around the magnetic field region develop into loops connecting regions of magnetic polarity at active region. The surface horizontal rotational flows could provide sufficient magnetic helicity and energy to produce solar flares [8]. Explosive events such as solar flares can be generated by changing the magnetic field causing by the magnetic reconnection caused by the transverse compressible wave [9].

Rotating sunspots are an extremely efficient way to inject energy into the magnetic field of the sun's atmosphere. Magneto hydrodynamic (MHD) waves play an important role because waves are natural carriers of energy, momentum and data 3-min oscillation over sunspots are one of the most distinct wave phenomena at chromospheric level and assumed to be related to slow magneto acoustic waves[10]. Class X flares are the strongest types of solar flares that can erupt from the sun. There are also two weaker categories: Class M flares, which are medium strength, but still powerful, and Class C flares, which are the weakest storms from the sun. Class A flares have purely thermal, compact sources while Class B flares are impulsive bursts which show double foot points in hard X-rays. It is widely accepted that solar flares are explosive phenomena that free the magnetic field energy kept in the solar corona mainly as thermal and kinetic energy of the plasma [11]. The typical spatial size, duration and released energy of a flare is about 10^9 cm, 10^3 sec and 10^{28} erg respectively.

Detailed analyzed of solar flare event on 2016 and 2017 based on e-CALLISTO, the higher the number of active region sunspots, the higher the class of flares produced and released. The sunspot number played an important role in the release of the solar flares. There are 85 numbers of sunspots on 5th May 2005 and on this day X class flare was released which is the highest class of flare. The cooler temperatures result in increased magnetic activity which allows the charged particles to escape from the sun's atmosphere because of their increased activity. The solar radio bursts are the results of solar flares take place on the surface of the Sun [12] .

3. Experimental

The CALLISTO (Compound Low Cost-Low Frequency for Transportable Observatories) system is one of the most outstanding project under ISWI with there are more than 80 instruments in more than 43 locations with users from more than 116 countries consist of indoor amplifier, CALLISTO

Receiver, and computer [13, 14]. This network systems are widely used for continuous data collection of solar activities every day through the internet connection and stored in the central database in the computer [15]. At the beginning, the CALLISTO spectrometer was covering a frequency range from 45MHz to 870MHz with the nearness of log-periodic antenna and has a full scope of frequency at Blein Radio Observatory around 50km west of Zurich [16]. The other frequency range can be seen by exchanging in a heterodyne up and down converter. The radiometric bandwidth around 300 KHz given by ceramic band pass channel, while the programmable step measure in frequency is 62.5 KHz. Callisto has constructed precisely for radio spectrometer and focusing on solar flare radio. Noted that all the sites are less suffering from local Radio Frequency Interference [17].

4. Result and Discussion

Four classes of solar flare were observed in categories of frequency, interplanetary magnetic field and sunspot number emitted by the solar. Such in solar flare class B which is the highest frequency is 900 MHz on 19/4/2016 which released energy value is 5.963 x 10^{-31} MeV while on 27/5/2017 the highest frequency is 70 MHz however it rising on 9/6/2017 to 360 MHz.

Table 1: Data energy emitted in solar flares

Date	Burst Duration (Minutes)	Lowest Frequency (MHz)	Highest Frequency (MHz)	Drift Rate	Energy at Lowest Frequency (MeV)	Energy at Highest Frequency (MeV)
19/4/2016	2	200	900	5.8	$1.325\text{x}10^{-31}$	$5.963\text{x}10^{-31}$
21/4/2016	1	65	160	1.6	$4.307\text{x}10^{-32}$	$1.060\text{x}10^{-32}$
27/5/2017	2	20	70	0.4	$1.325\text{x}10^{-32}$	$4.368\text{x}10^{-32}$
9/6/2017	2	110	360	2.1	$7.289\text{x}10^{-32}$	$2.385\text{x}10^{-31}$

Active Region 2529 just produced a moderately strong B9 solar flare peaking at 23:00 UTC (April 19th). Solar activity has been at very low levels for the past 24 hours. There are currently 2 numbered sunspot regions on the disk Solar activity is expected to be low with a slight chance of an M-class flare on day one (20th April) and expected to be very low with a chance for a C-class flares and a slight chance for an M-class flare on day two (21st April) and expected to be very low with a slight chance of a C-class flare on day three (22nd April).

For class B solar flare, the highest frequency which is 500 MHz with $3.31\text{x}10^{-31}$ MeV was started on 28th March 2016 and on 25th July 2016 stated the lower frequency among the highest frequency, 90 MHz with $4.64\text{x}10^{-32}$ MeV.

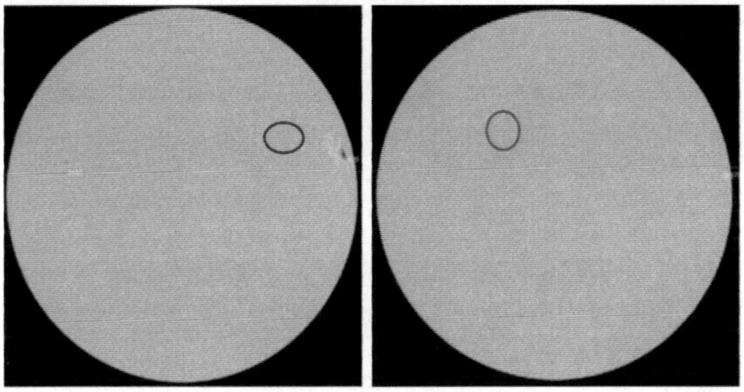

Figure 1: The location of the active region AR2529 which release the B9 flare The location of the active region AR2565 which release the C1

Table 2 Data energy emitted in solar flares

Date	Burst Duration (Minutes)	Lowest Freque ncy (MHz)	Highest Frequency (MHz)	Drift Rate	Energy at Lowest Frequency (MeV)	Energy at Highest Frequency (MeV)
19/4/2016	2	200	900	5.8	1.325×10^{-31}	5.963×10^{-31}
21/4/2016	1	65	160	1.6	4.307×10^{-32}	1.060×10^{-32}
27/5/2017	2	20	70	0.4	1.325×10^{-32}	4.368×10^{-32}
9/6/2017	2	110	360	2.1	7.289×10^{-32}	2.385×10^{-31}

The largest solar event of the period was a C1 event observed at Region 2567. There are currently no sunspot regions on the disk. The solar activity is expected to be very low with a chance for a C-class flares and a slight chance for an M-class flare on day one (26th July) and expected to be very low with a slight chance of a C-class flare on days two and three (27th July, 28th July).

By using the interplanetary magnetic field data, the magnetic energy contained in the active region sunspots has been calculated which this magnetic energy triggered the emission of solar flare. The highest data obtained was when 27th May 2016 and 12th February 2016 which the magnetic field was 21.9 nT and 17.4 nT respectively. By this, the magnetic energy contained in the active region sunspots valued were 6.85×10^{80} ergs on 12th February 2016 and 1.09×10^{81} erg on 27th May 2016 but this only determined the magnetic energy stored in active region sunspots and does not relate heavily to the triggered of solar flares.

5. Conclusion

As a conclusion, solar flares are produced from the active region sunspots, based on a few factors such as the magnetic field energy contained in the sunspots, level of energy released which can be calculated from the frequency data obtained from the e-CALLISTO software which is updated daily and can be observed from the software and its websites. And the most important factor that triggers the solar flares based on their level of strongest are from the origin which is the number of sunspots. The higher number of sunspots on the surface of the Sun, the higher the class of flares. The explosion of solar flares has potential to interrupt GPS system, the power grid and also other telecommunication systems. We have highlighted the knowing the importance active region sunspots in the energy release in the explosion of solar flares.

6. Acknowledgment

We are grateful to CALLISTO network; STEREO, LASCO, SDO/AIA,NOAA and SWPC make their data available online. This work was partially supported by the Ministry of Higher Education (MOHE) (600-RMI/FRGS 5/3 (0077/2016).

REFERENCES

1. Yousif, M.E., *The electromagnetic radiation mechanism.* International Journal of Fundamental Physical Sciences (IJFPS), 2014. **4**(3): p. 72.
2. Shariff, N., Z. Hamidi, and N. Zainol. *Analysis of Slow Partial Halo CME Events with Velocity of 100-200 km/s: An Observation through CALLISTO System.* in *Industrial Engineering, Management Science and Application (ICIMSA), 2017 International Conference on.* 2017. IEEE.
3. Parker, E.N., *The Solar-Flare Phenomenon and the Theory of Reconnection and Annihiliation of Magnetic Fields.* The Astrophysical Journal Supplement Series, 1963. **8**: p. 177.
4. Hamidi, Z. and N. Shariff. *The Non-thermal of a Large Solar Flares Associated with High Energy of Solar Burst Type III on 4th September 2017.* in *Journal of Physics: Conference Series.* 2018. IOP Publishing.
5. Hamidi, Z. and N. Shariff. *Automatic solar radio burst type II and III image processing tracking by using CALLISTO system.* in *Industrial Engineering, Management Science and Application (ICIMSA), 2017 International Conference on.* 2017. IEEE.
6. Hamidi, Z., et al. *Magnetic Reconnection of Solar Flare Detected by Solar Radio Burst Type III.* in *Journal of Physics: Conference Series.* 2014. IOP Publishing.
7. Rachmeler, L., *Modeling eruptive coronal magnetohydrodynamic systems with FLUX.* 2010, University of Colorado at Boulder.
8. Zhao, J. and A.G. Kosovichev, *Helioseismic observation of the structure and dynamics of a rotating sunspot beneath the solar surface.* The Astrophysical Journal, 2003. **591**(1): p. 446.
9. Nakariakov, V. and V. Melnikov, *Quasi-periodic pulsations in solar flares.* Space Science Reviews, 2009. **149**(1-4): p. 119-151.
10. Giordano, S., *The Connection between solar magnetic fields and photospheric dynamics.* 2008.
11. Heyvaerts, J., E.R. Priest, and D.M. Rust, *An emerging flux model for the solar flare phenomenon.* The Astrophysical Journal, 1977. **216**: p. 123-137.
12. Husien, N., et al. *Solar Radio Bursts Detected by CALLISTO System and Their Related Events.* in *Industrial Engineering, Management Science and Application (ICIMSA), 2016 International Conference on.* 2016. IEEE.
13. Hamidi, Z.S., et al., *The international cooperation of solar radio burst project using e-Callisto system network*, in *Information Science and Applications (ICISA) 2016.* 2016, Springer. p. 309-317.
14. Ali, M., et al. *e-CALLISTO network system and the observation of structure of solar radio burst Type III.* in *Industrial Engineering, Management Science and Application (ICIMSA), 2016 International Conference on.* 2016. IEEE.
15. Zainol, N., et al. *Effective data collection and analysis of solar radio burst type II event using automated CALLISTO network system.* in *Industrial Engineering, Management Science and Application (ICIMSA), 2016 International Conference on.* 2016. IEEE.
16. Sabri, S.N.U., et al., *The Significance of e-CALLISTO System and Construction of Log Periodic Dipole Antenna (LPDA) and CALLISTO System for Solar Radio Burst Study*, in *Information Science and Applications (ICISA) 2016.* 2016, Springer. p. 181-188.
17. Shariff, N., Z. Hamidi, and S. Sabri. *The Variance Formation of M9. 3 Solar Flare Associated with Solar Radio Burst Type V Occurred at Eight (8) Different Sites of the CALLISTO System on 4th August 2011.* in *Journal of Physics: Conference Series.* 2018. IOP Publishing.

IOP Conf. Series: Journal of Physics: Conf. Series **1298** (2019) 012018 doi:10.1088/1742-6596/1298/1/012018

An Analysis on the Formation of Solar Radio Burst Type II, III and IV by Using e-CALLISTO, IUGONET and Space Weather Data

N A M Norsham[1,2], Z S Hamidi[1,2]

[1] Institute of Science, Universiti Teknologi MARA Shah Alam Selangor Malaysia
[2] Faculty of Applied Sciences, Universiti Teknologi MARA Shah Alam Selangor Malaysia

E-mail: nuramira396@gmail.com

Abstract. The study of the radio solar burst data on 30th March 2018 and 31st March 2018 are essential for the understanding of type II, type III and type IV nature. Through this paper, the relationship of these three types of solar radio burst can be studied. The necessary data were collected from the free online websites which includes the Space Weather Website, e-CALLISTO and IUGONET metadata database. From this study, the occurrence of solar radio burst type II, III, and IV showed that there was the production of solar flares and Coronal Mass Ejections (CMEs) where it contributed to the geomagnetic disturbances to the Earth. On 30th March 2018, the Sun produced C-class flares with C4 value while the next day, the sun had B1 solar flares' event. The planetary K-index for both days were 2 (30th March 2018) and 3 (31st March 2018) respectively.

1.Introduction

The Sun is our nearest star to the Earth with the distance of 1AU. When the activity of the Sun is active, the Sun will produce a fast-moving solar wind that eventually will cause the effect to the surrounding interplanetary space. Basically, when the sun is very active, they will produce intense magnetic field which then will trigger the formation of solar flares or another powerful phenomenon which is known as Coronal Mass Ejection (CME).

Type II solar radio burst has slow drift bursts where they drift from high to low frequencies at the rates of about ¼ Mc per sec per sec [1] where the drift rate decreasing when the wavelength is decreasing [2, 3]. This kind of solar radio burst type are released in second harmonic and fundamental bands as proposed by Murray, Wild and Rowe that the drift of the frequency is caused by the excitation in the plasma oscillation of the coronal gas where the frequency decreases with time. It is due to the exciting activity travels to the region of decreasing electron density. The metric of type II burst is produced by the Coronal Mass Ejections, flares and also different types of ejecta that is considered as drivers of the MHD shocks [4-6]. Solar radio burst type II was identified and discovered by Payne Scott et al. [7] and Wild and McCready [8] respectively. They are categorized by the broadband and have time duration of about 20 minutes to a few hours.

Solar radio burst type III is a fast drift burst and most common to be found in meter wavelength bursts with the frequency range of 10 MHz to 500 MHz. This type of solar radio burst was introduced by Wild [9] and can be occurred singly, in group or in the storm. The intensity of the signal may vary for type III burst with the frequency. The duration of type III burst can exceed 30 minutes [10]. The electron plasma waves of type III burst are coalescence from the nonthermal distribution with the thermal distribution charge fluctuations [11, 12]. Type IV solar radio burst shows the formation of new active region [13, 14]. Some of type IV burst will come before type II burst at meter wavelength. This type of bursts is related to the production of the sunspot group. It is also well known to be followed by the geomagnetic disturbance [15].

2. Methodology and Observation

All the data for this study was collected from the Space Weather Website, IUGONET and e-CALLISTO. CALLISTO is the abbreviation for Compound Astronomical Low cost Low frequency Instrument for Spectroscopy and Transportable Observatory where currently there are more than 150 instruments at more than 90 locations worldwide. CALLISTO is design for solar radio burst and RFI monitoring for education and astronomical science purpose [16] and is a global network that is connected via the internet [17]. The data collected for solar radio burst were from BLENSW, GREENLAND and BLEN5M site. BLENSW and BLEN5M are the observatory from Switzerland while GREENLAND is from the ISR Observatory, Greenland. The magnetic field data was taken from the IUGONET metadata database and other parameters of the sun such as the speed, density, and radio sun were collected from the Space Weather Website. All data then were analysed to see their relationship to each other.

3.Discussion

The tables and figures provide the data and information about solar radio burst type II, III, and type IV.

Table 1. Sun's current conditions on 30th March 2018 (Credit: Space Weather)

Parameter	Value
Solar Wind	454 km/sec
X-ray Solar Flares	C4
Sunspot no.	0
Radio Sun	10.7 cm flux: 69 sfu
Planetary K-index	Kp = 2 (quiet)
Interplanetary Magnetic Field	B_{total}: 8 nT B_z: -4 nT South

Table 2. Probabilities of Geomagnetic Storms at High-latitudes (Credit: Space Weather)

	0-24 hr	24-48 hr
Active	20%	15%
Minor	25%	20%
Severe	20%	15%

Figure 1. The Sun surface for sunspot and coronal hole region on 30th March 2018.(Credit: SDO/HMI/AIA)

IOP Conf. Series: Journal of Physics: Conf. Series **1298** (2019) 012018 doi:10.1088/1742-6596/1298/1/012018

Table 1 above shows the condition on the Sun's surface on 30th March 2018. The solar wind was a bit high with the speed of 454 km/sec although the sunspot number was 0. But, as can be seen from Figure 1 of coronal holes structure, we can observe that solar wind stream was high as the temperature and densities of this region was very low. Thus, the area is thin and make it easy for solar wind to break through it. We are reaching the solar minimum as the spotless days of the Sun can be seen for the past 2 years. From the X-ray solar flares recorded, the flare was classified to be in the C-class with the indication of giving few visible consequences to the Earth. C-class flare ranging from $10^{-6} \leq 1 < 10^{-5}$.

Radio sun with 69 s.f.u. shows the correlation with the sunspot number and a number of UV and visible solar irradiance records. The K-index showed the value of 2 which means the sun is in calm condition. The chances for the occurrence of geomagnetic storm at high latitudes of severe rate was low with the reading of 15% for one to two days alert and only 1% at mid-latitudes regions as mentioned in Table 2.

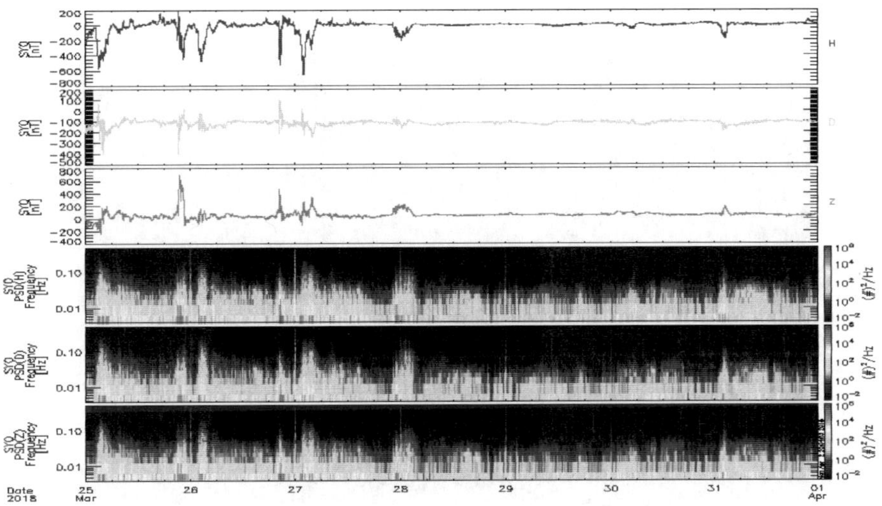

Figure 2. Magnetic field data with 1sec resolution from the fluxgate magnetometer at Syowa Station, Antarctica (Credited: IUGONET data; National Institute of Polar Research, Japan)

The interplanetary magnetic field of the Sun during that day (30th March 2018) was B_z: -4 nT South. It is known that when the value of this magnetic field is lower, the chances for the aurora activity is much higher. Figure 2 shows the magnetic field data from 25th March 2018 until 1st April 2018. The above magnetic field data shows that on 26th, 27th and 28th March 2018, the peak was a bit high in a term of their H, D and z components. This may indicate that the recorded solar bursts from the CALLISTO systems was caused by the sun activities on those days.

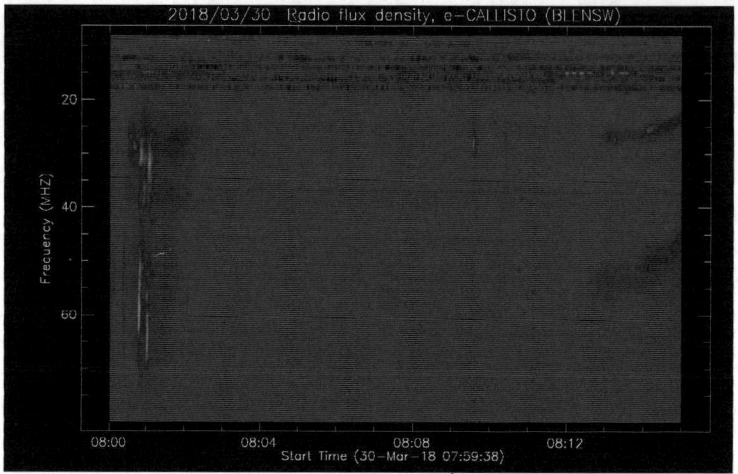

Figure 3. Solar radio burst type III and type II on 30th March 2018 (Credit: e-CALLISTO)

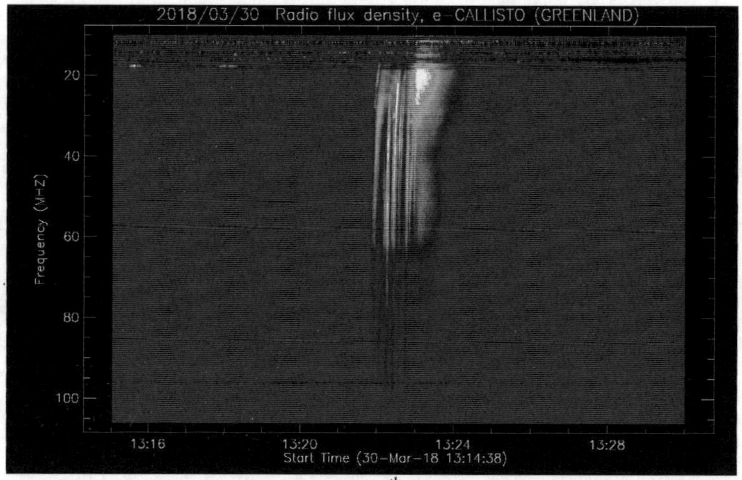

Figure 4. Type III solar radio burst on 30th March 2018 (Credit: e-CALLISTO)

There are two events of solar radio burst data recorded on 30th March 2018 (Figure 3) after a few quiet days of the Sun. The first event of solar radio burst occurred at 08:00 UT while the second event at 13:22 UT. The event on 08:00 UT was producing type III solar radio burst followed with type II burst. Type II burst showed the fundamental (occurs between 20 to 150 MHz) and first harmonic radiation which lasted for ~1 minute. Radio burst type III started at 20 MHz and ended at 70 MHz while type II started at 20 MHz until 60 MHz. From the burst shown, we knew that the sun was experienced solar flare and followed by the production of Coronal Mass Ejection (CME). The next event (Figure 4) at 13:22 UT showed small group of solar radio burst type III with duration of two minutes. It means that, there was another occurrence of solar flares after the event at 08:00 UT. The burst data was obtained at Greenland site of CALLISTO.

Table 3. The condition of the Sun on 31st March 2018 (Credit: Space Weather)

Parameter	Value
Solar Wind	469 km/sec
X-ray Solar Flares	B 7.1
Sunspot no.	11
Radio Sun	10.7 cm flux: 69 sfu
Planetary K-index	Kp = 3 (quiet)
Interplanetary Magnetic Field	B_{total}: 7 nT ; B_Z: -5 nT South

IOP Conf. Series: Journal of Physics: Conf. Series **1298** (2019) 012018 doi:10.1088/1742-6596/1298/1/012018

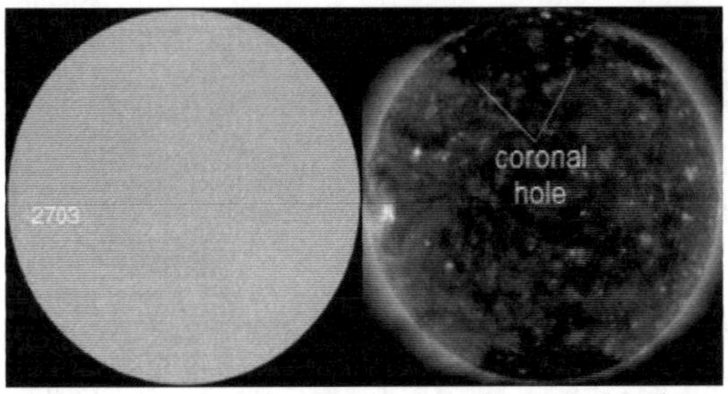

Figure 5. The Sun's surface showing the sunspot and coronal hole on 31st March 2018. (Credit: SDO/HMI/AIA)

Table 3 above shows the condition of the Sun on 31st March 2018. On that day, there was 11 sunspot numbers that appeared on the Sun's surface (new sunspot appeared, AR2703) with the radio flux of 69 s.f.u. That day solar flare was recorded and categorized under B1 class which was caused by the AR2703. The solar wind speed on that day was quite high with 469 km/sec. The Kp index was 3 (quiet). This Kp index was recorded by a magnetometer, an instrument that measure the disturbances of geomagnetic field. The data from magnetometer for this 31st March 2018 can be refer from Figure 2. Besides that, the B_z component value was -5.0 nT pointing to the South and the probability of geomagnetic occurrence was 10% severe for the coming one to two days starting on that day.

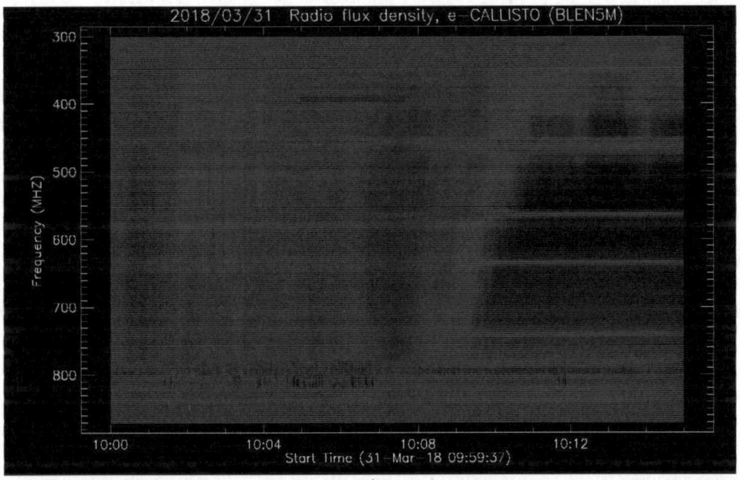

Figure 6. Radio burst type IV on 31st March 2018 (Credit: e-CALLISTO)

Figure 6 above displays the data of solar radio burst type IV which was recorded by the CALLISTO system at site Bleien, Switzerland. As can be seen from the data above, the frequency range were from 300 MHz to 800 MHz with duration of approximately 500 seconds. Usually, type IV occurs after a great outburst which related to solar flares or plasma emissions. Besides, in other word, type IV burst can also occur when there is the formation of the new active region and production of sunspot group where they might yield the geomagnetic disturbance [13-15]. The Sun's surface on 31st March 2018, showed the appearance of the new active region 2703. This can be a proof of the production of the solar radio burst type IV.

4. Conclusion

From the all the data collected on 30th March 2018 and 31st March 2018, it shows that even the Sun is not active (in terms of the production of the sunspot), we can still get the data which indicates that the sun, somehow, is active for the production of the flares and Coronal Mass Ejection. In addition, the activities from the past days before 30th March 2018 contributed to the production of the solar flares and Coronal Mass Ejections thus enhancing the geomagnetic disturbance.

Acknowledgments
We appreciate the efforts of the teams of CALLISTO network, SDO/AIA, NOAA, SWPC and IUGONET. This research was supported by UiTM under grant 600-RMI/FRGS 5/3 (0077/2016).

5. References

[1] M. R. Kundu, "Solar radio astronomy," *New York: Interscience Publication, 1965,* 1965.
[2] B. Vršnak, J. Magdalenić, and P. Zlobec, "Band-splitting of coronal and interplanetary type II bursts-III. Physical conditions in the upper corona and interplanetary space," *Astronomy & Astrophysics,* vol. 413, no. 2, pp. 753-763, 2004.
[3] N. Gopalswamy, "Coronal mass ejections and type II radio bursts," *GEOPHYSICAL MONOGRAPH-AMERICAN GEOPHYSICAL UNION,* vol. 165, p. 207, 2006.
[4] A. Nindos, H. Aurass, K.-L. Klein, and G. Trottet, "Radio emission of flares and coronal mass ejections," *Solar Physics,* vol. 253, no. 1-2, p. 3, 2008.
[5] M. Pick and N. Vilmer, "Sixty-five years of solar radioastronomy: flares, coronal mass ejections and Sun–Earth connection," *The Astronomy and Astrophysics Review,* vol. 16, no. 1-2, pp. 1-153, 2008.
[6] B. Vršnak and E. W. Cliver, "Origin of coronal shock waves," *Solar Physics,* vol. 253, no. 1-2, p. 215, 2008.
[7] R. Payne-Scott, D. Yabsley, and J. Bolton, "Relative times of arrival of bursts of solar noise on different radio frequencies," *Nature,* vol. 160, no. 4060, p. 256, 1947.
[8] J. P. Wild, McCready, L.L., *Aust. J. Sci Res,* pp. 3387-3398, 1950.
[9] J. Wild, S. Smerd, and A. Weiss, "Solar bursts," *Annual Review of Astronomy and Astrophysics,* vol. 1, no. 1, pp. 291-366, 1963.
[10] Z. S. Hamid, N. Shariff, Z. Ibrahim, and N. Ramli, "Investigation of the statistical properties of solar radio burst type II and III," in *Space Science and Communication (IconSpace), 2015 International Conference on*, 2015, pp. 335-338: IEEE.
[11] D. Melrose, "On the theory of type II and type III solar radio bursts. I. The impossibility of nonthermal emission due to combination scattering off thermal fluctuations," *Australian Journal of Physics,* vol. 23, no. 5, pp. 871-884, 1970.
[12] Z. Hamidi and N. Shariff, "Automatic solar radio burst type II and III image processing tracking by using CALLISTO system," in *Industrial Engineering, Management Science and Application (ICIMSA), 2017 International Conference on*, 2017, pp. 1-4: IEEE.
[13] Z. Hamidi, Z. Abidin, Z. Ibrahim, C. Monstein, N. Shariff, and M. Sabaghi, "The Beginning Impulsive of Solar Burst Type IV Radio Emission Detection Associated with M Type Solar Flare," *International Journal of Fundamental Physical Sciences,* vol. 2, pp. 32-34, 2012.
[14] Z. Hamidi, N. Shariff, and C. Monstein, "Disturbances of Solar Eruption From Active Region AR1613," *International Letters of Chemistry, Physics and Astronomy,* vol. 13, no. 1, p. 77, 2014.
[15] H. V. Cane, W. Erickson, and N. Prestage, "Solar flares, type III radio bursts, coronal mass ejections, and energetic particles," *Journal of Geophysical Research: Space Physics,* vol. 107, no. A10, pp. SSH 14-1-SSH 14-19, 2002.
[16] Z. Hamidi, Z. Abidin, Z. Ibrahim, and N. Shariff, "Indication of radio frequency interference (RFI) sources for solar burst monitoring in Malaysia," in *AIP Conference Proceedings*, 2012, vol. 1454, no. 1, pp. 43-46: AIP.
[17] N. Anim, Z. Hamidi, Z. Abidin, C. Monstein, and N. Rohizat, "Radio frequency interference affecting type III solar burst observations," in *AIP Conference Proceedings*, 2013, vol. 1528, no. 1, pp. 82-86: AIP.

IOP Conf. Series: Journal of Physics: Conf. Series **1298** (2019) 012019 doi:10.1088/1742-6596/1298/1/012019

The Impact on Climate Change Due to the Effect of Global Electromagnetic Waves of Solar Flare and Coronal Mass Ejections (CMEs) Phenomena

[1,2]N Mohamad Ansor, [1,2]Z S Hamidi, [1,3]N N M Shariff

[1]Institute of Science, Universiti Teknologi MARA, 40450 Shah Alam, Selangor, Malaysia
[2]Faculty of Science, Universiti Teknologi MARA, 40450 Shah Alam, Selangor, Malaysia
[3] Academy of Contemporary Islamic Studies, Universiti Teknologi MARA, 40450 Shah Alam, Selangor, Malaysia

Email address: nurain_ansor@yahoo.com

Abstract. Climate change has a close relation with the variability of the Sun's energy radiated, which is associated with solar activity and this issue has been the top debate topic among climate scientists in finding out the main source of the changing in Earth's climate. Previous studies were discussing on human activities and formation of ^{14}C and ^{10}Be isotopes which are likely contributing to the temperature increment of Earth's surface. However, the pattern of climate on this planet is always depending on the condition of its main source of energy; the Sun. In this paper, we briefly review the impact of production of ^{14}C and ^{10}Be isotopes to Earth's temperature. We also discussed in this article on how the variability of solar activity contributes to climate change. The main method in carrying out this study is by solar observation. We present the summary of several sets of significant solar events in 2017 and its relationship with Earth's temperature. It is believed that Earth's climate is likely to be affected by the variability of solar activity, in the sense that solar minimum in 2017 causes it to be the second highest annual temperature with 0.89°C warmer than normal temperature.

1. Introduction

Climate change basically refers to the concentration of carbon dioxide produced in Earth's atmosphere. This phenomenon has been happening for the last 600,000 years and scientists believed that solar activity, which we are focusing on in this paper are solar flares and Coronal Mass Ejections have a crucial role in changing this climate. Solar flares and Coronal Mass Ejections (CMEs) emit radiation, which covers across the electromagnetic (EM) waves spectrum from radio waves up to gamma rays. High energy flares can disrupt our power grids, communications, Global Positioning System (GPS) signal and cover the Earth's upper atmosphere with hazardous radiation [1]. Meanwhile, CME can trigger geomagnetic storm when it interupts the Earth's magnetic field. If worse, this storm can cause blackout in electrical power and damage the communications satellite [2]. Starting in 1860 until 1990, the global mean annual surface temperature increased by 0.55°C [3]. Hypothesis has been made on the warmer temperature that is caused by ongoing release of carbon dioxide gas from the industries. However, analysis of climate records propose that the cause of the warming is more than the influence of increasing carbon dioxide gases as the records expose outstanding inter-annual and inter-decadal variability [4]. Variability of solar activity modifies the climate in so many ways. Solar activity can be minimized or maximum depends on the number of sunspots present on Sun's surface. The long-term significant event in climate study is the Maunder Minimum (1645-1715) which is happened to coincide with the Little Ice Age, where our Earth was somehow cooler than present [5]. A study conducted by [6] has successfully shown the relation between the number of sunspots observed during the Maunder Minimum and Earth's climate during the period. Maunder Minimum is described by its minimum sunspots observed by astronomers. As [7] discovered in their research, the Sun during Maunder Minimum was 0.25% less bright than it was during previous solar minimum in 1985-1986. It is estimated that the solar irradiance during the period is 0.24%, lesser from present that should contribute to decreasing global temperature by 0.2°C [6].

Global climate changes through total radiation which varies with solar activity, specifically in the ultraviolet (UV) range. At the first place, scientists assumed that the Total Solar Irradiance (TSI) was constant and unvarying when satellites have yet to be established in the space. Starting in 1978, when satellites have been launched, TSI could be measured by direct measurement by the satellites [8]. Later, [9] come up with their combined record to prove that TSI is not constant and is varied ~0.1% from minimum to maximum of solar cycle. A strong solar flare does not raise the TSI more than a hundredth percent, however they temporarily amplify the solar output at UV, radio and X-ray wavelengths [8].

2. Production of cosmogenic isotopes ^{14}C and ^{10}Be at Earth's atmosphere

The formation of these isotopes are the consequences of cosmic rays of both galactic and solar origin and are formed in the upper atmosphere. They are produced from the change in cosmic ray flux fluctuations that intrude the Earth in which the fluctuations are caused by solar wind [10]. Solar wind includes the ejction of particles from the Sun through the interplanetary medium and can strongly influence the Earth's magnetic field. The production of ^{14}C is low as the Earth is shielded from cosmic rays due to prolonged solar magnetic field during high solar activity. Hence, low solar activity leads to more cosmic rays penetrate through Earth's atmosphere and producing more ^{14}C. Sudden increase in ^{14}C concentration indicates a change towards colder or wetter climate, which shows that solar does play a role in climate change [10]. Therefore, data of ^{14}C concentration can be a good measure for solar output [11].

However, records of ^{10}Be are also taken into account as changes in ^{14}C concentration is too weak to prove this assumption. The production of ^{10}Be is the consequence of nuclear reactions induced by cosmic rays and happens in two lowest layers of the Earth's atmosphere which are stratosphere and upper troposphere. Moreover, [12] found out that ^{14}C and ^{10}Be are alongside each other from 8000 to 5000 BP. This shows the variations of ^{10}Be is also a good indicator in relating changing solar activity and cosmic ray intensity. A data record of ^{10}Be concentration and $\delta^{18}O$ which was produced by [9] shows strong fluctuations between ^{10}Be concentration data of 40,000-11,000 years ago and $\delta^{18}O$ record of the same ice core in which they are parallel to each other. This implies warm phases give low values of ^{10}Be concentration and peaks in ^{10}Be concentration happen during cold phases. Moreover, the difference in ^{10}Be is believed to indicate changes in snow accumulation rate [13].

3. Solar Observation

In this study, we observed and analyzed several selected solar events based on class of solar flares that were occurred recently in 2017. Since 2017 was at solar minimum which implies solar activity was at inactive phase, we will correlate this event to the temperature changes in 2017. During solar minimum, smaller number of sunspots present on Sun's surface. A sunspot is known as a coldest region on the Sun's surface with a temperature of approximately 4000K and is appearing as black due to other surrounding of Sun's surface has a temperature of about 6000K. Data for solar cycle progression with their respective sunspot number can be referred to Figure 1 which includes solar cycle 23 and 24. Every data is taken from [14] and is updated every month and later they are compiled and arranged by [15].

IOP Conf. Series: Journal of Physics: Conf. Series **1298** (2019) 012019 doi:10.1088/1742-6596/1298/1/012019

Focusing on 2017, sunspot number decreases which complies the condition of solar minimum with average sunspot number of 19 [14]. Decrease in sunspot number gives low chance information of many solar flares. However, there are still strong solar events recorded throughout the year, in which the strongest flare is X9.3. We summarized the top 10 strongest flares in Chart 1 with respective to their region, sunspot number and Kp-index adapted from [14]. Based on the chart, region 2673 has been the most active region throughout the year as it has always produced most of the strongest flares because it is a complex region with β-γ-δ classification. The highest flare was recorded on 6th September with X9.3 magnitude.

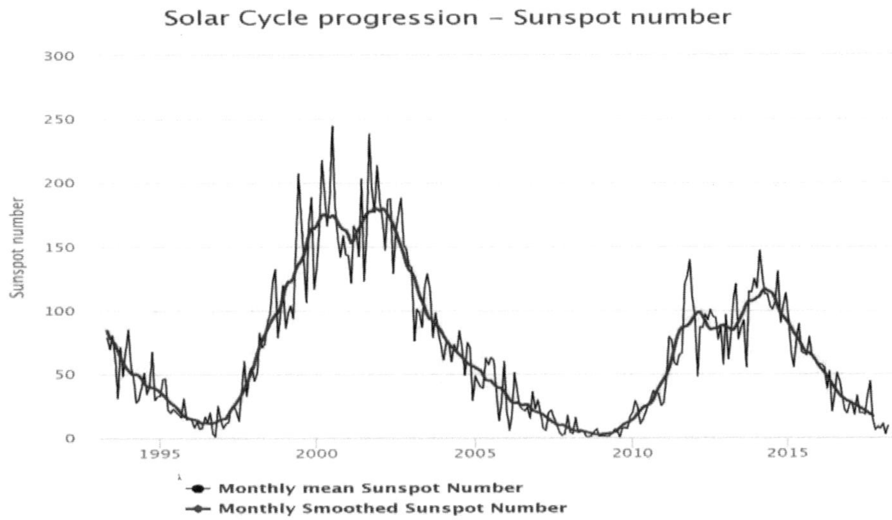

Figure 1. Solar cycle progression with respective to sunspot number [15]

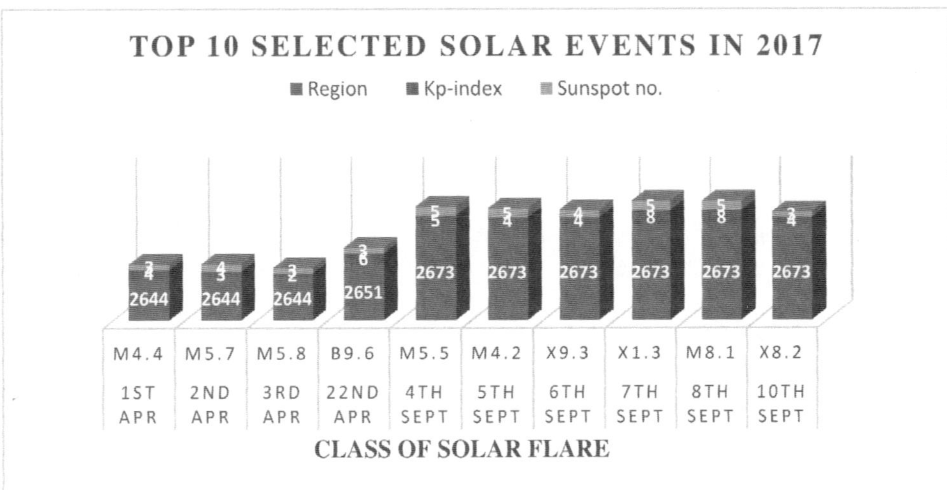

Chart 1. Summary of 10 selected solar events as in 2017

We are concerned on global annual mean surface air temperature based on land and ocean data and also annual mean temperature change for the hemisphere. Generally, 2017 has been the second hottest year on record since 1880 [16][17]. Figure 2 displays the global mean temperature based on land and ocean data adapted from NASA GISS. As shown in the figure, the second highest annual temperature is marked at 2017 to 0.89°C warmer than normal temperature, while 2016 holds the highest annual mean temperature by 0.99°C hotter.

Meanwhile, data of annual mean temperature change in North and South hemispheres also prove that 2017 is the second hottest year this Earth has been. It can be referred in Figure 3 which shows hemispheric temperature change since 1880. For the North Hemisphere, 2017 and 2015 have slightly the same annual mean temperature as they differ by only 0.01°C, in which the temperature anomaly of 2017 is 1.11°C and 2015 is 1.12°C. However, for South Hemisphere, 2017 still ranks the second highest temperature with 0.68°C warmer.

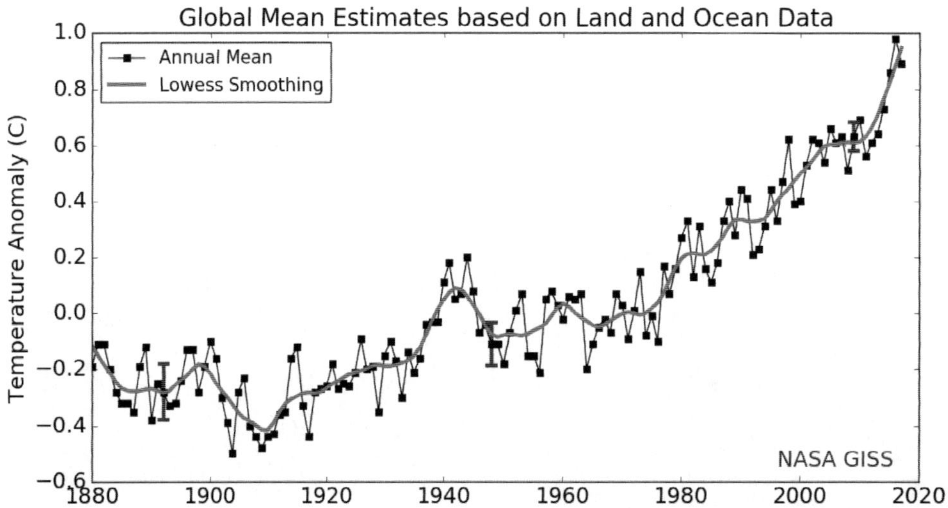

Figure 2. Global mean temperature based on land and ocean data since 1880 [16]

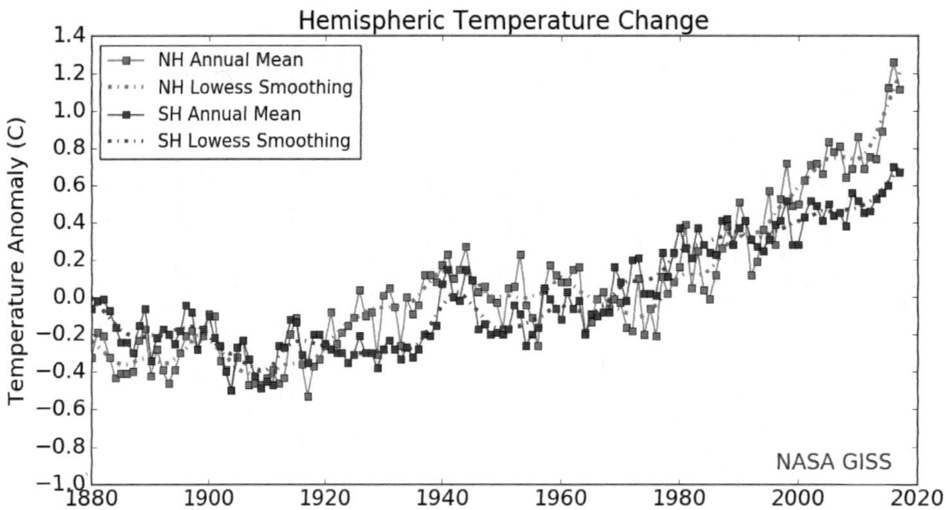

Figure 3. Hemisphere temperature change since 1880 [16]

4. Correlation between solar activity and Earth's temperature

In order to relate the variations of solar activity to Earth's temperature, we apply the equation for the luminosity of the Sun which is given as

$$L = 4\pi r^2 \sigma T^4 \quad (1)$$

where, r is radius of Sun, σ is Stefan-Boltzmann constant and T is temperature of Sun.

From the equation, Sun's luminosity is directly proportional to Sun's temperature, which means high temperature produces high energy radiated by the Sun. The Earth's climate is strongly depending on solar radiation that is provided by the Sun. Some of solar radiation is absorbed by the Earth and some of it will be reflected back to space mostly by clouds and atmospheric particles to balance Earth's energy and temperature. The amount of solar radiation absorbed and reflected determine the rising or declining in Earth's climate. Generally, Earth's system absorbs 71% of total incoming solar radiation and 29% is reflected back. When atoms and molecules absorb energy, the matter become excited and moves quickly and randomly and leads to increase in the matter's temperature. If more energy is absorbed, more particles increase in temperature and eventually Earth's atmosphere and surface start to warm up. The origin of this result goes back to Sun's temperature, where, the hotter the Sun is, more solar energy is produced leads to high incoming energy absorbed by Earth. Meanwhile, the temperature of the Sun is based on the presence of sunspots; less sunspot number, more hotter the Sun will be.

5. Conclusion

Overall, this article reviews some evidences from previous studies and we present solar activity data throughout 2017 during solar minimum to prove the possibility of solar radiation emitted by solar flares and coronal mass ejections in changing Earth's climate pattern. By gathering the data records, we believe climate change is contributed by the variability of solar activity, in which, solar minimum results in high solar radiation produced and received by the Earth. Hence, less amount of energy absorbed results in low of the Earth's temperature.

Acknowledgments

We are grateful to CALLISTO network, STEREO, LASCO, SDO/AIA, NOAA and SWPC make their data available online. This work was partially supported by the UITM internal BESTARI grant, 600-IRMI/DANA 5/3/BESTARI (067/2016) from the Kementerian Pengajian Tinggi Malaysia. Special thanks to the National Space Agency and the National Space Centre for giving us a site to set up this project and support this project. Solar burst monitoring is a project of cooperation between the Institute of Astronomy, ETH Zurich, and FHNW Windisch, Switzerland, Universiti Teknologi MARA and University of Malaya. This paper also used the NOAA Space Weather Prediction Centre (SWPC) for the sunspot, radio flux and solar flare data for comparison purpose. The research has made use of the National Space Centre Facility and a part of an initiative of the International Space Weather Initiative (ISWI) program.

6. References

[1] Ramli, N., Hamidi, Z. S., Abidin, Z. Z., & Shahar, S. N. (2015). The Relation between Solar Radio Burst Types II, III and IV due to Solar Activities. International Conference on Space Source and Communication 2015, 123-127

[2] Shariff, N. N., Hamidi, Z. S., & Zainol, N. H. (2017). Analysis of slow partial halo CME events with velocity of 100-200 km/s: An observation through CALLISTO system. International Conference on Industrial Engineering, Management Science and Application 2017

[3] Parker, D. E., Jones, P. D., Folland, C. K., & Bevan, A. (1994). Interdecadal changes of surface temperature since the late nineteenth century. J Geophys Res 99 (D7), 14373-14399.

[4] Hady, A. A. (2013). Deep solar minimum and global climate changes. Journal of Advance Research 4, 209-214.

[5] Eddy, J. A. (1976). The Maunder Minimum. Science, 192, 1189-1202.

[6] Guinan, E. F., & Ribas, I. (2002). Our changing Sun: The role of solar nuclear evolution and magnetic activity on Earth's atmospere and climate. ASP Conference Series 269, 85-104.

[7] Lean, J., Skumanich, A., & White, O. (1992). Estimating the Sun's radiative output during the Maunder Minimum. Geophysical Research Letters 19, 1591-1594.

[8] Tsiropoula, G. (2003). Signatures of solar activity variability in meteorogical parameters. Journal of Atmospheric and Solar-Terrestrial Physics Vol. 65, 469-482.
[9] Frohlich, C., & Lean, J. (1998). Total solar irradiance variations. In: Deubner, F.L. (Ed), Proceedings of the IAU Symposium Vol. 185. Kluwer Academic Publisher, Dordrecht, 89-102.
[10] van Geel, B., Raspopov, O., Renssen, H., Plicht, V. D., Johannes, D. V., & Meijer, H. (1999). The Role of Solar Forcing Upon Climate Change. Quartenary Science Reviews, 331-338.
[11] Bard, E., Raisbeck, G. M., Yiou, F., & Jouzel, J. (1997). Solar Modulation of Cosmogenic Nuclide Production over The Last Millenium; Comparison between C-14 and Be-10 records. Earth and Planetary Science Letters 150, 453-462.
[12] Finkel, R. C., & Nishiizumi, K. (1997). Beryllium 10 concentrations in the Greenland Ice Sheet Project 2 ice core from 3-40ka. Journal of Geophysical Research 102, C12, 26699-26706.
[13] Yiou, F., Raisbeck, G. M., Baumgartner, S., Beer, J., Hammer, C., Johnsen, S., . . . Yiou, P. (1996). Berylium 10 in the Greenland Ice Core Project ice core at Summit, Greenland. Journal of Geophysical Research, 102(C12), 26783-26794.
[14] SpaceWeatherLive. (2017). Retrieved from SpaceWeatherLive web site: https://www.spaceweatherlive.com/
[15] Space Weather Prediction Center National Oceanic and Atmospheric Administration. (2017). Retrieved from National Oceanic and Atmospheric Administration web site: https//www.swpc.noaa.gov/
[16] National Aeronautics and Space Administration Goddard Institute of Space Studies. (2018). Retrieved from National Aeronautics and Space Administration web site: https://data.giss.nasa.gov/gistemp
[17] National Aeronautics and Space Administration. (2017). Retrieved from National Aeronautics and Space Administration web site: https//www.nasa.gov/

IOP Conf. Series: Journal of Physics: Conf. Series **1298** (2019) 012020 doi:10.1088/1742-6596/1298/1/012020

A prediction model for *superconductor critical temperature* using stepwise discriminant analysis based on feature extraction

P C Guo, Wei Li[1], Z Y Su

School of Science, Dalian Ocean University, Dalian, 116023, China.

[1] Email: wwli@dlou.edu.cn

Abstract. Although the critical temperature is a very important step for the extensive application of superconductors, it's difficult to find the correlations of many kinds of physical superconductor properties and guarantee the accuracy in predicting the critical temperature. In this paper, an efficient prediction model using stepwise discriminant analysis based on feature extraction is provided to give the reduction of superconductor physical properties and predict the superconductor critical temperature. Firstly principal component analysis and clustering analysis are implemented to reduce the data dimension of superconductor physical properties and give the reductive clusters to complete the feature extraction. The 71 physical properties data of 1300 superconductors is efficiently reduced to 3 main components and 7 clusters. According to the extracted features and improved stepwise discriminant analysis based on the principle of binary search, bayesian discriminant function of each layer is established. At last, Python programming is designed to input the characteristic values and output the predicted the range of critical temperature to finish the efficient computer implementation of this model. Its result that only to the fifth layer, the superconductor critical temperature accuracy of 20,000×71 data matrix is 3. 125 has verified the efficiency of this model.

1. Introduction

In the era of big data, although large data can accurately reflect the overall situation, it is easy to suffer from "dimension disaster". In the existing data, there are more than 70 kinds of physical properties of superconductors and the correlations between many variables can't be ignored. It increases the difficulties of data analysis. Recently, several properties with the greatest impact on the critical temperature are generally used to approximately replace all of properties according to the previous experience, and then the prediction is completed by multiple linear regression analysis[1]. Firstly, the selection of properties in the above process is not scientific enough, and its correlation is unknown, which is easy to cause the loss of the information of the original data. Secondly in the multiple linear regression analysis, the model established can't guarantee the accuracy in predicting the critical temperature and its margin of error can't be calculated. It is easy to lead to the prediction results inconsistent with the facts. In this paper, an efficient prediction model using stepwise discriminant analysis based on feature extraction is provided to give the reduction of physical properties data of superconductors and predict the superconductor critical temperature to overcome the above limitations.

Superconductor is a conductive material with the properties of diamagnetism and resistance approaching 0 when approaching the critical temperature. Superconductivity was first discovered in 1911. During that time, the scientists found that at extremely low temperature, the resistance of some materials completely disappeared, showing a superconducting state. The temperature at which the resistance of a superconductor is zero is called the critical temperature of a superconductor[2]. Now the superconductor materials have extremely wide applications in the application of strong electricity, weak electricity and diamagnetism. Therefore, it is necessary to study its critical temperature for the more applications of superconductors.

This paper consists of four sections. Section 1 is the introduction, including the research purpose and significance; section 2 is the theoretical introduction of this prediction model including principal component analysis (PCA), cluster analysis and stepwise discriminant analysis; in Section 3, based on the theoretical guidance in section 2, the superconductor critical temperature is efficiently analyzed using this prediction model; Section 4 is the summary and discussion.

2. Methods

An efficient prediction model using stepwise discriminant analysis based on feature extraction is provided to give the reduction of physical properties data of superconductors and predict the superconductor critical temperature. Firstly PCA is implemented to reduce the data dimension of superconductor physical properties. Based on the principal component data, clustering analysis is used to give the reductive clusters to complete the feature extraction. Secondly, according to the extracted features and improved stepwise discriminant analysis, the bayesian discriminant function of each layer is established. Finally, Python programming is used to finish the efficient computer implementation of this model to obtain the predicted value of critical temperature.

2.1. Principal component analysis

PCA is a multivariate statistical method, which reduces the dimension of multivariate variables through linear transformation and selects several comprehensive indexes[4]. The above comprehensive indicators are the principal components after the linear transformation and combination of the original variables. Furthermore, there is no correlation or the correlation among the indicators is extremely low. It not only preserves the information contained in the original indicators, but also simplifies the problems in data analysis. Therefore it greatly reduces the workload and improves the analysis efficiency[5-6]. The algorithm steps are as follows:

(1) standardize the original data, unify its dimensions and order of magnitude, and obtain the standardized matrix X.

(2) use X to calculate the correlation coefficient matrix R.

(3) find the eigenvalue and eigenvector of R matrix by Jacobian algorithm.

(4) when the characteristic value is greater than 1 and the cumulative contribution rate is up to 80-95%, the principal component is determined and the principal component expression is obtained.

(5) establish the initial factor loading matrix and explain the principal components.

2.2. Clustering analysis

The main principle of clustering analysis is to divide data sources into several categories based on their similarity size, so that the objects in the same category can be approximately regarded as one object, greatly improving the efficiency of data analysis[7-8]. In this paper, the system clustering method is used to analyze the main components obtained by PCA. The algorithm steps are as follows:

(1) divide n samples into n categories and form a distance matrix by calculating the distance between each data point.

(2) combine the nearest two classes into one class, leaving n-1 class. Repeat (1) operation to get a new distance matrix.

(3) repeat the operation (2) until the artificially specified number of classes is collected[9].

2.3. Stepwise discriminant analysis

Traditional Discriminant Analysis[3] is a Multivariate Statistical Analysis method to identify the type attribution of a research object based on various eigenvalues under the condition of classification determination. Stepwise discriminant analysis is a method with the bayesian discriminant function as criterion and the ability to variables selection. The main implementation principle is to introduce gradually the variables according to their importance and eliminate the relevant variables according to their correlations until all variables are important. The filtering is completed step by step. In this paper, the principle of binary search in programming[10] is used to improve the analysis method. The algorithm steps are as follows:

(1) determine the grouping variables, divide the samples into two groups according to the grouping criteria and establish the first-layer bayesian discriminant function.

(2) continue to classify one group downward and do the same for the other group to establish the second bayesian discriminant function.

(3) repeat the operation (2) until the desired effect is achieved, and establish the fifth layer bayesian discriminant function, as follows in figure 1:

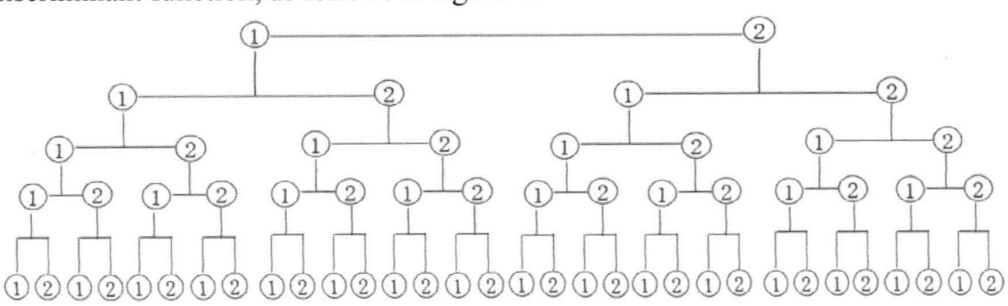

Figure 1. Structure of stepwise discriminant analysis.

3. Results

Firstly, PCA is implemented to reduce the data dimension of superconductor physical properties, with SPSS software. Then clustering analysis is used to classify the main components in PCA to complete the feature extraction. Secondly, according to the extracted features and improved stepwise discriminant analysis, the bayesian discriminant function of each layer is established. Finally, Python programming is used to finish the efficient computer implementation of this model to obtain the predicted value of critical temperature.

3.1 PCA Result

This part is the verification that the physical properties data of superconductors is suiTable for PCA. The original data is from paper[1]. It is 71 physical properties of 21, 263 superconductors as table 1.

Table 1. Physical attribute data in superconductors.

material	La1. 85Sr0. 15Cu0. 982Fe0. 018O4	La1. 85Sr0. 15Cu0. 975Ga0. 025O4	...	Y0. 83Ca0.02Pr0. 15Ba2Cu3O6. 95	Y1Ba2Cu4O8
number_of_elements	5	5	...	6	4
mean_atomic_mass	72. 383174	75. 158774	...	81. 12731667	76. 4445625
wtd_mean_atomic_ mass	56. 7890145	56. 83087779	...	51. 908679	49. 71593667
...	...	...	...	...	...
gmean_Valence	2. 352158045	2. 352158045	...	2. 40187391	2. 213363839
wtd_gmean_Valence	2. 228543963	2. 229447743	...	2. 069202467	2. 054799318
entropy_Valence	1. 589026915	1. 589026915	...	1. 748970751	1. 368922361

We select the first 20,000 of the above data as samples and use the remaining data as the comparison samples for the discriminant analysis to obtain the 20,000 ×71 data matrix.

3.1.1 Factor extraction.

Solving factor solution is chosen as the main component analysis method. The basis for determining the number of factors is that the cumulative contribution rate of variance is over 85%, as shown in table 2 and figure 2:

Table 2. Total variance interpretation.

	Total	Variance percentage	Accumulation %	Total	Variance percentage	Accumulation %
1	1303. 073	98. 568	98. 568	1303. 073	98. 568	98. 568

IOP Conf. Series: Journal of Physics: Conf. Series **1298** (2019) 012020 doi:10.1088/1742-6596/1298/1/012020

2	15. 464	1. 17	99. 738	15. 464	1. 17	99. 738
3	1. 07	0. 081	99. 819	1.07	0.081	99.819
4	0. 766	0. 058	99. 877			
. . .	. . .	. . .	. . .			
19995	0. 002	0	99. 999			
20000	0. 001	0	100			

In order to simplify the analysis without affecting the analysis results, 1300 samples of superconductors are randomly selected here.

3.1.2 Scree plot

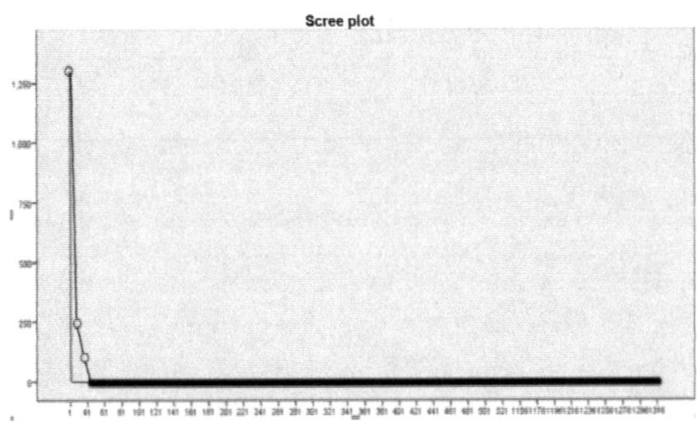

Figure 2. Scree plot of the main component.

According to figure 2, the first principal component is significantly distinguished from other eigenvalues. The first three principal component eigenvalues account for 99.819% of all information and the majority of all information. From the fourth principal component, it's basically flat. Therefore, we obtain 3 principal components to express the vast majority of the information that can be expressed by all the original indicators[1].

3.1.3 Score of each component

Calculating the score matrix of each component with SPSS yields the following results in table 3.

Table 3. Factor scoring coefficient matrix.

	composition		
	1	2	3
La1. 85Sr0. 15Cu0. 982Fe0. 018O4	. 001	-. 005	-.007

La1. 85Sr0. 15Cu0. 975Ga0. 025O4	. 001	-. 005	-.006
La1. 85Sr0. 15Cu0. 977Ga0. 023O4	. 001	-. 005	-.006
. . .	. . .	. . .	...
La1. 87Ba0. 13Cu1O4	. 001	-. 004	-.002
La1. 65Ba0. 35Cu1O4	. 001	-. 004	.000

3.2 Clustering analysis of Table 3

Based on the principal component results, with SPSS software, clustering analysis is implemented to further analyze the data. The number of cluster for the data of 3 principal components of 1300 superconductors is 7, as table 4:

Table 4. Cluster analysis based on PCA.

Case	7 clusters	Case	7clusters
1	1	13	4
. . .	. . .	14	3
8	2	. . .	. . .
9	7	68	6
. . .	. . .	71	1

As table 4, based on the three principal components, 71 physical attribute features in the data are successfully classified into 7 categories. Based on the characteristics of the association between various attributes of objects, one item of each category can be selected as the indicator to represent the class, which is respectively: Range_atomic_mass, mean_fie, wtd_mean_fie, wtd_range_fie, wtd_range_atomic_radius, wtd_mean_ThermalConductivity, range_ThermalConductivity. Now the feature extraction operation is completed.

3.3 Stepwise Discriminant Analysis

Based on the seven features extracted by PCA and clustering analysis, the stepwise discriminant analysis is performed. We use the above seven characteristics as independent variables to predict the dependent variable(critical temperature T). The symbols of independent variables are a, b, c, d, e, f, and g, respectively. Firstly, the samples are divided into the analysis samples and the comparison samples for the following analysis. The former is to take the critical temperature as a known data sample to construct the bayesian discriminant function model [11], and the latter is to take the critical temperature as an unknown data sample to test whether the model is accurate. Secondly, to analyze multiple binary classification for the critical temperature T of sample, namely (1) and (2). Because of the selected data in T within 0 ~ 100 °C, the first layer of the interval length is 50 °C, namely 0 ~ 50 °C for (1) class, 50 ~ 100 °C for (2) class; the second layer is 25 °C, namely 0 ~ 25 °C for (1) class, 25 ~ 50 °C for (1) class, 50 ~ 75 °C for (2) class;75 ~ 100 °C for (2) class. . . and so on, until the fifth layer interval length by 3. 125 °C, namely 0 ~ 3. 125 °C for (1) class, 3. 125 ~ 6. 25 °C for (2) class. . . 93. 625 ~ 96. 875 °C for (1) class, 96. 875 ~ 100 °C for (2) class. Until this point, the classification of data is completed. The algorithm is as follows:

Table 5. Discriminant analysis data table.

ıic_mass	mean_fie	...	range_ThermalConductivity	critical_temp	group1	...	group5
)607	781. 46	...	399. 97342	0. 2	0	...	0
)607	745. 38	...	399. 97342	0. 7	0	...	0
)81	748. 0666	...	399. 97342	1. 5	0	...	0
.	...	...	...	...	...	...	...
ı006	781. 9	...	399. 97342	99. 25	1	...	1
)81	769. 1	...	399. 97342	100	1	...	1

Based on table 5, SPSS is used to construct the bayesian discriminant function layer by layer. First, the typical discriminant function of the first layer is tested:

Table 6. Typical discriminant function test table.

Werker Lambda				
Function test	Werker Lambda	chi-square	Degrees of freedom	significance
1	. 673	772. 129	7	. 000
...	...	...	...	...

As shown in table 6 significance test, the effect of discriminant function is significant.
The classification function coefficient of the first layer:

Table 7. Coefficient of classification function

	Group1	
	0	1
range_atomic_mass	2. 244	2. 277
mean_fie	. 293	. 290

wtd_mean_fie	4. 373	4. 416
wtd_range_fie	-1. 468	-1. 477
wtd_range_atomic_radius	6. 871	6. 881
wtd_mean_ThermalConductivity	3. 391	3. 484
range_ThermalConductivity	-. 378	-. 382
(constant)	-2121. 348	-2166. 998
Fisher linear discriminant function		

As the table 5, table 6, table 7, the first bayesian discriminant function is constructed, namely:
T=2. 244*a+0. 293*b+4. 373*c-1. 468*d+6. 871*e+3. 391*f-0. 378*g-2121. 348
T=2. 277*a+0. 290*b+4. 416*c-1. 477*d+6. 881*e+3. 484*f-0. 382*g-2166. 998
The bayesian discriminant function established layer by layer. Repeating the above operation yeilds the five levels of classification, a total of 62 bayesian discriminant functions:
1:T=2. 720*a-0. 464*b+6. 340*c-2. 607*d+10. 015*e+3. 213*f-0. 899*g-2394. 941
2:T=2. 710*a-0. 469*b+6. 331*c-2. 606*d+10. 020*e+3. 203*f-0. 896*g-2382. 983
1:T=3. 883*a-0. 059*b+12. 264*c-5. 069*d+15. 878*e+6. 065*f-0*g-5202. 347
2:T=3. 899*a-0. 075*b+12. 229*c-5. 049*d+15. 855*e+6. 084*f-0*g-5197. 540
Until this point, all the bayesian discriminant functions are constructed.

3.4 Python Programming

Using Python to achieve binary search, the part of the original code is shown as figure 3:

```
# calculation
for lis in lis3:
    # gives [1,1] and [2,2], respectively.
    for func in lis1:
        #get   a, b, and c
        lis4.append(func(lis[0]. lis[1]. lis[2].lis[3]. lis[4]. lis[5].lis[6]))
        num_of_index += 1
        if num_of_index == num_of_lis1:
            lis2.append(lis4)
            lis4 = []
            num_of_index = 0

# compare
for lastlis in lis2:
    for index, value in enumerate(lastlis).
        if value == max(lastlis).
            print('Index:', index, 'Value:', value)
```

Figure 3. Implementation Code

The above seven values can be input to obtain the estimated range of the critical temperature. Since the purpose of this paper is to verify the feasibility of this model, only the fifth layer is achieved, that is, the interval length is 3. 125. Therefore the accuracy is 3. 125. If you keep going down, the length of the interval shrinks exponentially and the accuracy increases correspondingly. The results verifies that the prediction data of this model is more effective and accurate.

4. Conclusions

In this paper, an efficient prediction model using stepwise discriminant analysis based on features extraction is provided to give the reduction of superconductor physical properties and predict the superconductor critical temperature. With PCA and cluster analysis, the 71 physical properties data of 1300 superconductors is efficiently reduced to 3 main components and 7 clusters. Based on the above extracted features, a improved stepwise discriminant analysis is provided to find the superconductor critical temperature accuracy. The principle of binary search is used to improve stepwise discriminant analysis to complete the classification in the form of a binary tree and set up the Bayes discriminant function of each layer. Finally, Python programming is designed to input the characteristic values and output the predicted the range of critical temperature to complete the efficient computer implementation of this model. Using this model, the superconductor critical temperature accuracy 20,000×71 data matrix is 3. 125. The results verify that this model is more efficient because it overcomes the limitations of the traditional prediction. The model is simpler and more accurate tool for the prediction of superconductor critical temperature and provides the potential analytical value for the application and promotion of superconductors.

5. Acknowledgments

Natural Science Foundation of Liaoning Province of China(20170540103), National Natural Science Foundation of China(11501076).

6. References

[1] Kam H 2018 *Comp. Mater. Sci.* **154** 346
[2] Zhang L Y 2016 *Beyond Freedom: the Magic Superconductor* (Beijing: Science Press)
[3] Zhao D F 2007 *Mathematical model and computation* (Beijing: Science Press)
[4] He X Q 2015 *Multivariate Statistical Analysis* (Renmin University of China Press)
[5] Su W H 2000 *Study on the theory and method of multi-index comprehensive evaluation* (Xiamen University)
[6] Guo Y J 2002 *Comprehensive evaluation theory and method* (Beijing: Science Press)
[7] Du Q, Jia L Y and Yan X F 2014 *Spss statistical analysis from introduction to mastery* (People's Post Office)
[8] Fang K T and Pan E P 1982 *Clustering analysis* (Beijing: Geological Publishing House)
[9] Li P Z 2008 *The application of cluster analysis and principal component analysis in the evaluation of regional comprehensive consumption level* (Nankai University)
[10] Kernighan B W, Pike R , etal 2000 *The Practice of Programming* (Beijing: Machinery Industry Press)
[11] Yang S Y 2011 *Pattern recognition and intelligent computing–matlab technology implementation* (Electronic Industry Press)

IOP Conf. Series: Journal of Physics: Conf. Series **1298** (2019) 012021 doi:10.1088/1742-6596/1298/1/012021

Stress intensity factor for cracks in bonded dissimilar materials

K B Hamzah[1,4], N M A NikLong[1,2,5], N Senu[1,2] and Z K Eshkuvatov[3]

[1]Laboratory of Computational Sciences and Mathematical Physics, Institute for Mathematical Research, Universiti Putra Malaysia, 43400 Serdang, Selangor, Malaysia
[2]Mathematics Department, Faculty of Science, Universiti Putra Malaysia, 43400 Serdang, Selangor, Malaysia
[3]Faculty of Science and Technology, UniversitiSains Islam Malaysia, 71800 Negeri Sembilan, Malaysia
[4]Fakulti Teknologi Kejuruteraan Mekanikal dan Pembuatan, Universiti Teknikal Malaysia Melaka, Hang Tuah Jaya, 76100 Durian Tunggal, Melaka, Malaysia

[5]E-mail: nmasri@upm.edu.my

Abstract. In this paper, an inclined and a semi circular arc cracks lying in the upper half of bonded dissimilar materials are considered.The modified complex variable function method is used to formulate the problem into the hypersingular integral equations (HSIEs) with the help of the continuity conditions of the traction and displacement.The traction along the crack serves as the right hand side of the integral equation. The HSIEs is solved numerically for the unknown crack opening displacement function using the appropriate quadrature formulas. Numerical results presented the behaviour of the nondimensional stress intensity factorsat the crack tips which are depends on the elastic constants ratio, the position of the crack and the distance between the crack and the boundary.

1. Introduction

The analysis of stress intensity factors (SIF) at the crack tips is very important to determine the stability and safety of the materials. In previous years, many researchers discussed on the nondimensionalSIFbehaviour for the crack problems in an infinite plane [1, 2], finite plane [3, 4] half plane [5, 6, 7] and bonded dissimilar materials [8, 9, 10, 11]. Chen [8] presented Fredholm integral equations with density distributions as undetermined functions to calculate the nondimensionalSIF for multiple crack problems in bonded dissimilar materials. Chen and Hasebe [9] obtained the logarithmic singular integral equation for a circular arc crack lies in the upper part of bonded dissimilar materials. Yang and Wang [10] utilised the combinations of Chebyshev polynomials and collocation methods for solving the nondimensionalSIF of a perpendicular crack to the interface of bonded dissimilar materials. Ghajar*et al.*[11] used the displacement method to analyze the effect of Poisson's ratio on the nondimensionalSIF for the interface crack in bonded dissimilar materials.The aim of this paper is to analyze the behaviour of nondimensionalSIF for a crack lies in the upper part of bonded dissimilar materials subjected to the remote stress $\sigma_x = p$.

IOP Conf. Series: Journal of Physics: Conf. Series **1298** (2019) 012021 doi:10.1088/1742-6596/1298/1/012021

2. Problem formulation

The complex variable function method plays an important role in solving the cracks problem in plane elasticity [12]. In this method, the stress components $(\sigma_x, \sigma_y, \sigma_{xy})$, resultant force function $f(X,Y)$ and displacements (u,v) are related to the two complex potentials $\phi(z)$ and $\omega(z)$ as follows

$$\sigma_x+\sigma_y=4\operatorname{Re}\Phi(z),\quad \sigma_y-i\sigma_{xy}=\Phi(z)+(z-\bar{z})\overline{\Phi'(z)}+\Omega(\bar{z}) \tag{1}$$

$$f=-Y+iX=\phi(z)+(z-\bar{z})\overline{\Phi'(z)}+\omega(\bar{z}) \tag{2}$$

$$2G(u+iv)=\kappa\phi(z)-(z-\bar{z})\overline{\Phi(z)}-\omega(\bar{z}) \tag{3}$$

where$G$ is shear modulus of elasticity, $\kappa=(3-v)/(1+v)$ for plane stress, $\kappa=3-4v$ for plane strain and v is Poisson's ratio.Thefunction $\omega(z)$, $\Phi(z)$, $\Psi(z)$ and $\Omega(z)$ are defined as

$$\omega(z)=z\bar{\Phi}(z)+\bar{\psi}(z),\quad \Phi(z)=\phi'(z),\quad \Psi(z)=\psi'(z),\quad \Omega(z)=\omega'(z) \tag{4}$$

where $\bar{\phi}(z)=\overline{\phi(\bar{z})}$. The derivative in a specified direction of equation (2) using equation (4) is defined as

$$\frac{d}{dz}\{-Y+iX\}=\Phi(z)+\overline{\Phi(z)}+\frac{d\bar{z}}{dz}\left[z\overline{\Phi'(z)}+\overline{\Psi(z)}\right]=N+iT \tag{5}$$

Where $N+iT$ denotes the normal and tangential traction along the crack segment $\overline{z, z+dz}$.

The complex potentials for a crack L in an infinite plane can be obtained by[1]

$$\phi(z)=\frac{1}{2\pi}\int_L\frac{g(t)dt}{t-z},\quad \psi(z)=\frac{1}{2\pi}\int_L\frac{\overline{g(t)d\bar{t}}}{t-z}+\frac{1}{2\pi}\int_L\frac{g(t)d\bar{t}}{t-z}-\frac{1}{2\pi}\int_L\frac{\bar{t}g(t)dt}{(t-z)^2} \tag{6}$$

where $g(t)$ is crack opening displacement (COD) function defined by

$$g(t)=\frac{2G}{i(\kappa+1)}\left[(u(t)+iv(t))^+-(u(t)+iv(t))^-\right] \tag{7}$$

$(u(t)+iv(t))^+$ and $(u(t)+iv(t))^-$ denote the displacements at point t of the upper and lower crack faces, respectively.The modified complex potentials for the crack lies in the upper part of bonded dissimilar materials are defined as

$$\phi_1(z)=\phi_{1p}(z)+\phi_{1c}(z),\quad \psi_1(z)=\psi_{1p}(z)+\psi_{1c}(z) \tag{8}$$

where subscript p and c represent the principal and complementary parts, respectively. Whereas, for the crack lies in the lower part, the complex potentials are represented by $\phi_2(z)$ and $\psi_2(z)$.

The traction and displacement continuity conditions are defined as

$$\left[\phi_1(t)+(t-\bar{t})\overline{\Phi'_1(t)}+\omega_1(\bar{t})\right]^+=\left[\phi_2(t)+(t-\bar{t})\overline{\Phi'_2(t)}+\omega_2(\bar{t})\right]^-, t\in L \tag{9}$$

$$G_2\left[\kappa_1\phi_1(t)-(t-\bar{t})\overline{\Phi_1(t)}-\omega_1(\bar{t})\right]^+=G_1\left[\kappa_2\phi_2(t)-(t-\bar{t})\overline{\Phi_2(t)}-\omega_2(\bar{t})\right]^-, t\in L. \tag{10}$$

Applying equation (8) into equations (9) and (10), the following complex potentials are obtainable:

$$\phi_{1c}(z)=\frac{G_2-G_1}{G_1+\kappa_1 G_2}\left[z\overline{\Phi_{1p}}(z)+\overline{\psi_{1p}}(z)\right] \tag{11}$$

$$\psi_{1c}(z)=\frac{\kappa_1 G_2-\kappa_2 G_1}{G_2+\kappa_2 G_1}\overline{\phi_{1p}}(z)-\frac{G_2-G_1}{G_1+\kappa_1 G_2}\left[z\overline{\Phi_{1p}}(z)+z^2\overline{\Phi'_{1p}}(z)+z\overline{\Psi_{1p}}(z)\right] \tag{12}$$

$$\phi_2(z)=\left(1+\frac{\kappa_1 G_2-\kappa_2 G_1}{G_2+\kappa_2 G_1}\right)\phi_{1p}(z) \tag{13}$$

$$\psi_2(z)=\left(\frac{G_2-G_1}{G_1+\kappa_1 G_2}-\frac{\kappa_1 G_2-\kappa_2 G_1}{G_2+\kappa_2 G_1}\right)z\Phi_{1p}(z)+\left(1+\frac{G_2-G_1}{G_1+\kappa_1 G_2}\right)\psi_{1p}(z). \tag{14}$$

where the principal parts of complex potentials are referred to the complex potentials for the cracks in an infinite plane.

The HSIEs for a crack lies in the upper part of bonded dissimilar materials can be obtained by substituting equation (8) into (5) and applying equations (6), (11) and (12), then letting point z approaches t_0 on the crack and changing $d\bar{z}/dz$ into $d\bar{t}_0/dt_0$, yields

$$N(t_0)+iT(t_0)=\frac{1}{\pi}h.p\int_L\frac{g(t)dt}{(t-t_0)^2}+\frac{1}{2\pi}\int_L A_1(t,t_0)g(t)dt+\frac{1}{2\pi}\int_L A_2(t,t_0)\overline{g(t)dt} \tag{15}$$

where

$$A_1(t,t_0)=\frac{1}{(t-t_0)^2}\left[\frac{(t-t_0)^2}{(\bar{t}-\bar{t}_0)^2}\frac{d\bar{t}}{dt}\frac{d\bar{t}_0}{dt_0}-1\right]+\frac{G_2-G_1}{G_1+\kappa_1 G_2}\left[\frac{1}{(t-\bar{t}_0)^2}+\frac{2(\bar{t}_0-\bar{t})}{(t-\bar{t}_0)^3}\right.$$

$$\left.+\frac{d\bar{t}_0}{dt_0}\left(\frac{2(2t_0-3\bar{t}_0+\bar{t})}{(t-\bar{t}_0)^3}-\frac{6(\bar{t}_0-\bar{t})(\bar{t}_0-t_0)}{(t-\bar{t}_0)^4}-\frac{1}{(t-\bar{t}_0)^2}\right)\right]+\frac{\kappa_1 G_2-\kappa_2 G_1}{G_2+\kappa_2 G_1}\frac{d\bar{t}_0}{dt_0}\frac{1}{(t-\bar{t}_0)^2}$$

$$+\frac{G_2-G_1}{G_1+\kappa_1 G_2}\left[\frac{1}{(\bar{t}-t_0)^2}+\frac{1}{(t-\bar{t}_0)^2}+\frac{d\bar{t}_0}{dt_0}\left(-\frac{1}{(t-\bar{t}_0)^2}-\frac{2(\bar{t}_0-t_0)}{(t-\bar{t}_0)^3}\right)\right]\frac{d\bar{t}}{dt}$$

$$A_2(t,t_0)=\frac{t-t_0}{(\bar{t}-\bar{t}_0)^3}\left[\frac{(\bar{t}-\bar{t}_0)}{(t-t_0)}\left(\frac{d\bar{t}}{dt}+\frac{d\bar{t}_0}{dt_0}\right)-2\frac{d\bar{t}}{dt}\frac{d\bar{t}_0}{dt_0}\right]+\frac{G_2-G_1}{G_1+\kappa_1 G_2}\left[\frac{1}{(\bar{t}-t_0)^2}+\frac{1}{(t-\bar{t}_0)^2}\right.$$

$$\left.+\frac{d\bar{t}_0}{dt_0}\left(-\frac{1}{(t-\bar{t}_0)^2}-\frac{2(\bar{t}_0-t_0)}{(t-\bar{t}_0)^3}\right)\right]+\frac{G_2-G_1}{G_1+\kappa_1 G_2}\left(\frac{1}{(\bar{t}-t_0)^2}+\frac{2(t_0-t)}{(\bar{t}-t_0)^3}\right)\frac{d\bar{t}}{dt}$$

If $G_2=0$ and $G_1=G_2$, then equation (15) reduce to the HSIEs for a single crack in a half plane [6] and an infinite plane elasticity [2], respectively.It is well known that to solve the HSIEs (15) numerically, we used the curved length coordinate method[1,6].

3. Results and discussions

The stress intensity factors (SIF) at the tip A_j of crack L is defined as

$$\left(K_1 - iK_2\right)_{A_j} = \sqrt{2\pi} \lim_{t \to t_{A_j}} \sqrt{\left|t - t_{A_j}\right|} g'_j(t), \quad j = 1, 2.$$

Consider an inclined crack (Figure 1(a)) and a semi circular arc crack (Figure 1(b)) lies in the upper part of bonded dissimilar materials subjected to remote stress $\sigma_x = p$ .Table 1 shows the nondimensional SIF at the crack tips A_1 and A_2 when $\alpha = 90^0$ and R/h varies for elastic constant ratios $G_2/G_1 = 0.0$ (Figure 1(a)). Our numerical results agree with those of Tada and Paris [5].

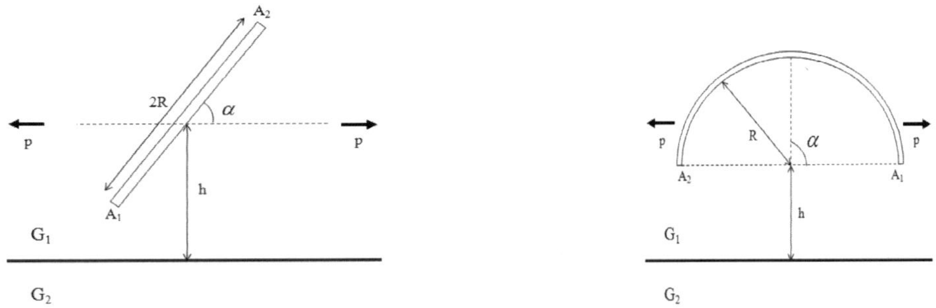

(a) (b)

Figure 1. A single crack lies in the upper part of bonded dissimilar materials.

Table 1. SIF for an inclined crack when $\alpha = 90^0$ and $G_2/G_1 = 0.0$ (Figure 1(a)).

SIF	*R/h*						
	0.3	0.4	0.5	0.6	0.7	0.8	0.9
F_{IA1}	1.0271	1.0527	1.0913	1.1490	1.2379	1.3879	1.7077
F_{IA1}[5]	1.0330	1.0520	1.0940	1.1480	1.2430	1.3850	1.6880
F_{IA2}	1.0201	1.0349	1.0539	1.0776	1.1074	1.1464	1.2038
F_{IA2}[5]	1.0180	1.0310	1.0500	1.0720	1.1060	1.1350	1.1940

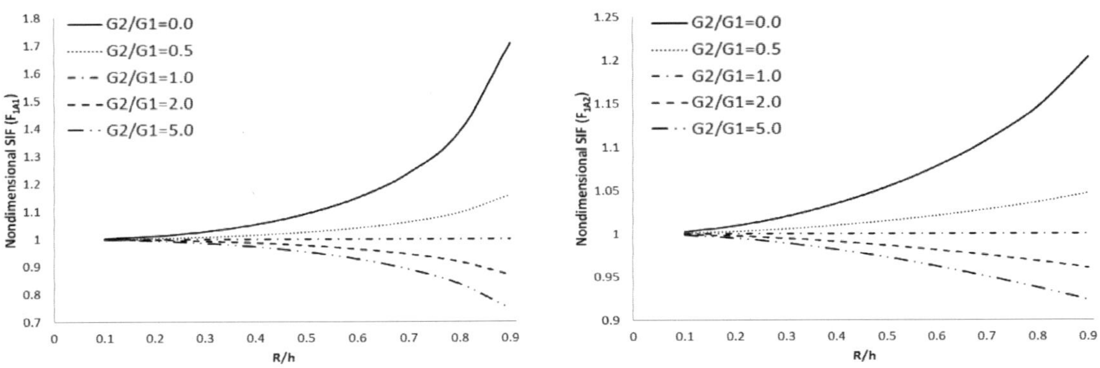

(a) SIF at the crack tip A_1 (b) SIF at the crack tip A_2

Figure 2. SIF against R/h for an inclined crack when $\alpha = 90^0$ (Figure 1(a)).

Figure 2 shows the nondimensionalSIF against R/h for different values of G_2/G_1 where R and h are defined as in Figure 1(a). It is observed that as the ratio G_2/G_1 increases, the nondimensional SIF decreases at all crack tips. For $G_2/G_1 = 1.0$ the nondimensionalSIF is constant as R/h varies due to the fact that the plane is filled up with one material only.However for $G_2/G_1 < 1.0$ and $G_2/G_1 > 1.0$, as R/h increases the nondimensional SIF increases and decreases, respectively.Figure 3 presented the nondimensionalSIF when α varies and $R/h = 0.9$ (Figure 1(a)). It is found that as the angle α and the ratio G_2/G_1 increases the nondimensional SIF increases and decreases, respectively.

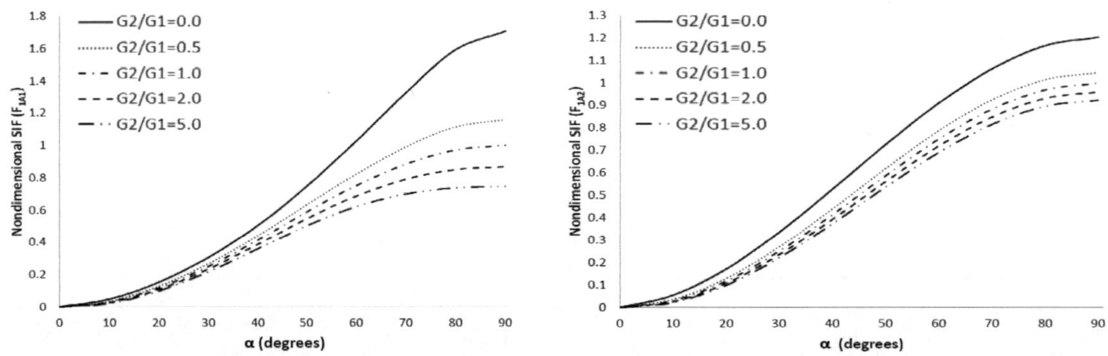

(a) SIF at the crack tip A_1 (b) SIF at the crack tip A_2

Figure 3.SIF for an inclined crack when α varies and $R/h = 0.9$ (Figure 1(a)).

Table 2. SIF for a semi circular arc crack when h/R varies and $G_2/G_1 = 0.0$ (Figure 1(b)).

SIF	*h/R*								
	0.2	0.4	0.6	0.8	1.0	1.2	1.4	1.6	2.0
F_{IA1}	0.7613	0.7207	0.6982	0.6833	0.6733	0.6674	0.6632	0.6605	0.6574
F_{IA1}[7]	0.7611	0.7205	0.6981	0.6831	0.6733	0.6672	0.6632	0.6605	0.6572

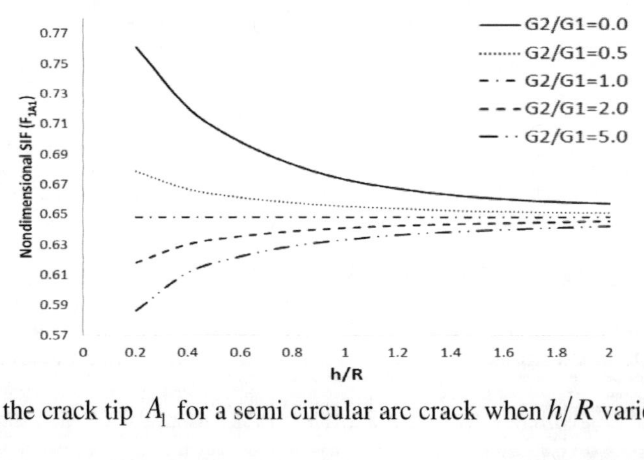

Figure 4.SIF at the crack tip A_1 for a semi circular arc crack when h/R varies (Figure 1(b)).

Table 2 shows the nondimensional SIF for a semi circular arc crack lies in the upper part of bonded dissimilar materials when h/R varies for $G_2/G_1 = 0.0$ (Figure 1(b)). Our results are in good agreement with those of Elfakhakhre *et al.* [7]. It is observed that the nondimensional SIF at crack tip A_1 is equal to SIF at tip A_2. For different values of G_2/G_1, the nondimensionalSIF at crack tip A_1 against h/R are presented in Figure 4. It is found that the nondimensional SIF decreases for $G_2/G_1 < 1.0$ and increases for $G_2/G_1 > 1.0$ as h/R increases. Whereas for $G_2/G_1 = 1.0$ the nondimensionalSIF does not change.

4. Conclusion

In this paper, HSIEs for the problems of an inclined and a semi circular arc crack lies in the upper part of bonded dissimilar materials with different elastic constants G_1 and G_2 are formulated using modified complex variable function method. The elementary solution of HSIEs reduces to the half plane and infinite plane elasticity problems when $G_2 = 0$ and $G_1 = G_2$, respectively. Numerical results showed the nondimensional SIFs at the crack tips for a single crack are influenced by the elastic constants ratio, the position of the crack and the distance between the crack and the boundary. The nondimensional SIF does not change for $G_1 = G_2$, increases for $G_2/G_1 < 1.0$ and decreases for $G_2/G_1 > 1.0$ as the distance between the crack and the boundary decreases. Whereas the nondimensional SIF at the crack tips decreases as G_2/G_1 increases for the constant distance between the cracks and the boundary. For the multiple cracks problems and the cracks lie in both upper and lower parts of bonded dissimilar materials will be treated later.

5. Acknowledgment

The first author would like to express his very great appreciation to the Ministry of Education Malaysia, Universiti Putra Malaysia (UPM) and Universiti Teknikal Malaysia Melaka for the financial support. The second author would like to thanks the UPM for Putra Grant, project no: 9567900.

6. References

[1] Nik Long NMA and Eshkuvatov ZK 2009 *International Journal of Solids and Structures* **46** 2611
[2] Rafar RA, Nik Long NMA, Senu N and Noda NA 2017 *ZAMM Journal of Applied Mathematics and Mechanics* **97**(11) 1482
[3] Chen YH 1987 *Computer Methods in Applied Mechanics and Engineering* **62** 1
[4] Lai J and Schijve J 1990 *Engineering Fracture Mechanics* **36**(4) 619
[5] Tada H, Paris PC and Irwin GR 1973 *The stress analysis of cracks handbook* (Pennsylvania: Del Research Corporation)
[6] Chen YZ, Lin XY and Wang XZ 2009 *Philosophical Magazine* **89**(26) 2239
[7] Elfakhakhre NRF, Nik Long NMA and Eshkuvatov ZK 2018 *Applied Mathematical Modelling* **60** 540
[8] Chen YZ 1986 *Engineering Fracture Mechanics* **25**(1) 1
[9] Chen YZ and Hasebe N 1992 *Theoretical and Applied Fracture Mechanics* **17** 189
[10] Yang M and Wang X 2018 *Acta Mechanica* **229**(5) 2063
[11] Ghajar R, Peyman S, Sheikhi J and Poorjamshidian M 2017 *Journal of Solid Mechanics* **9**(1) 172
[12] Muskhelishvili NI 1953 *Some Basic Problems of the Mathematical Theory of Elasticity* (Leyden: Noordhoff International Publishing)

Electron transport mechanisms in CdS_xTe_{1-x} solid solution thin films prepared by thermal evaporation method

T. Gaewdang* **and Ng. Wongcharoen**

Physics Department, King Mongkut's Institute of Technology Ladkrabang, 1 Soi Chalongkrung 1, Ladkrabang District, Bangkok 10520, Thailand

*Email: thitinai.ga@kmitl.ac.th

Abstract. In this paper, CdS_xTe_{1-x} ($0 \leq x \leq 1.0$) thin films were prepared by thermal evaporation method in vacuum. X-ray diffraction (XRD) revealed that the films exhibited a zincblende structure with the preferred orientation of (111) plane when $x \leq 0.2$. However, when $x \geq 0.8$, they had a wurtzite structure with the preferred orientation of (002) plane. For the composition $0.2 \leq x \leq 0.6$, the zincblende and wurtzite phases coexisted in the system and the films became less preferentially oriented. Atomic force microscopy (AFM) was used to study the morphological features of the samples. Energy gap values in the range of 1.45-2.39 eV were deduced from optical transmission spectra. The temperature dependence of electrical conductivity from room temperature down to liquid-helium temperature of CdS_xTe_{1-x} thin films was performed in order to identify the dominant conduction mechanism. It was shown that three types of conduction mechanisms can be expected. In the high temperature range about 220-300 K, Seto's model of thermionic emission dominated, whereas, in the low temperature range (about 145-185 K), the dominant hopping conduction was Mott variable range hopping. However, the lower temperature region (about 70-105 K) Efros-Shklovskii variable range hopping was observed. The set of parameters describing the properties of localized electrons in each conduction behavior were examined.

1. Introduction

In the past years CdS/CdTe heterostructure has been regarded as one of the most promising structure for high efficiency and low cost solar cells. Theoretical calculations suggest a maximum efficiency around 30 % while the report efficiency is around 22.1% [1]. Light absorption and charge separation in CdTe thin film solar cells largely occur in the vicinity of the CdS/CdTe interface [2]. An interdiffusion process taking place at the CdS-CdTe interface during the cell production has been pointed out as one of the efficiency limiting factors through the resulting change in the band gap [3]. Thus, it is important to have a full understanding of the optoelectronic properties of such ternary compound CdS_xTe_{1-x}. The films of CdS_xTe_{1-x} were prepared by variety of methods, such as close spaced sublimation, magnetron sputtering, chemical bath deposition, spray pyrolysis and thermal evaporation methods. In this study, we prepared CdS_xTe_{1-x} thin films of the whole range of composition (x) by a single source thermal evaporation in vacuum. The crystal structure of the films

IOP Conf. Series: Journal of Physics: Conf. Series **1298** (2019) 012022 doi:10.1088/1742-6596/1298/1/012022

was analyzed by XRD. Surface morphology was observed by AFM. The energy gap values of the films were evaluated from the optical transmission spectra. Electrical conduction mechanisms of all the films were investigated at the temperature down from 300 to 20 K. We have studied these solid solution ternary thin films in order to improve the performance of CdS/CdTe solar cell.

2. Materials and methods

Cadmium sulfotelluride (CdS_xTe_{1-x}) thin films deposited on slide glass substrates by a single source thermal evaporation in vacuum of 5×10^{-5} mbar by keeping the substrate temperature at 100°C. CdS and CdTe starting materials with 99.999 % purity were purchased from Sigma-Aldrich. The starting materials were weighed regarding to the stoichiometric composition, further mixed and ground together; then pressed in a pellet form with a pressure about 6 GPa for using as the precursor for thermal evaporation in vacuum. The as-deposited films were annealed at 200°C for 30 min in a stream of slowly flowing pure N_2 gas. The crystal structure of these films was checked by X-ray diffraction technique with a Brucker D 8 diffractometer using CuK_α radiation. The crystallite size (L) was calculated using Scherrer's formula, neglecting peak broadening due to residual stresses in the films, $L= 0.9\lambda/(\beta\cos\theta)$ where β is the broadening of diffraction line measured at half its maximum intensity in radian and λ is the wavelength of X-rays (0.1540 nm). Surface morphology was examined by Park XE-100 AFM. Optical transmission measurements were performed with UV-Vis-NIR LAMDA 750 spectrophotometer in the wavelength range of 350-2500 nm. The temperature-dependent conductivity of CdS_xTe_{1-x} thin films was measured by two probe method. Two silver electrodes with area 0.05 cm^2 were fabricated on the sample surface. The cryogenic system with a He closed cycle allows us to measure electrical conductivity of the films from 300 to 20 K using a cryostat model RDK 10-320 supplied by Leybold vacuum.

3. Results and discussion

3.1 Structural analysis

Figure 1 shows XRD patterns of CdS_xTe_{1-x} thin films deposited on slide glass substrates by a single source thermal evaporation. For $x<0.2$, XRD revealed that CdS_xTe_{1-x} thin films exhibited a zincblende structure with the preferred orientation of (111) plane. However, for $x \geq 0.8$, the films existed in the wurtzite structure with the preferred orientation of (002) plane. For the composition $0.2 \leq x \leq 0.6$, the zincblende and the wurtzite phases coexisted in the system and the films became less preferentially oriented. The coexistence of the zincblende and wurtzite phases often occurring in films with composition (x) in the range 0.2-0.8 has been previously reported [4-8]. With increasing composition (x), the strongest peak intensity shifted to a higher 2θ value; this is due to the partial replacement of Te atoms by S atoms. Figure 2 shows AFM images of the films of different compositions. AFM images of the films with composition $x = 0$ showed a polycrystalline character with pyramidal shaped crystallites, typical of cubic crystallites oriented in (111) plane. The grain size increased from 105 nm to 260 nm, as the value of x increased from 0 to 1. As composition (x) increased up to 0.4, the coexistence of flattening and pyramidal shaped crystallites was observed; this was consistent with the mixed phases seen in the XRD results. Then, a well-defined prism geometry was observed when x=1. The surface roughness decreased from 5.10 nm (for x = 0) to 3.12 nm (for x = 0.4), and then increased to 6.43 nm (for x=1).

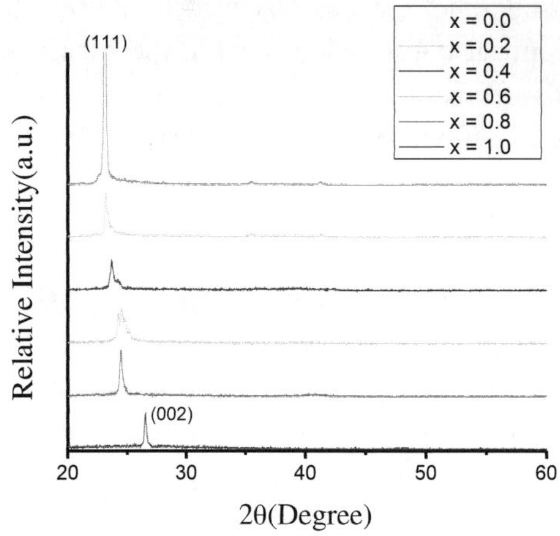

Figure 1. XRD patterns of CdS_xTe_{1-x} thin films.

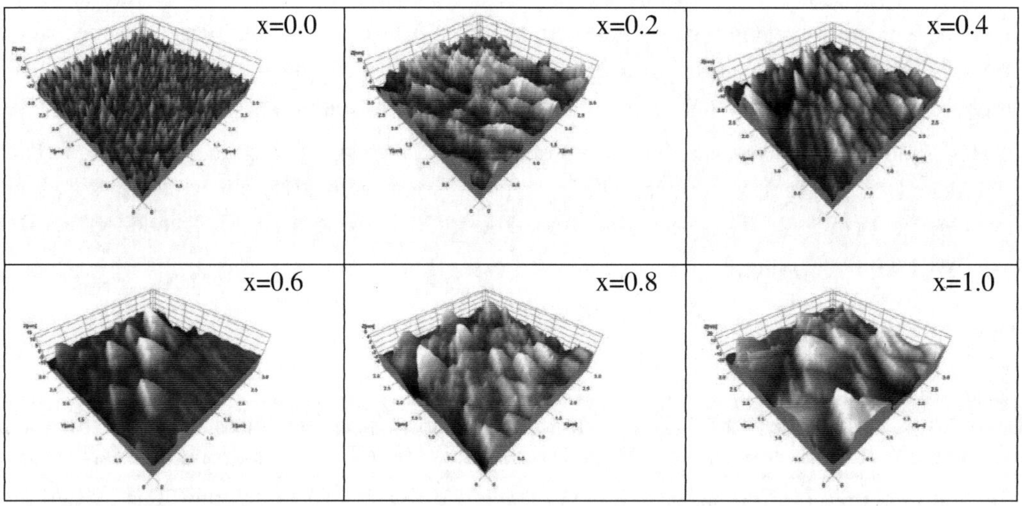

Figure 2. AFM micrographs of CdS_xTe_{1-x} thin films.

3.2 Optical properties

Figure 3 shows the optical transmission spectra of CdS_xTe_{1-x} thin films in the wavelength range of 350-2500 nm. The value of energy gap is given by the intercept of the straight line with the energy axis. Figure 4 shows the plots of $(\alpha h\nu)^2$ vs. hν of the films with different compositions approximated a straight line, indicating that the optical transition in the films corresponded to the model for direct allowed transition. The value of the energy gap value is given by the intercept of the straight line with the energy axis. Energy gap values were found to be 1.51, 1.47, 1.45, 1.54, 2.00, 2.39 eV corresponding to composition x = 0, 0.2, 0.4, 0.6, 0.8, 1.0, respectively.

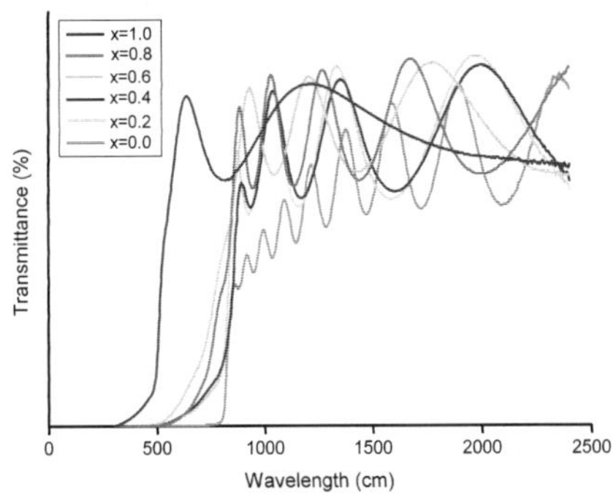

Figure 3. Transmission spectra of CdS_xTe_{1-x} thin films.

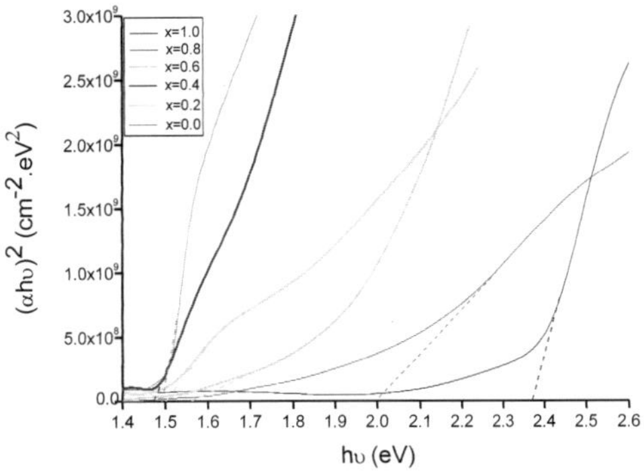

Figure 4. $(\alpha h\nu)^2$ vs. hν plot of CdS_xTe_{1-x} thin films.

3.3 Electrical properties

The variation of electrical conductivity of the films with composition (x) at room temperature is shown in Figure 5. The lowest conductivity value about $1.6\times 10^{-9}\ (\Omega\cdot cm)^{-1}$ was observed in the films with x=0.6. However, the highest conductivity value around $4.5\times 10^{-3}\ (\Omega\cdot cm)^{-1}$ as obtained in the films with x=0.8. Normally, CdS films exhibit n-type semiconductor while CdTe films exhibit p-type semiconductor. From the thermoelectric test, the films with $x \leq 0.4$ show p-type, but the films with $x \geq 0.6$ show n-type. Our results are in good agreement with the earlier report [9]. The conductivity type started to change from p-type to n-type when x reached 0.6. It indicated that the amount of charge carrier between the electrons and holes is approximately the same order or refute. The Arrhenius plots of $\ln(\sigma)$ vs. 1000/T of CdS_xTe_{1-x} thin films for the whole composition ($0\leq x\leq 1.0$) are shown in Figure 6.

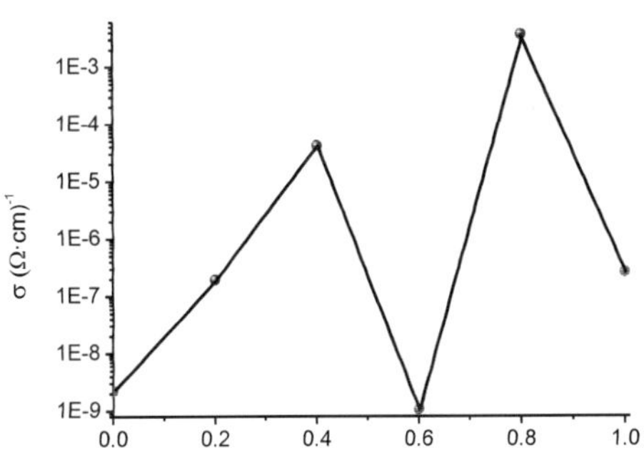

Figure 5. Variation of conductivity with composition (x) of CdS_xTe_{1-x} thin films at 300 K.

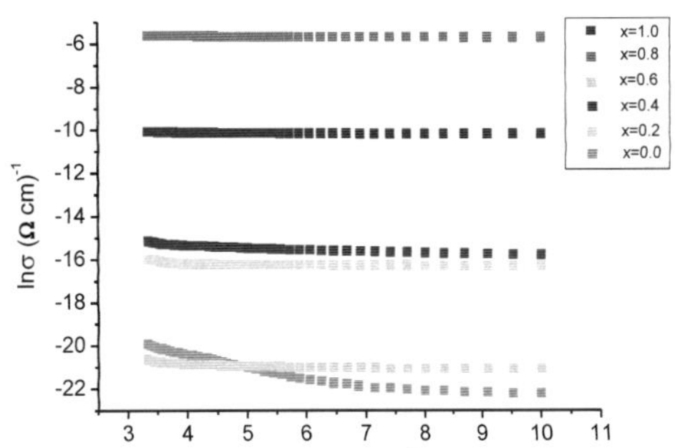

Figure 6. ln(σ) vs. 1000/T plot of CdS_xTe_{1-x} thin films.

Figure 7 (a) shows the plot of $\ln(\sigma)$ and 1000/T for the films with x=0.8. Two regimes are observed in the curve. At high temperature, lnσ exhibits an Arrhenius like behavior, while notably deviates from it at lower temperature. For high temperature range, the electrical conductivity of the sample is dominated by thermal activation passing through grain boundaries. The temperature dependence of the

IOP Conf. Series: Journal of Physics: Conf. Series **1298** (2019) 012022 doi:10.1088/1742-6596/1298/1/012022

electrical conductivity of the films was firstly analyzed using Seto's model as the following relation [10]:

$$\ln\left(\sigma T^{1/2}\right) = -\frac{E_a}{k}\left(\frac{1}{T}\right) + \ln\sigma_0 \tag{1}$$

where σ_0 is pre-exponential factor, k is Boltzmann's constant and E_a is the activation energy. This model related to grain boundaries which trap carriers and the barrier height value appeared. As shown in Figure 7 (b), a good linear relationship can be observed in the Arrhenius plot in the high temperature range (220-300 K). The important parameters deduced from Seto's model were listed in Table 1: L is crystallite size, E_b is barrier height, N_d is donor concentration, E_t is trap energy level, N_t is trap concentration, L_D is Debye length and ε_r is dielectric constant taken from the literature [11]. Figure 7 (c) shows the plot of $\ln(\sigma T^{1/2})$ vs $T^{-1/4}$. The plot was found to be linear at temperature range of 145-185 K. This suggested that the conduction is controlled by Mott variable-range hopping (Mott-VRH) instead of the Seto's model at high temperature. The Mott's parameters can be extracted from the relation [12]:

$$\ln\left(\sigma T^{1/2}\right) = -\left(\frac{T_{0,Mott}}{T}\right)^{1/4} + \ln\sigma_{0,Mott} \tag{2}$$

The important parameters deduced from Mott-VRH model were tabulated in Table 2: $T_{0,Mott}$ is a characteristic temperature coefficient, $\sigma_{0,Mott}$ is pre-exponential factor, $N(E_F)$ is density of state at the Fermi level, $R_{hop, Mott}$ is average hopping distance, and $W_{hop, Mott}$ is average hopping energy. At very low temperature range (70-105K), the measured data were satisfied the Efros-Shklovskii variable-range hopping (ES-VRH). In this electrical conduction mechanism region, the electron interaction caused the density of state at Fermi level is separated and became to the coulomb gap (Δ_C). The conductivity in ES-VRH model is given as [13]

$$\ln\left(\sigma T\right) = -\left(\frac{T_{0,ES}}{T}\right)^{1/2} + \ln\sigma_{0,ES} \tag{3}$$

Figure 7 (d) shows the plot of $\ln(\sigma T)$ vs $T^{-1/2}$. The important parameters deduced from ES-VRH model were listed in Table 3: $T_{0,ES}$ is a characteristic temperature coefficient, $\sigma_{0,ES}$ is pre-exponential factor, $R_{hop, ES}$ is average distance, ξ_{ES} is localization length and $W_{hop, ES}$ is average hopping energy.

IOP Conf. Series: Journal of Physics: Conf. Series **1298** (2019) 012022 doi:10.1088/1742-6596/1298/1/012022

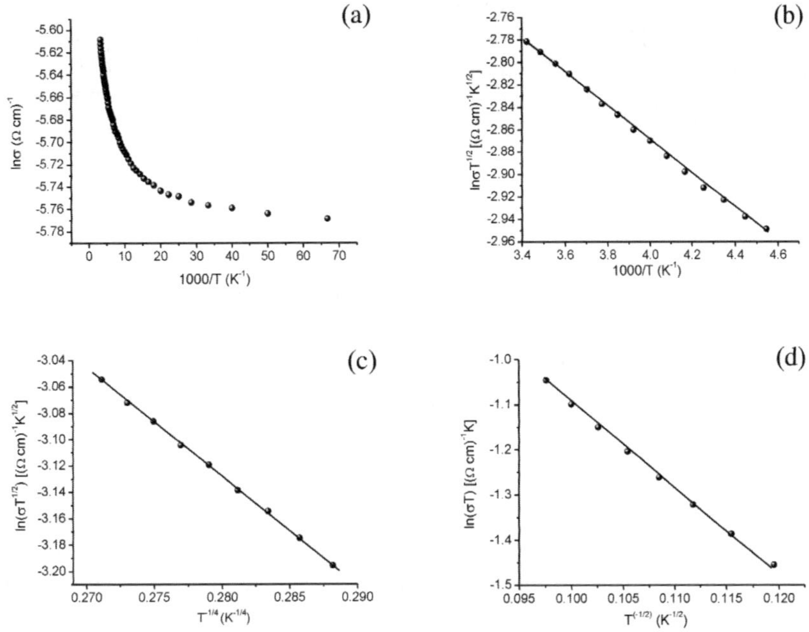

Figure 7. (a) ln(σ) vs 1000/T plot, (b) $\ln(\sigma T^{1/2})$ vs 1000/T plot, (c) $\ln(\sigma T^{1/2})$ vs $T^{-1/4}$ plot and (d) ln(σT) vs $T^{-1/2}$ plot of CdS_xTe_{1-x} thin films for x=0.8.

Table 1. Seto's parameters at high temperature of CdS_xTe_{1-x} thin films for x=0.8.

Composition (x)	E_g (eV)	ε_r	L (nm)	E_b (meV)	N_d (cm^{-3})	E_t (eV)	N_t (10^{11} cm^{-2})	L_D (nm)
0.0	1.51	12.62	77.7	59.9	4.18×10^{16}	0.694	3.20	16.53
0.2	1.47	12.24	39.9	13.5	3.50×10^{16}	0.721	1.41	17.81
0.4	1.45	11.74	15.1	17.3	3.16×10^{17}	0.703	4.77	5.95
0.6	1.54	11.51	3.4	23.5	1.45×10^{18}	0.746	28.72	1.15
0.8	2.00	11.13	5.9	12.9	8.54×10^{18}	0.907	9.09	2.69
1.0	2.39	10.56	34.2	25.4	9.0×10^{16}	1.174	3.08	11.24

Table 2. Mott's parameters at low temperature of CdS_xTe_{1-x} thin films for x=0.8.

Composition (x)	$\sigma_{0,Mott}$ $(\Omega\cdot cm)^{-1}$	$T_{0,Mott}$ (K)	$N(E_F)$ ($cm^{-3}\cdot eV^{-1}$)	$R_{hop,Mott}$ (nm)	$\frac{R_{hop,Mott}}{\xi_{Mott}}$ (@160 K)	$W_{hop,Mott}$ (meV)
0.0	1.57×10^{-3}	5.58×10^{6}	9.69×10^{19}	3.81	4.72	110.49
0.2	2.21×10^{-5}	5.44×10^{6}	6.44×10^{22}	0.85	1.05	42.91
0.4	4.90×10^{-3}	5.13×10^{3}	8.73×10^{22}	0.74	0.92	36.85
0.6	2.75×10^{-7}	1.21×10^{4}	1.98×10^{22}	1.03	1.28	77.48
0.8	4.34×10^{-1}	4.57×10^{3}	8.84×10^{22}	0.79	0.91	36.64
1.0	6.01×10^{-3}	3.65×10^{4}	2.92×10^{21}	1.66	2.06	104.23

Table 3. ES's parameters at very low temperature of CdS_xTe_{1-x} thin films for x=0.8.

Composition (x)	$\sigma_{0,ES}$ $(\Omega\cdot cm)^{-1}$	$T_{0,ES}$ (10^3 K)	$R_{hop,ES}$ (nm)	ξ_{ES} (nm)	$\frac{R_{hop,ES}}{\xi_{ES}}$	$W_{hop,ES}$ (meV)	Δ_C (μeV)

					(@100 K)		
0.0	4.31×10^{-7}	9.01	48.12	64.13	0.75	12.90	1.28
0.2	8.86×10^{-5}	3.51	77.03	164.04	0.47	8.16	20.6
0.4	3.15×10^{-2}	4.53	67.81	127.38	0.53	9.18	27.2
0.6	5.20×10^{-7}	4.05	71.72	142.87	0.50	8.69	12.3
0.8	2.14×10^{0}	3.45	77.64	167.73	0.46	8.02	23.9
1.0	9.57×10^{-5}	3.37	112.03	171.14	0.65	5.59	4.29

4. Conclusions

CdS_xTe_{1-x} thin films in the entire range ($0\leq x\leq 1.0$) were prepared by vacuum thermal evaporation method on glass substrate using mixed powders of CdS and CdTe compounds. From XRD results, CdTe-like phase belonging to zincble phase was observed when $x \leq 0.2$. In contrast, CdS-like phase belonging to wurtzite structure was obtained when $x \geq 0.8$. For the composition $0.2 \leq x \leq 0.6$, the zincblende and wurtzite phases coexisted in the system and the films became less preferentially oriented. The crystallite size in the range 105-260 nm was elucidated by AFM. The optical transition on CdS_xTe_{1-x} thin films was found to be direct allowed transition type. The variation of energy gap value with composition (x) in the range of 1.45-2.39 eV was observed. From the thermoelectric test, the conductivity type started to change from p-type to n-type when composition (x) reached x=0.6 due to auto-compensation of charge carrier between the electrons and holes. The temperature range of 20-300 K, it was shown that three types of electrical conduction mechanisms can be expected. In the high temperature range (>220 K), thermionic emission across grain boundary dominates, whereas in the low temperature range (145-185 K) the dominant hopping conduction is Mott's variable range hopping. At very low temperature range (70-105 K), the Efros-Shklovskii variable-range hopping is observed. The films with x=0.8 showed the highest conductivity and also high energy gap value which may suitable to solar cell application.

5. References

[1] Chander S and Dhaka M S 2018 *Mater. Res. Bull.* 97 128
[2] Kuciauskas D, Kanevec A, Duenow J N, Dippo P, Young M, Li J V, Levi D H and Gessert T A 2013 *Appl. Phys. Lett.* 102 173902
[3] Ikhmayies S J 2016 *Energy* 9 234
[4] Al-Shakarchi E M 2010 *J. Korean Phys. Soc.* 57 1421
[5] Wood D A, Roger K D, Lane D W, Conibeer G J and Parton D. 1998 *J. Mater. Sci. Lett.* 17 1511
[6] El-Nahass M M, Sallam M M, Afifi M A and Zedan I T 2007 *Mater. Res. Bull* 42 371
[7] Adachi S 2009 Properties of semiconductor alloys group-IV, III-V and II-VI semiconductor (John Wiley & Sons, New York) p 210
[8] Gaewdang T and Wongcharoen 2017 *CMUJ Nat. Sci.* 19 263
[9] Jacome C, Florez M, Gurevich Y G, Giraldo J and Gordillo G 2001 *J. Phys. D: Appl. Phys.* 34 1862
[10] Tomakin M, Altunbas M, Bacaksiz E and Celik S 2012 *Thin Solid Films* 520 2532
[11] Ameri M, Mesbah S, Al-Douri Y, Bouhafs B, Vashney, D and Ameri I 2014 *Acta Phys. Pol. A* 125 1110
[12] Sharma K, Al-Kabbi A S, Saini G S S and Tripathi S K 2012 *Appl. Phys. A* 108 911
[13] Ansari M Z and Khare N 2015 *J. Appl. Phys.* 117 025706

Performed Of Filter Nano Textile For Purification Of Aerosol Water By Electrospinning Technique

[1,1]K H Jawad, [2,2]D B Mohammed and [3,3]M R Alsaffar

[1]University of Babylon, Department of Poly. Eng. ,Iraq
[2]University of Technology , Department of Applied Science, Iraq
[3]Oil ministry / Iraq/

[1]hanajawad21@yahoo.com;[2]balkees1957@yahoo.com; [3]rusulrusul83@yahoo.com

Abstract. Poly styrene (PS) Nano filter prepare by electrospinning technique for cleaning the dust water from dumps and maintain the properties of the resulting water. 18% con. of (PS + dimethyl forma amide DMF) solution prepared for electrospinning pump on Aluminium surface for (10hr.), 4μm thickness filter resulted. Some test performed on this filter involve: morphology by using scanning electron microscopy (SEM) , thickness by digital micrometre, contact angle by contact angle shape drop analyser .In addition that, some tests performed on the water sample after dumps separating from water by preparing nano filter, these tests involve: viscosity, surface tension, density, PH number, and weight of the sample, Results show the prepared filter have hydrophobic behavior and it has (108°) θ (contact angle), the prepared filter has nano structured with beads through its morphology.

1. Introduction

Fast population expansion generate the huge problem of limited pure water supply in the world and the dust water treatment is the only purification process that can beat on this problem. Nanotechnology is one of solutions and it has extensive domain for dust water treatment. [1]

Filtration is a process of separation of particles and impurities from water by coercion the water through a porous media. There are some types of porous media involve: natural, as sand, gravel and clay, or nanotextile wall made of various materials [2].The development of Nano science and technology has presented new chances for using smaller and more steady structures for porous films. [3]

A number of methods have been employed to fabricate super hydrophobic surfaces, such as electrochemical sedimentation [4], electrophoresis precipitation [5], chemical inscription [6], layer after layer self-assembly [7], chemical vapour deposition (CVD) [8], electrospinning technique [9]. Electrospinning technique is a novel method for preparation various types of nanofibers textile as 'having a diameter of less than one micron for different application such as medical textile, personal care, filtration, barrier, composite, insulation and energy storage [10]

2. Experimental

2.1 Materials
Polystyrene (PS) pellets, with molecular weight (MW = 35,000) , density (1.06 g/cm3), from Sigma Aldrich. PS is, organic , Hydrophobic Polymers N, N - Di methyl forma amide (DMF) (CH3)2NCH(OCH3)2 , boiling point (65oC), as a solvent was used

2.2 Electrospinning set up: Electro spinning set up
Electrospinning solution placed in a syringe connected to a syringe pump to regulate the fluid flow, while the positive high voltage power supply connected to metallic needle , and the negative power supply connected to earthly metallic collector

2.3 Fabrication of Porous Hydrophobic Textile:
15ml of (PS/ DMF) pumped by electrospinning system for (10hr) , and collected on aluminum foil (20cm* 15cm2), then it cut according to required test as in table (1).

2.4 dust water System Preparation:
(1g) soil mixed with (49g) distil water with constantly moving for (5min).

2.5 Soil / water (Dust water) System Separation:
Soil/ water system separation by using prepared nanofilter was performed, by using the following system, fig (2) :
2 mbar pressure used for pumping the dust water from tank of dust water to clean water tank passing through the nanofilter.

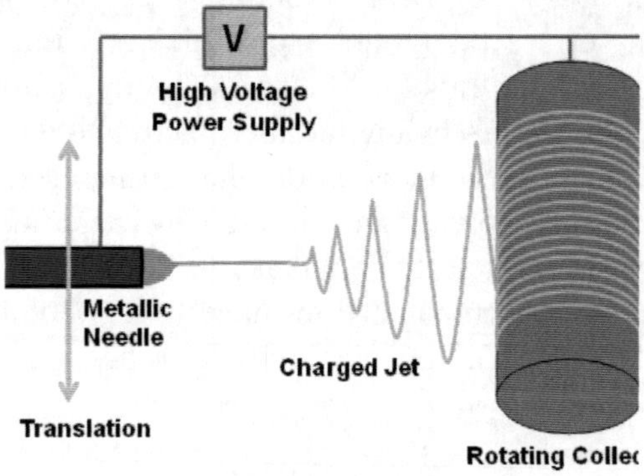

Figure 1 electrospinning set up

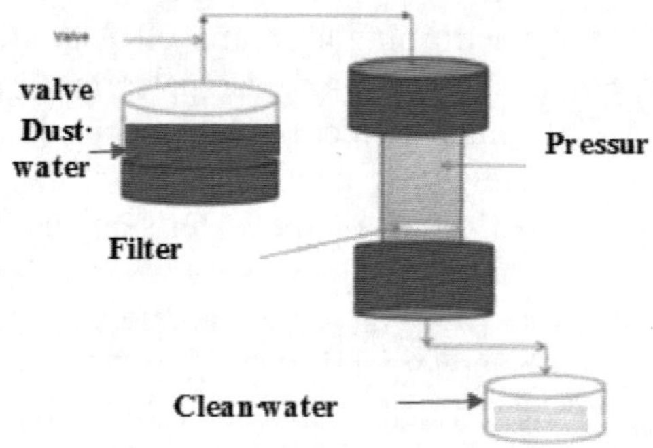

Figure 2 water purification system

3. Result and Discussion

3.1 Morphology of Nanofilter by SEM :

Fig (3) show the SEM image of prepared nano filter: We show from fig (3 a) there are an beads in morphology were showed, these because instability through electrospinning operation. These beads are very important at this time, because it lead to increase of roughness of nanofiber surface which leads to increase of hydrophobic and decrease the wettability of nonfilter. These beads collected the dust and impurities from water, then it permit water to pass through it by its porous. Fig (3 b) show the SEM image of filter after removal the dust from water which agglomerates on the surface of water by hydrophobicity of this filter

IOP Conf. Series: Journal of Physics: Conf. Series **1298** (2019) 012023 doi:10.1088/1742-6596/1298/1/012023

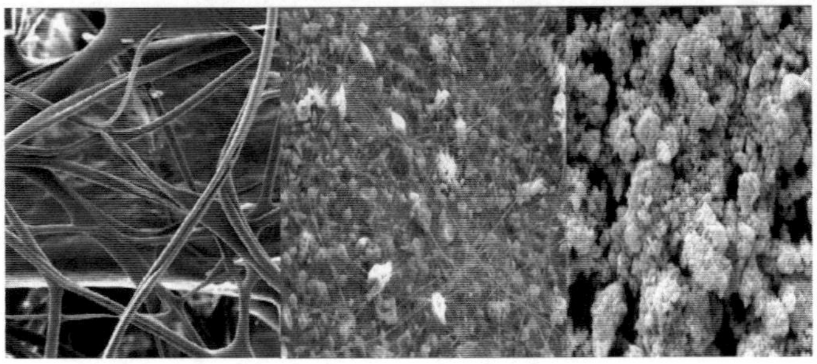

a b

Figure 3 (a) before cleaning the water **Figure 3 (b) after cleanining the water**

Figure (3 a, b) SEM images of PS nanofilter

3.2 Contact angle analysis :

Fig (4) show the contact angle of (PS) nanofilter, we noticed the contact angle of nonofilter is (108 ^{o}) , and this leads to high hydrophobic properties, and this nanofilter is able to separate the water from the dust and soil high efficiency. [11]

Figure 4. contact angle of PS

3.3 Wettability :

Figs (5 - 6) show the change of contact angle with time for constant periodic time (55 min) with constant increasing ratio (5min) Fig (7) referees to there are an small change of contact angle with increasing of remaining time on the textile surface , this is because the roughness due to the beads which prevents water from spreading and penetrate through the of fabric texture hydrophobic prepare textile surface. [12] Also from fig (6) notice the contact angle decrease to 66 ^{o} after the water drop passing through the filter by porous effecting

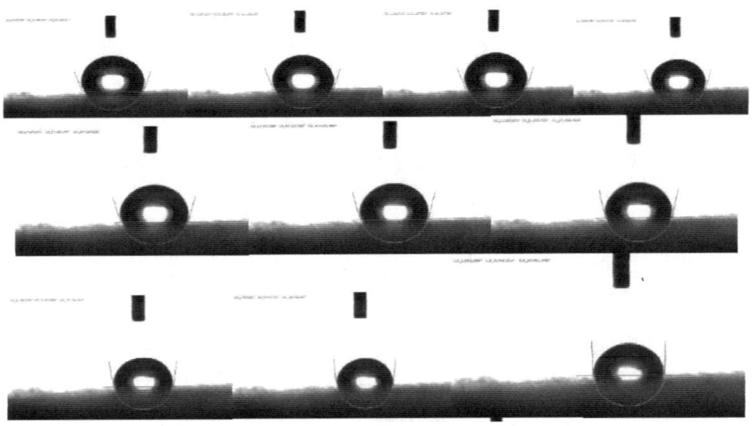

Figure 5 contact angle of nano textile

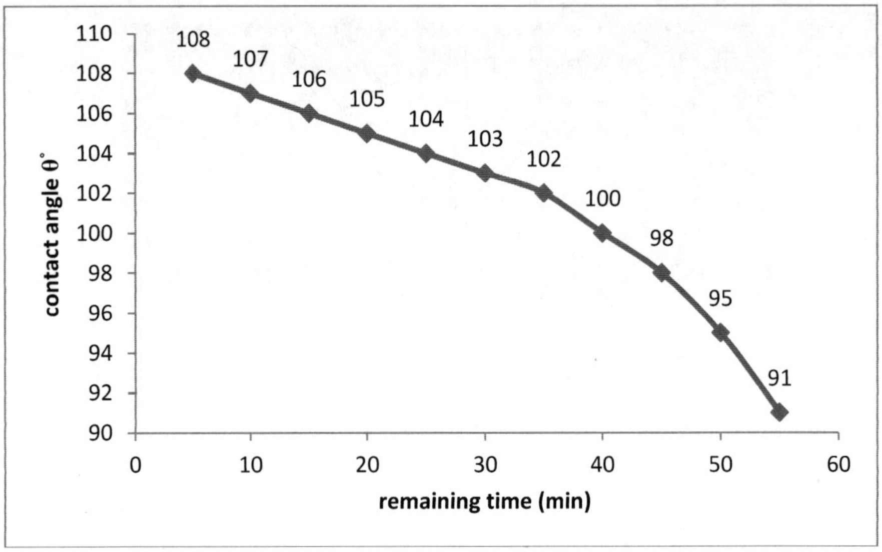

Figure 6 Relationship between contact angle and remaining time on the surface of Textile

3.4 water characteristic before and after dust separation:

Table 1. Dimension of Samples According to Tests

Sample No.	Dimension cm^2	Test
1	1	SEM (Scan Electron Microscopy)
2	4 (2*2)	Contact angle
3	5 cm diameter	Cleaning of dust water

Table 2 characteristic of water after & before separation from dust

Characteristic	Before separation from dust	After separation from dust
Weight g	50.001	50.09
Density g/cm^3	1	1.09
Viscosity Cp	0.98	1
Surface tension mN/m	6.8	6.9
PH	7	7.2

Table (2) show the characteristic of water before and after the separation from dust: From table (2) observed there are an agreement between the results of characteristic of water (before & after) its separation from dust, this refers to high efficiency of prepare textile for obtain on normal pure water with very little steps, and this prepared textile have superior properties for its separation from dust. This is because two factors, involve hydrophobic surface property and winding depth filter property. [13]

4. Conclusion:

1- The electrospinning technique is very active method for preparation the hydrophobic textile for separation dust/water system.

2- this textile has two filtration properties, surface filtration by its hydrophobic characteristic, and depth filtration by winding and beads of nanofibers characteristic.

5. Acknowledgements

The authors extend their appreciation to the Technology university /Applied science department /materials branch , Nano technology research center and Babylon University / Material engineering collage for perform this search

6. References:

[1] Anil K., Akansha G.&Bhagwan N., "A review: Nano Membrane and Application" Inter. Jour. of Innovative Res. in Sci., Eng. and Tech.", V.3, Iss.1, (2014)

[2] Vikas Sh. and Akhilesh Sh., " Nanotechnology : An Emerging Future Trend in Wastewater Treatment with its Innovative Products and Processes ", Inter. Jour. of Enhanced Research in Sci. Tech. and Eng. VOL. 2 ISSUE 1, DEC-(2013)

[3] Y. Han "Molecular sieving using nanofilters: past, present and future". Lab. Chip 8, 23–33 (2008).

[4] L. Feng, Z. Zhang, Z. Mai, Y. Ma, B. Liu, L. Jiang, D. Zhu, A. Super-Hydrophobic, Super-oleophilic coating mesh film for the separation of oil and water, Angewandte Chemie International Edition 43, (2014)

[5] Xufeng W. and Gaoquan Shi, " Production and Characterization of Stable Superhydrophobic Surfaces Based on Copper Hydroxide Nanoneedles Mimicking the Legs of Water Striders",J. Phys.Chem. B, v.110 No.23. (2006)

[6] Zuxin S., Qing Li, Zhongwei Wang, Longqin Li, Funan Chen, and Juncen Zhou," Novel Method for Controllable Fabrication of a Super hydrophobic CuO Surface on AZ91D Magnesium Alloy", ACS Appl. Mater. Interfaces, 4 (8), pp 4348–4356(2012)

[7] Zhijian Ch. and Wei Li, " Introduction to the special issue of the International Symposium on Crystal Engineering & Drug Delivery Systems", Frontiers of Chem. Eng. in China, ,V. 4, Iss. 1, (2010)

[8] Shunsuke N., Hitomi S., Xintong Z., Zhaoyue Liu, Kazuya N., Taketoshi M., Yoshihiro K. and Akira F., "Assembly of Self-Assembled Monolayer-Coated Al_2O_3 on TiO_2 Thin Films for the Fabrication of Renewable Super hydrophobic−Superhydrophilic Structures", *Langmuir*, 25 (13), (2009)

[9] Al dabbagh B. M. & Alshimary H. J. " Sio2/ Polyamide Nanocomposite textile for super hydrophobic coating By Electrospinning technique ", international journal of research in science,Vol.(2), No.(4) 2016

[10] Al dabbagh B. M. & Alshimary H. J. "Polyamide Nanofibers Coating by Electrospinning Technique for Anti Corrosion Behavior", Engineering and Technology Journal, Vol. 35, Part A, No. 10, 2017

[11] Haiyan S., Zhen Xu Ch., " Multifunctional, Ultra-Flyweight, Synergistically Assembled Carbon Aerogels", Advanced materials, V.25, Iss.18 . (2013)

[12] Chao-Hua X., Peng-Ting J. , P. Z., Ya-Ru L. , Shun-T. J., " Fabrication of superhydrophobic and superoleophilic textiles for oil–water separation ", Applied Surface Science. (2013). DOI: 10.1016 /j.apsusc. 2013.07.120

[13] K. Sutherland, Filters and filtration handbook, 5th , Elsevier, USA, (2008).

IOP Conf. Series: Journal of Physics: Conf. Series **1298** (2019) 012024 doi:10.1088/1742-6596/1298/1/012024

Effect of Soret and Dufour numbers on double diffusive mixed convection boundary layer flow induced by a shrinking sheet

Siti Suzilliana Putri Mohamed Isa[1,2*], Norihan Md. Arifin[1,3], and Umer Farooq[4,5]

[1]Institute for Mathematical Research, Universiti Putra Malaysia, 43400 UPM Serdang, Selangor Darul Ehsan, MALAYSIA.
[2]Centre of Foundation Studies for Agricultural Science, Universiti Putra Malaysia, 43400 UPM Serdang, Selangor Darul Ehsan, MALAYSIA.
[3]Department of Mathematics, Universiti Putra Malaysia, 43400 UPM Serdang, Selangor Darul Ehsan, MALAYSIA.
[4]Department of Mathematics, Faculty of Science, Jiangsu University, 212013 Zhenjiang, CHINA.
[5]Department of Mathematics, COMSATS University Islamabad, Park Road, Tarlai Kalan Islamabad 44000, PAKISTAN.

Email: ctsuzilliana@upm.edu.my

Abstract. The numerical study of double diffusive mixed convection boundary layer flow over a shrinking sheet are presented in the presence of Soret and Dufour effects. Shrinking velocity, wall temperature and wall concentration are assumed to have exponential function forms. Non-similarity method is applied for the governing basic equations (flow, momentum, energy and concentration equations) before they are solved numerically using bvp4c program in Matlab software. The numerical results of velocity, temperature and concentration are reported graphically for various values of suction parameter, dimensionless coordinate along the plate parameter, shrinking parameter, mixed convection parameter, buoyancy ratio parameter, Soret parameter and Dufour parameter. Since dual solutions is obtained, stability analysis is performed to select the stable solution.

1. Introduction

The study of thermodynamics is beneficial as its concepts are being applied to engineering [1,2]. In engineering applications, the mixed convection in fluid flow due to a stretching/shrinking sheet will produce desired findings. In the sectors of science and engineering, mixed convection flows (combination of forced and natural convection) are applied in spray and flash dying, crystal growth, float gas production, combustion of atomized liquid fuels, thermal–hydraulics of nuclear reactors, the cooling of sophisticated electronics and heat exchanger devices, and dehydration operations in chemical processing [3]. On the other hand, the flow due to a stretching or shrinking sheet plays a role in cooling of metallic plate in a bath, glass fibre and paper production, extrusion of polymer sheet from a die, and continuous casting [4]. Therefore, a lot of reports regarding to the exponential variations of stretching/shrinking velocity and temperature distributions have been published [5–11].

In the process in fluid flow induced by temperature and concentration differences, Soret and Dufour effects must not be neglected. Mass transfer induced by temperature differences is known as Soret effect, whereas Dufour effect refers to the heat transfer caused by concentration gradient in the fluid. From experimental reports, the diffusion of matter and heat due to Soret and Dufour effects become effective when the gradients of these two elements are very large. When the gradient of the temperature and the concentration are large, the coupled interaction between Soret and Dufour effects are sufficiently great [12]. As a results, many researchers explored the impacts of these to parameters in their model of the fluid flow in the presence of stretching/shrinking sheet [13–18]. They considered linear, nonlinear and exponential variations of stretching/shrinking velocity, temperature and concentration distributions.

In numerical solutions, the existence of multiple solutions are able to be obtained. These multiple solutions always occur in the case of convection occurred due to stretching/shrinking sheet. Stability analysis is used to determine the most stable solution, and the other is declared as unstable. This analysis can be performed by using the bvp4c programme from MATLAB. This analysis calculates the growth of disturbances for dual solutions, where the solution with initial decay of disturbance labeled as stable solution (first solution). Besides, unstable solution (second solution) owns initial growth of disturbance. The application of stability analysis in the related cases (Soret and Dufour impacts) have been done by some researchers [19–20] for unsteady state. This method also have been applied in the case of fluid flow occurred by the movement of exponential stretching or shrinking sheet [21].

This paper applies the problem formulation developed by Srinivasacharya and RamReddy [13], and stability analysis is applied in the numerical solutions. The previous investigators [13] studied the case of the sheet is stretched and impermeable, whereas this paper focuses on the sheet when it is shrieked and the fluid is allowed to pass through the sheet. The variations of velocity, temperature and concentration in this problem are exponential [13–14,15]. To the best of the authors' knowledge, the stability analysis of the impacts of Soret and Dufour parameters in the case of mixed convection over an exponentially flat stretching/shrinking sheet is not yet to be reported in the literature. The governing equations are transformed to ordinary differential equations using dimensionless similarity transformation parameter and are solved numerically by Matlab. Computed numerical results are performed through graphs of velocity temperature and concentration profiles.

2. Mathematical formulation and stability analysis

2.1. Mathematical formulation

Consider the two-dimensional incompressible, viscous and electrically conducting fluid over an exponentially permeable stretching/shrinking surface with the effect of Soret and Dufour parameters (Figure 1). The x-axis runs along the shrinking surface in the direction opposite to the sheet motion and the y-axis is perpendicular to it. The viscous fluid is placed at $y > 0$. Under the assumption of Boussinesq and boundary layer approximations, the problem is governed by the following equations [13]:

$$\frac{\partial u}{\partial x} + \frac{\partial v}{\partial y} = 0 \tag{1}$$

$$u\frac{\partial u}{\partial x} + v\frac{\partial u}{\partial y} = \upsilon\frac{\partial^2 u}{\partial y^2} + g\beta_T\left(T - T_\infty\right) + g\beta_C\left(C - C_\infty\right) \tag{2}$$

$$u\frac{\partial T}{\partial x}+v\frac{\partial T}{\partial y} = \alpha\frac{\partial^2 T}{\partial y^2} + \frac{DK_T}{C_s C_P}\frac{\partial^2 C}{\partial y^2} \tag{3}$$

$$u\frac{\partial C}{\partial x}+v\frac{\partial C}{\partial y} = D\frac{\partial^2 C}{\partial y^2} + \frac{DK_T}{T_m}\frac{\partial^2 T}{\partial y^2} \tag{4}$$

where u and v are the components of velocity in the x and y directions, $\upsilon=\mu/\rho$ is the kinematic viscosity, μ is the viscosity, ρ is the fluid density, g is the acceleration due to gravity, β_T is the coefficient of thermal expansion, β_C is the coefficient of solutal expansions, T is the temperature of the fluid, C is the concentration of the fluid, α is the thermal diffusivity, D is the solutal diffusivity of the medium, K_T is the thermal diffusion ratio, C_s is the concentration susceptibility, C_p is the specific heat at constant pressure and T_m is the mean fluid temperature. The subscripts of w and ∞ at parameters T and C indicate the conditions at the wall and at the outer edge of the boundary layer.

The appropriate boundary conditions are

$$\begin{aligned} &u=u_w(x)=\lambda U_0\exp(x/L), \quad v=v_w(x), \\ &T_w(x)=T_\infty + T_0\exp(x/2L), \quad C_w(x)=C_\infty + C_0\exp(x/2L), \quad \text{at } y=0, \\ &u\to 0, \quad T\to T_\infty, \quad C\to C_\infty \quad \text{as } y\to\infty, \end{aligned} \tag{5}$$

where $\lambda<0$ is the shrinking parameter and $v_w(x)<0$ is the wall mass suction velocity. The term $\exp(x/2L)$ in exponential temperature distribution T_w and exponential concentration distribution C_w are used [13-14,15] in this problem for the case of two-dimensional flow with the presence of Soret-Dufour effects and when concentration equation included in governing equations. The assumption of temperature and concentration functions are to make sure the numerical results will satisfy final boundary condition after substituting similarity variables.

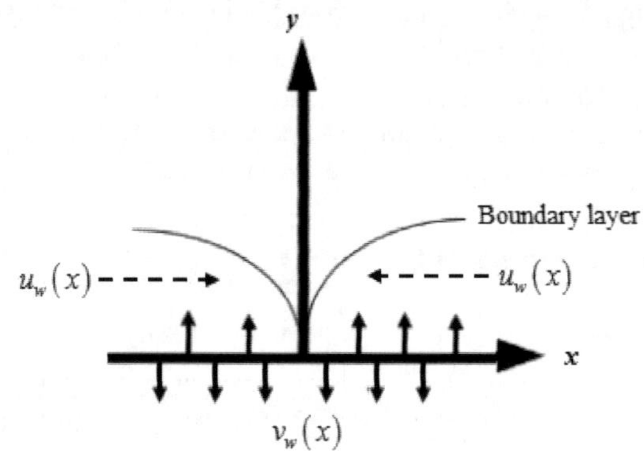

Figure 1. Physical model and coordinate system

In this problem, the stream function is stated as $\psi(x,y)=(2\upsilon LU_0)^{1/2}\exp(x/2L)\, f(\eta)$, where

$$u=\frac{\partial\psi}{\partial y} \text{ and } v=-\frac{\partial\psi}{\partial x} \tag{6}$$

Introducing new similarity variables:

$$\theta(\eta)=\frac{T-T_\infty}{T_w-T_\infty},\qquad \phi(\eta)=\frac{C-C_\infty}{C_w-C_\infty},\qquad \eta=y\left(\frac{U_0}{2\upsilon L}\right)^{1/2}\exp(x/2L),$$
$$u=U_0\exp(x/L)\,f'(\eta),\qquad v=-\left(\frac{\upsilon U_0}{2L}\right)^{1/2}\exp(x/2L)\left[f(\eta)+\eta f'(\eta)\right], \tag{7}$$

where prime denotes differentiation with respect to η.

When the governing equations (2)–(4) and boundary conditions (5) are substituted by similarity variables, it will be transform to the following form:

$$f'''+ff''-2(f')^2+2Ri\left[\exp\left(\frac{-3X}{2}\right)\right](\theta+N\phi)=0. \tag{8}$$

$$\frac{1}{\Pr}\theta''+f\theta'-f'\theta+Db\phi''=0. \tag{9}$$

$$\frac{1}{Sc}\phi''+f\phi'-f'\phi+Sr\theta''=0. \tag{10}$$

$$\begin{aligned} &f'(\eta)=\lambda,\quad f(\eta)=S,\quad \theta(\eta)=1,\quad \phi(\eta)=1 \quad \text{at } \eta=0,\\ &f'(\eta)\to 0,\quad \theta(\eta)\to 0,\quad \phi(\eta)\to 0 \quad \text{as } \eta\to\infty, \end{aligned} \tag{11}$$

where $Ri=Gr/\mathrm{Re}^2$ is the mixed convection parameter, $Gr=g\beta_T(T_0-T_\infty)L^3/v^2$ is the thermal Grashof number, $\mathrm{Re}=u_0L/v$ is the Reynolds number, $X=x/L$ dimensionless coordinate along the plate parameter, L is the characteristic length of the plate, $N=\beta_C(C_0-C_\infty)/\beta_T(T_0-T_\infty)$ is the buoyancy ratio, $\Pr=\upsilon/\alpha$ is the Prandtl number, $Sc=\upsilon/D$ is the Schmidt number, $Sr=DK_T(T_0-T_\infty)/T_m\upsilon(C_0-C_\infty)$ is the Soret number, $Db=DK_T(C_0-C_\infty)/C_sC_p\upsilon(T_0-T_\infty)$ is the Dufour number and $S=(v_w(x)/\exp(x/2L))\,(2L/\upsilon U_0)^{1/2}>0$ is the suction parameter.

2.2. *Stability analysis*

Since the numerical results obtain 2 solutions, stability analysis is performed to select the stable one. The first step for performing stability analysis is to consider unsteady state for governing equations (2)–(4). Then, the new dimensionless time variable τ is introduced in similarity variables for exponential variation of sheet velocity [21]. Consequently, the following equations are obtained:

$$\frac{\partial u}{\partial t}+u\frac{\partial u}{\partial x}+v\frac{\partial u}{\partial y}=\upsilon\frac{\partial^2 u}{\partial y^2}+g\beta_T(T-T_\infty)+g\beta_C(C-C_\infty) \tag{12}$$

$$\frac{\partial T}{\partial t}+u\frac{\partial T}{\partial x}+v\frac{\partial T}{\partial y}=\alpha\frac{\partial^2 T}{\partial y^2}+\frac{DK_T}{C_sC_p}\frac{\partial^2 C}{\partial y^2} \tag{13}$$

$$\frac{\partial C}{\partial t}+u\frac{\partial C}{\partial x}+v\frac{\partial C}{\partial y}=D\frac{\partial^2 C}{\partial y^2}+\frac{DK_T}{T_m}\frac{\partial^2 T}{\partial y^2} \tag{14}$$

$$\theta(\eta,\tau)=\frac{T-T_\infty}{T_w-T_\infty}, \qquad \phi(\eta,\tau)=\frac{C-C_\infty}{C_w-C_\infty}, \qquad \tau=\frac{U_0}{2L}\left(\exp(x/L)\right)t,$$
$$\eta=y\left(\frac{U_0}{2\upsilon L}\right)^{1/2}\exp(x/2L), \qquad u=U_0\exp(x/L)\frac{\partial}{\partial\eta}f(\eta,\tau),$$
$$v=-\frac{\eta}{2L}\sqrt{2\upsilon LU_0}\exp(x/2L)\frac{\partial}{\partial\eta}f(\eta,\tau)-\frac{U_0}{2L^2}\exp(3x/L)t\sqrt{2\upsilon LU_0}\frac{\partial}{\partial\tau}f(\eta,\tau)$$
$$-\frac{1}{2L}\sqrt{2\upsilon LU_0}\exp(x/2L)f(\eta,\tau), \tag{15}$$

By substituting equation (15) into equations (12)–(14), then these equations are transformed as below:

$$\frac{\partial^3 f}{\partial\eta^3}+f\frac{\partial^2 f}{\partial\eta^2}-2\left(\frac{\partial f}{\partial\eta}\right)^2+2Ri\left(\exp\left(\frac{-3X}{2}\right)\right)\theta+2Ri\,N\left(\exp\left(\frac{-3X}{2}\right)\right)\phi-\frac{\partial^2 f}{\partial\eta\partial\tau}$$
$$+2\tau\frac{\partial f}{\partial\tau}\frac{\partial^2 f}{\partial\eta^2}-2\tau\frac{\partial f}{\partial\eta}\frac{\partial^2 f}{\partial\eta\partial\tau}=0 \tag{16}$$

$$\frac{1}{\Pr}\frac{\partial^2\theta}{\partial\eta^2}+f\frac{\partial\theta}{\partial\eta}-\frac{\partial f}{\partial\eta}\theta+Db\frac{\partial^2\phi}{\partial\eta^2}-\frac{\partial\theta}{\partial\tau}-2\tau\frac{\partial\theta}{\partial\tau}\frac{\partial f}{\partial\eta}+2\tau\left(\exp\left(\frac{-3X}{2}\right)\right)\frac{\partial f}{\partial\tau}\frac{\partial\theta}{\partial\eta}=0 \tag{17}$$

$$\frac{1}{Sc}\frac{\partial^2\phi}{\partial\eta^2}+f\frac{\partial\phi}{\partial\eta}-\frac{\partial f}{\partial\eta}\phi+Sr\frac{\partial^2\theta}{\partial\eta^2}-\frac{\partial\phi}{\partial\tau}-2\tau\frac{\partial\phi}{\partial\tau}\frac{\partial f}{\partial\eta}+2\tau\left(\exp\left(\frac{-3X}{2}\right)\right)\frac{\partial f}{\partial\tau}\frac{\partial\phi}{\partial\eta}=0 \tag{18}$$

subject to boundary conditions with the existence of τ:

$$\frac{\partial f}{\partial\eta}(\eta,\tau)=\lambda, \quad f(\eta,\tau)=S, \quad \theta(\eta,\tau)=1, \quad \phi(\eta,\tau)=1 \quad \text{at } \eta=0,$$
$$\frac{\partial f}{\partial\eta}(\eta,\tau)\to 0, \quad \theta(\eta,\tau)\to 0, \quad \phi(\eta,\tau)\to 0 \quad \text{as } \eta\to\infty, \tag{19}$$

The analysis developed by Weidman et al. [22] are stated as below, to determine the stability of the solution

$$f(\eta,\tau)=f_0(\eta)+\left(\exp(-\gamma\tau)\right)F(\eta,\tau),$$
$$\theta(\eta,\tau)=\theta_0(\eta)+\left(\exp(-\gamma\tau)\right)G(\eta,\tau),$$
$$\phi(\eta,\tau)=\phi_0(\eta)+\left(\exp(-\gamma\tau)\right)H(\eta,\tau), \tag{20}$$

where γ denotes an unknown eigenvalue while $F(\eta,\tau)$, $G(\eta,\tau)$ and $H(\eta,\tau)$ are small relative to $f_0(\eta)$, $\theta_0(\eta)$ and $\phi_0(\eta)$. Substituting equation (20) into equations (16)–(18), and take $\tau=0$, the following linear eigenvalue problem are obtained:

$$F_0'''+f_0F_0''+f_0''F_0-4f_0'F_0'+2Ri\left(\exp\left(\frac{-3X}{2}\right)\right)(G_0+NH_0)+\gamma F_0'=0 \tag{21}$$

$$\frac{1}{\Pr}G_0''+f_0G_0'+\theta_0'F_0-\theta_0F_0'-G_0f_0'+Db\,H_0''+\gamma G_0'=0 \tag{22}$$

$$\frac{1}{Sc}H_0'' + f_0H_0' + \phi_0'F_0 - \phi_0F_0' - H_0f_0' + Sr\,G_0'' + \gamma H_0' = 0 \tag{23}$$

along with the boundary conditions:

$$\begin{aligned} &F_0'(\eta)=0, \quad F_0(\eta)=0, \quad G_0(\eta)=0, \quad H_0(\eta)=0 \quad &&\text{at } \eta=0, \\ &F_0'(\eta)\to 0, \quad G_0(\eta)\to 0, \quad H_0(\eta)\to 0 &&\text{as } \eta\to\infty, \end{aligned} \tag{24}$$

According to Harris et al. [23], the range of possible eigenvalues can be determined by relaxing a boundary condition on $F_0(\eta)$, $G_0(\eta)$ and $H_0(\eta)$. For the present problem, we relax the boundary condition $F_0'(\eta)\to 0$ as $\eta\to\infty$. Finally, the system of equations (21)–(23) subject to equation (24) along with the new boundary condition $F_0''(\eta)=1$ is solved.

3. Results and discussion

Numerical computations have been performed for the velocity profile $f'(\eta)$, temperature profile $\theta(\eta)$ and concentration profile $\phi(\eta)$ for for various values of physical parameters, such as suction *S*, dimensionless coordinate along the plate *X*, shrinking λ, mixed convection *Ri*, buoyancy ratio parameter *N*, Soret *Sr* and Dufour *Db*. In the present study, the following values are adopted for the numerical computations: $S=3.1$, $X=1$, $\lambda=-0.4$, $Ri=1$, $N=0.5$, $Sr=1.8$ and $Db=0.1$. These values are used throughout the calculations unless otherwise mentioned. In addition, the appropriate values of *S*, *X*, λ, *Ri*, *N*, *Sr* and *Db* are used in MATLAB routine bvp4c, to obtain the profiles which satisfy the boundary conditions. First solution is presented by solid line, whereas dashed line denoted as second solution.

In order to validate the present numerical method used in this paper, we have compared the present results with those obtained by Magyari and Keller [5] and Srinivasacharya and RamReddy [13] for values of wall-temperature gradient $-\theta'(0)$. This physical parameter is calculated by using the formulation $Nu_x/\sqrt{\text{Re}_x} = -\sqrt{X/2}\;\theta'(0)$. The comparison is tabulated in Table 1. The bvp4c programme is used to calculate the values of wall-temperature gradient. As a conclusion, the present values of $-\theta'(0)$ are in good agreement with those reported by the previous investigators. Therefore, the good comparison proves that our bvp4c MATLAB is applicable to present our numerical computation.

Figures 2–8 shows the velocity profiles $f'(\eta)$ for different values of *S*, *X*, λ, *Ri*, *N*, *Sr* and *Db*. From Figures 4–8, the magnitude of velocity is increased due to the increment of λ, *Ri*, *N*, *Sr* and *Db* for first solution. The instantaneous velocity is reduced by the increment of *S* and *X*, as shown in Figures 2–3. For second solution and for high parameters *S* and *X*, the instantaneous negative velocity decreases and reaches minimum peak, then it increases until it reaches zero value. The decrement of negative velocity denotes the increment of velocity magnitude. After that, velocity magnitude is slowing down in opposite direction, until it reaches stationary state. In addition, the velocity values decrease at the point far from shrinking sheet for the addition of parameters λ, *Ri*, *N*, *Sr* and *Db* (Figures 4–8).

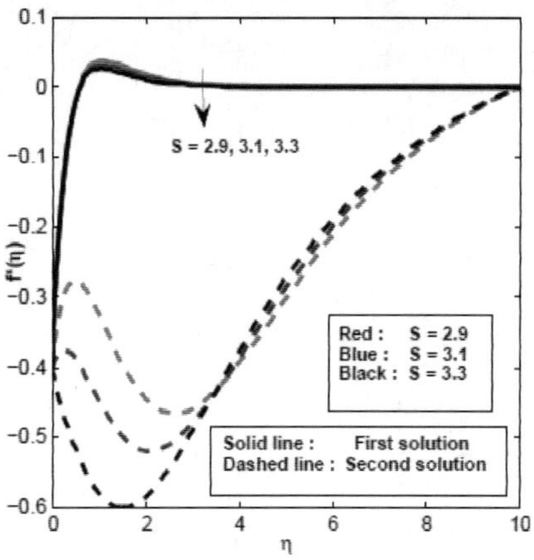

Figure 2. Variation of velocity for various S

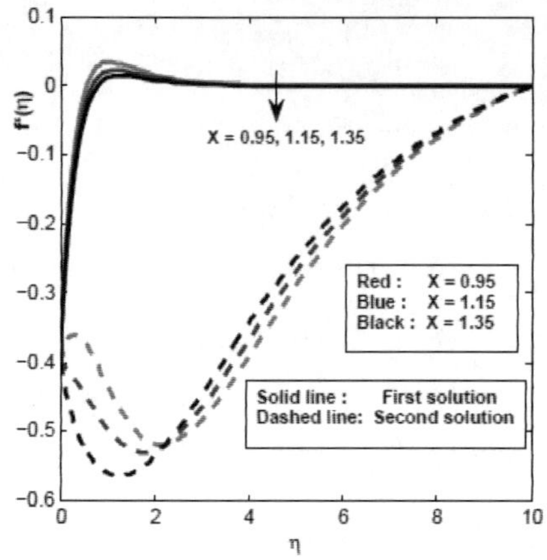

Figure 3. Variation of velocity for various X.

The temperature profiles $\theta(\eta)$ for various values of S, X, λ, Ri, N, Sr and Db are presented by Figures 9–15, respectively. For variation of temperature of first solution is continuously decrease until it reaches zero value, controlled by the distance from the shrinking sheet. Moreover, the temperature of second solution increases until it achieves maximum temperature. The maximum temperature is gained at the near point. Subsequently, temperature is measured as zero at the point far from shrinking sheet. The decrement of temperature for first solution is contributed by an augmentation of S, Ri, N and Sr (Figures 9, 12–14). The effect of Dufour parameter Db on temperature, which showed in Figure 15 causes the pattern of temperature variation is non-uniform (decreases at close point and subsequently increases at far point).

Table 1. Comparison between wall-temperature gradient $-\theta'(0)$ calculated by the present method and that of Magyari and Keller [5] and Srinivasacharya and RamReddy [13] for $S = 0$, $\lambda = 1$ and $Ri = Sc = Sr = Db = N = X = 0$.

Pr	Magyari and Keller (1999)	Srinivasacharya and RamReddy (2011)	Present
0.5	-0.59434	-0.59438	-0.59466
1.0	-0.95478	-0.95478	-0.95478
3.0	-1.86908	-1.86908	-1.86907
5.0	-2.50014	-2.50015	-2.50013
8.0	-3.24213	-3.24218	-3.24212
10.0	-3.66038	-3.66043	- 3.66037

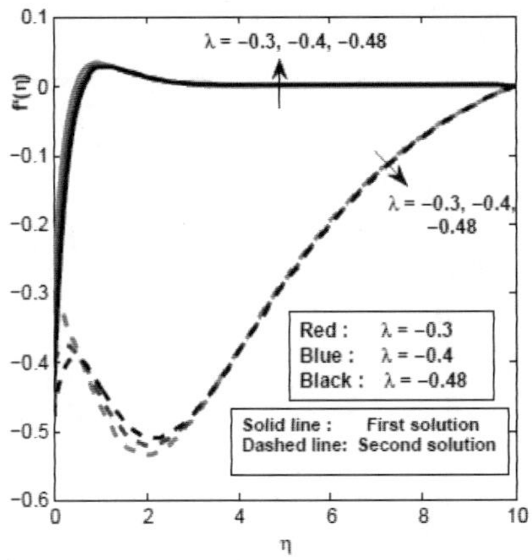

Figure 4. Variation of velocity for various λ.

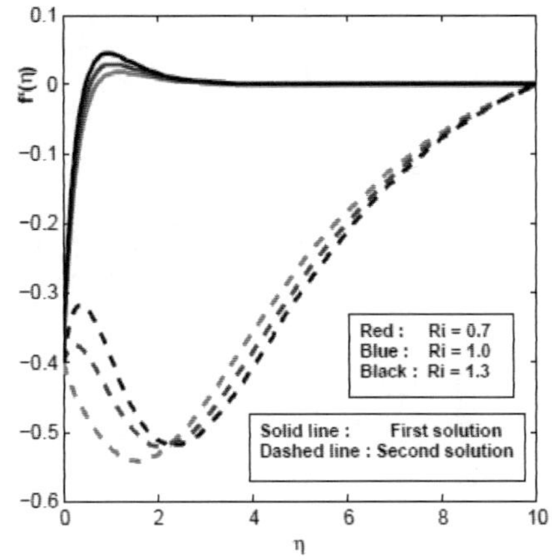

Figure 5. Variation of velocity for various *Ri*.

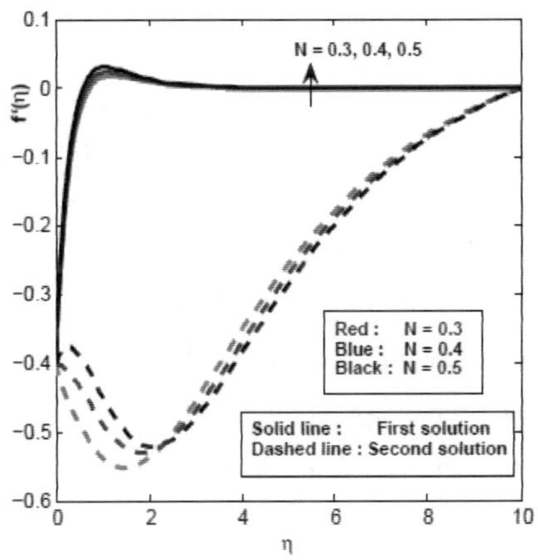

Figure 6. Variation of velocity for various *N*.

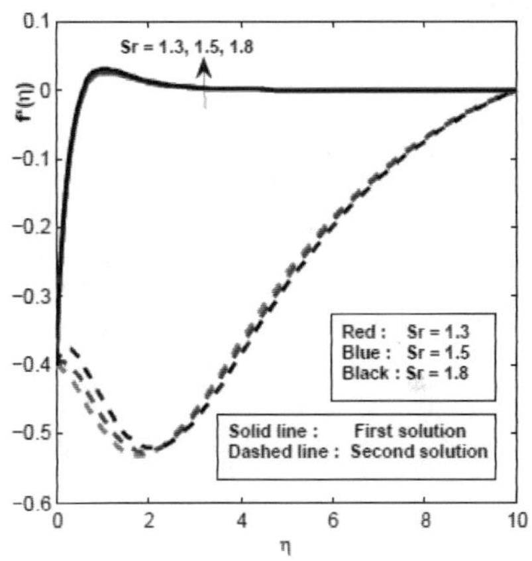

Figure 7. Variation of velocity for various *Sr*.

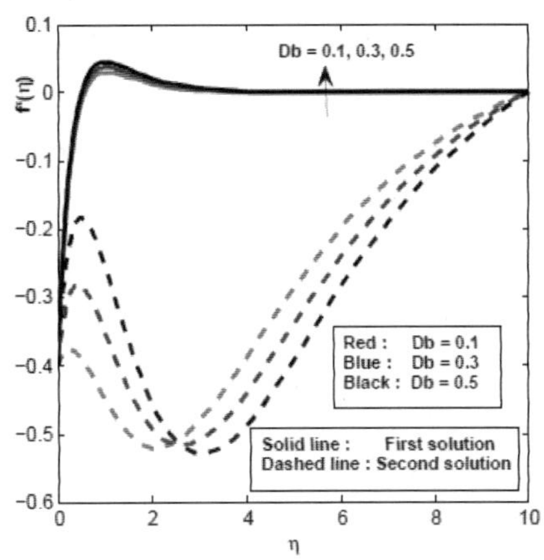

Figure 8. Variation of velocity for various *Db*.

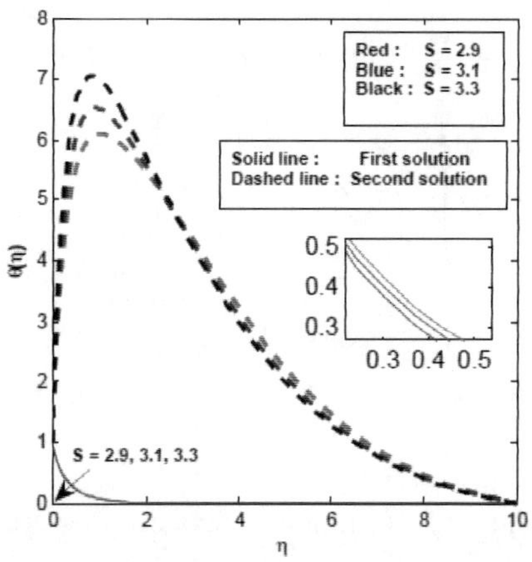

Figure 9. Variation of temperature for various *S*.

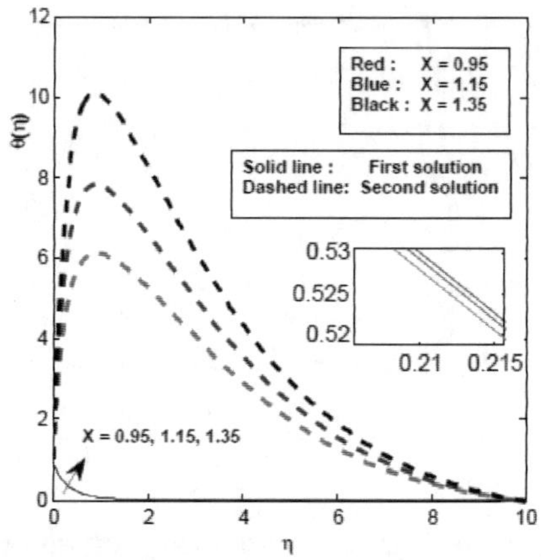

Figure 10. Variation of temperature for various *X*.

The graphs of the concentration profiles $\phi(\eta)$ are depicted by Figures 16–22, respective to the controlling parameters *S*, *X*, λ, *Ri*, *N*, *Sr* and *Db*. From Figures 17, 18 and 21, the concentration values of first solution are ascended due to the improvement of the following: *X*, λ and *Sr*. On the other hand, the concentration values of first solution become lessen due to the impact of *Ri* and *N* (Figures 19 and 20). However, the effect of *S* and *Db* are to increase the concentration $\phi(\eta)$ of first solution at the point close to the sheet (Figures 16 and 22). Furthermore, concentration profiles of second solution consistently increase due to the effect of *Ri* and *Db* (Figures 19 and 22). Moreover, the impact of *X*, *N* and *Sr* are to reduce the concentration value of second solution. These unchanging pattern of decrement can be seen in Figures 17, 20 and 21. Refer to the Figure 16, the concentration values of second solution are decrease when parameter suction *S* is increased. However, these feature changes when the distance of the concentration point is far from the shrinking sheet ($\phi(\eta)$ is increased due to addition of *S*). The impact of shrinking parameter λ on concentration values second solution (Figure 18) shows a reverse trend, compared to the impact of suction parameter *S* (Figure 16).

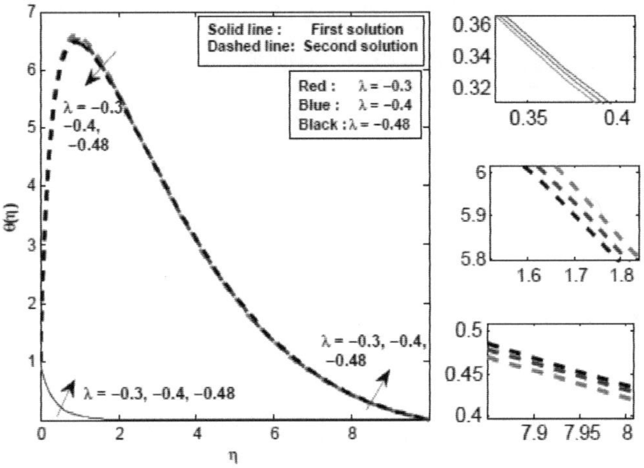

Figure 11. Variation of temperature for various λ.

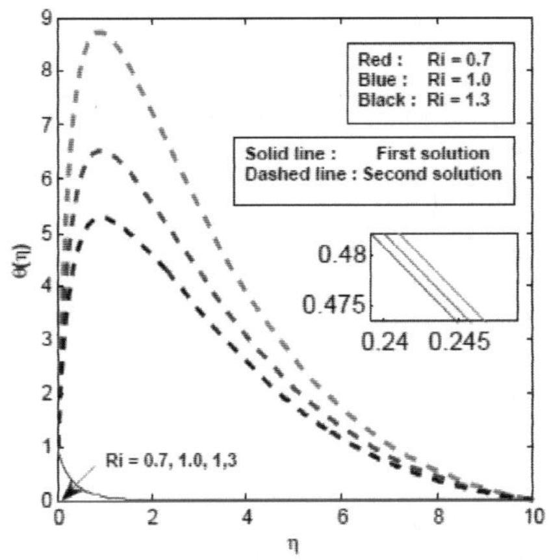

Figure 12. Variation of temperature for various *Ri.*

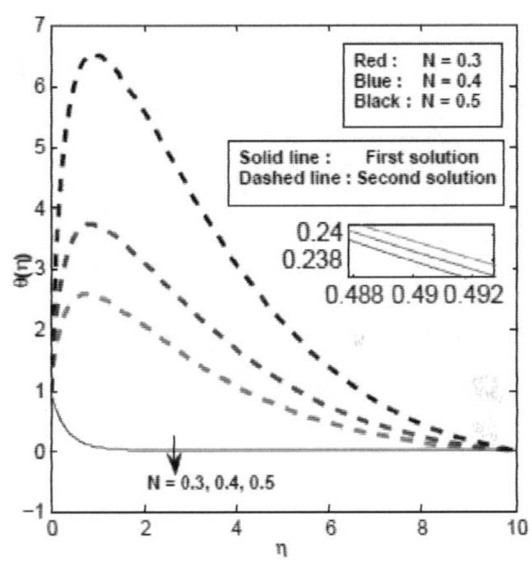

Figure 13. Variation of temperature for various *N*.

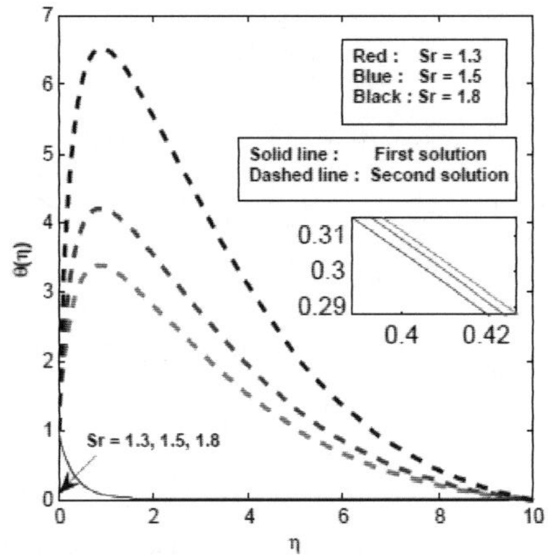

Figure 14. Variation of temperature for various *Sr.*

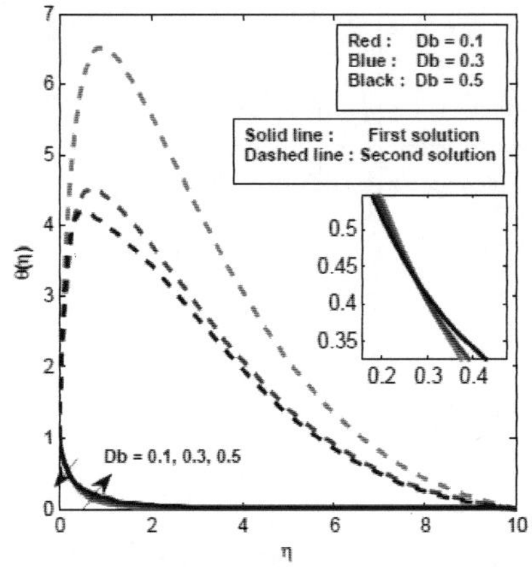

Figure 15. Variation of temperature for various *Db.*

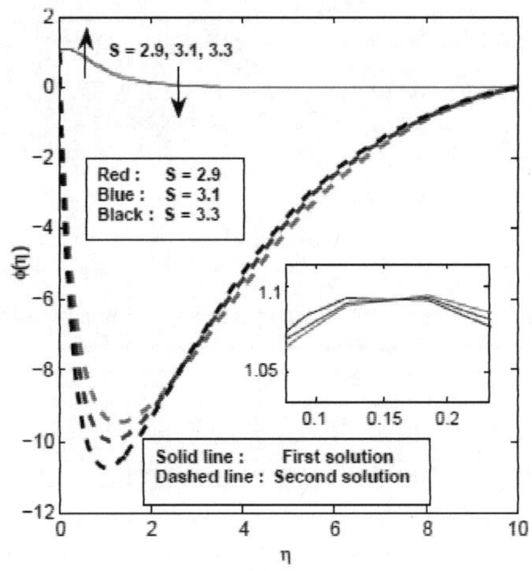

Figure 16. Variation of concentration for various *S.*

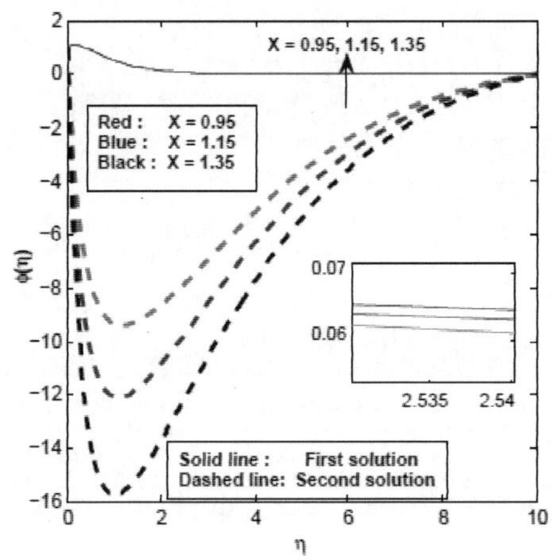

Figure 17. Variation of concentration for various *X.*

In the case of second solution for all profiles, the effects of controlling parameters in this problem mostly depends on the distance from the shrinking sheet. For example: the impact of suction parameter *S* is to reduce concentration rate, at the point close to the sheet (Figure 16). However, the condition at the point far from sheet is not supposed to follow the previous decrement of concentration affected by the same parameter. Furthermore, inconsistent pattern can be seen for all velocity profiles (Figures 2–8), temperature profiles (Figures 9, 11) and concentration profiles (Figures 18).

IOP Conf. Series: Journal of Physics: Conf. Series **1298** (2019) 012024 doi:10.1088/1742-6596/1298/1/012024

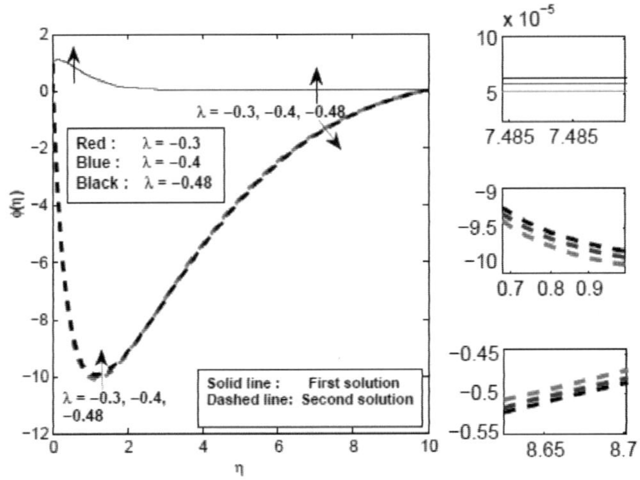

Figure 18. Variation of concentration for various λ.

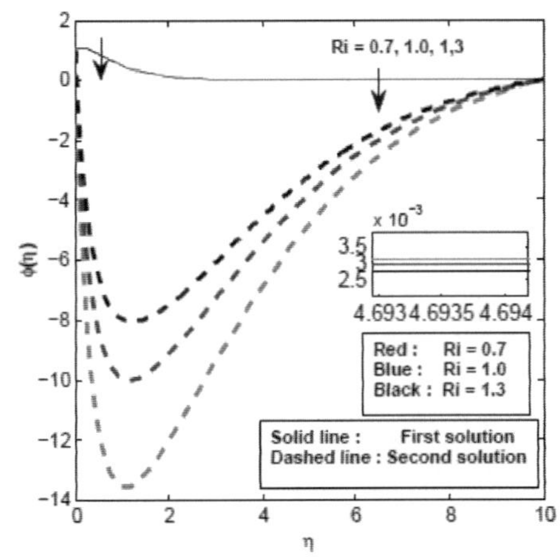

Figure 19. Variation of concentration for various *Ri*.

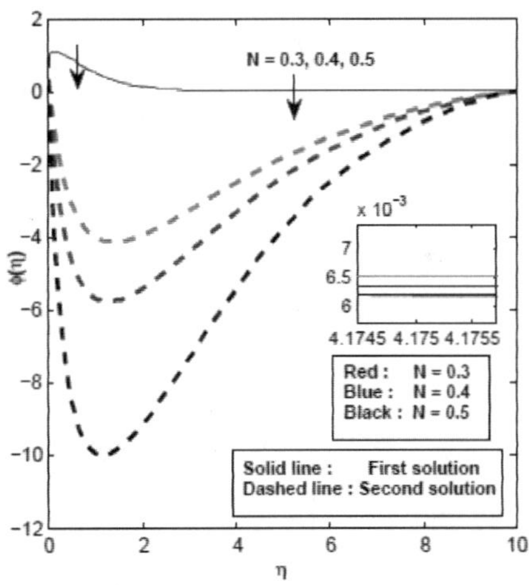

Figure 20. Variation of concentration for various *N*.

Stability analysis is tabulated in Table 2, which reports the smallest eigenvalues for first solution (positive eigenvalues) and second solution (negative eigenvalues). Therefore, the first solutions with positive smallest eigenvalues indicate that the solution is stable and physically reliable. Moreover, second solution absolutely has reverse characteristics compared to the first one.

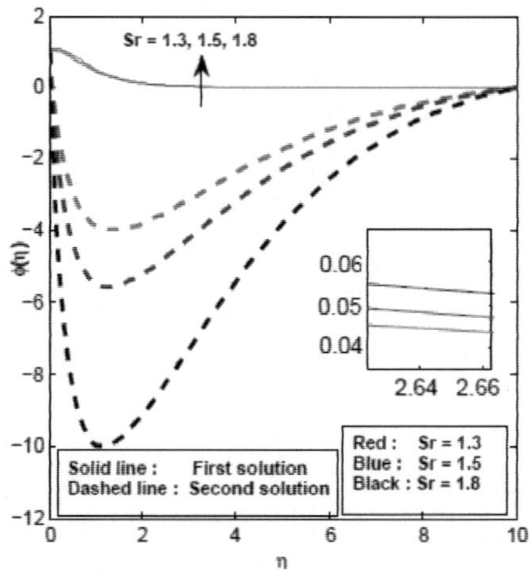

Figure 21. Variation of concentration for various *Sr*.

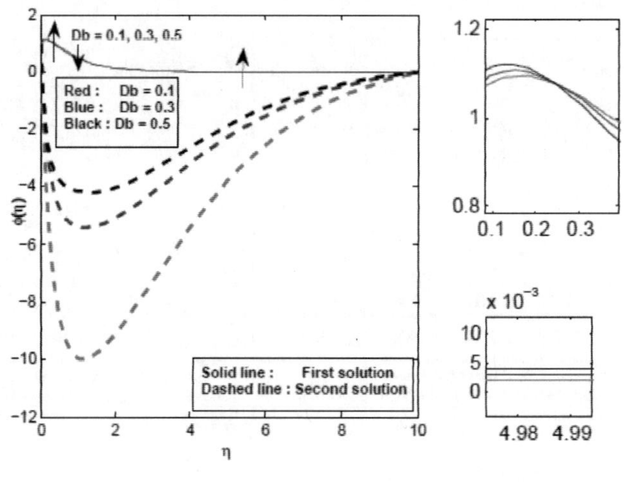

Figure 22. Variation of concentration for various *Db*.

Table 2. Smallest eigenvalues γ for several values of S and Ri.

S	Ri	γ	
		First solution	Second solution
	-0.01001	0.07766	-0.07915
1.75	-0.01	0.07955	-0.08097
	-0.00999	0.08180	-0.08341
	-0.01001	0.28926	-0.31038
1.80	-0.01	0.28978	-0.31098
	-0.00999	0.29031	-0.31157

4. Conclusion

The characteristics of the fluid flow due to a shrinking sheet with the exponential variation of temperature and concentration are reported, by the presence of Soret and Dufour effects. From the findings in the form of table and graphs, the main conclusions are as follows:

a) The effect of increasing values of parameters λ, Ri, N, Sr and Db are to increase velocity distribution for the first solution.
b) The temperature distribution by first solution in flow region increases with increasing the magnitude of parameters X, λ and Db. The effect of Dufour parameter Db to increase the temperature rate is effective when the instantaneous point is far from shrinking sheet.
c) It can be seen that larger X, λ and Sr values always imply higher values of concentration of first solution.
d) The variation of velocity, temperature and concentration values of second solution which affected by governed parameters in this problem show inconsistent pattern (increment and decrement). This pattern is depends on the location of the measured point from the shrinking sheet.
e) The calculation of positive eigenvalues for the first solution proved that it is stable and reliable.

5. Acknowledgement

We would like to express an appreciation to the Putra Grant of Universiti Putra Malaysia (Project code: GP-IPM/2018/9596900) for the financial support received.

6. References

[1] De Groot S R, Mazur P 1962 *Non-Equilibrium Thermodynamics* (Amsterdam: North-Holland Publishing Company)
[2] Mcdougall T J 1983 *J. Fluid Mech.* **126** 379–397.
[3] Mojumder S, Saha S, Rahman M R, Rahman M M, Rabbi K M, Ibrahim T A 2017 *Eng. Sci. Technol. Int J.* **20** 272–282.
[4] Hayat T, Shafiq A, Alsaedi, A 2014 *PLoS One* **9** e83153.
[5] Magyari E, Keller B 1999 *J. Phys. D: Appl. Phys.***32** 577–585.
[6] Isa S S P M, Arifin N M, Nazar R, Bachok N, Ali F M, Pop I 2014 *International Journal of Mechanical and Mechatronics Engineering* **8** 1548–1553.
[7] Hayat T, Ashraf B, Shehzad S A, Alsaedi A, Bayomi N 2015 *Int. J. Numer. Method H.* **25** 333-357.
[8] Ahmad K, Hanouf Z, Ishak A 2016 *AIP Adv.* **6** 035024.
[9] Isa S S P M, Arifin N M, Nazar R, Bachok N, Ali F M, Pop I 2017 *Sci. Iran.* 24 637–647.
[10] Isa S S P M, Arifin N M, Nazar R, Bachok N, Ali F M, Pop I 2017 *J. Phys. Conf. Ser***949** 012016.
[11] Besthapu P, Haq R U, Bandaria S, Al-Mdallal Q M 2017 *J. Taiwan Inst. Chem. Eng.* **71** 307–314.
[12] Pal D, Mondal H 2013 *Nucl Eng Des* **256** 350–357.
[13] Srinivasacharya D, RamReddy Ch 2011 *IJNS* **12** 60–68.
[14] Sreenivasulu P, Redddy N B 2012 *Elixir Applied Mathematics* **51** 10809–10816.
[15] Hayat T, Muhammad T, Shehzad S A, Alsaedi A 2015 *Int. J. Numer. Method H.* **25** 762-781.
[16] Sharada K, Shankar B 2015 *WJM* **5** 165–177.
[17] Sravanthi 2016 *J. Nigerian Math. Soc.* **35** 208–226.
[18] Chandra Sekhar K V, Manjula V 2018. *IJCIET* **9** IJCIET_09_03_042

[19] Dzulkifli N F, Bachok N, Pop I, Yacob N A, Arifin N M, Rosali H 2017 *J. Phys. Conf. Ser* **890** 012031
[20] Dzulkifli N F, Bachok N, Pop I, Yacob N A, Arifin N M, Rosali H 2017 *J. Chem. Eng. Process Technol.* **8** 1000336.
[21] Adnan N S M, Arifin N M 2017 *J. Phys. Conf. Ser***890** 012046.
[22] Weidman P D, Kubitschek D G, Davis A M H J 2006 *Int. J. Eng. Sci.* 44 730-737.
[23] Harris S D, Ingham D B, Pop I 2009 *Transport Porous Med.* **77** 267-285.

IOP Conf. Series: Journal of Physics: Conf. Series **1298** (2019) 012025 doi:10.1088/1742-6596/1298/1/012025

Molecular Engineering of Quinodimethane-based Dyes with Different Anchor Groups for Dye Sensitized Solar Cells (DSSCs)

Jianhao Tang[a], Yun Gong[b, *]

[a] Chongqing Foreign Language School, No.1 Hongyupo Shiqiaopu, Jiulongpo District, ChongQing, 400039, P.R.China.
[b] Cavendish Laboratory, University of Cambridge, J. J. Thomson Avenue, Cambridge CB3 0HE, United Kingdom

* Corresponding author: yun.gong.cambridge@gmail.com

Abstract. This paper mainly works on changing different anchoring groups of quinodimethane-based dyes, which represent a twisting phenyl ring during the excitation process, for dye-sensitized solar cells (DSSCs). Five different groups, (1) cyanoacrylic acid (-CHC(CN)COOH), (2) carboxylic acid (-COOH), (3) pyridine (-C5H5N), (4) sulfonic acid (-SO3H), (5) phosphoric acid (-PO3H2), were adopted as anchors on the quinodimethane moieties forming dyes 1-5, to further explore their effects on optoelectronic properties via molecular engineering methods. Density functional theory (DFT) and time-dependent DFT (TD-DFT) were used to calculate their electronic structures and electro-optical properties. Computational results show that all dyes show suitable molecular orbital energy levels between the conductive band of TiO2 and the potentials I-/I3- redox couple so that all of 1-5 can be potential candidates for DSSCs application. Also, TDDFT results exhibit that dye 1 has the widest absorption band in UV-visible region, with the first excitation energies of 2.782 eV, while other dyes show similar first excitation energies about 0.2 eV higher than that of dye 1. In particular, dye 3 with pyridine as anchor group shows a very excellent charge transfer properties from the highest occupied molecular orbitals (HOMO) to the lowest unoccupied molecular orbitals (LUMO), with the highest HOMO level of -5.16 eV, which offers a strong driving force for dye regeneration, and a LUMO level of -2.36 eV, which enables excited electrons successfully to be injected in to titania film. Also, this pyridine anchor group offers a larger π conjugation, which may improve the ability of electron injecting and further strengthen the photovoltaic performance in DSSCs.

1. Introduction

In 1991, Grätzel published a paper about dye sensitized nanocrystals solar cells on Nature,1 and the photoelectric conversion efficiency reached 7% with a low cost, and this work exploits a new age of the history of solar cells. Since then, people are working on improving the performances of DSSCs. So far, the most efficient metal-free DSSCs have reached power-conversion efficiency (PCE, η) values of up to 14.3%.2

When exposed under illumination, dye molecules will be motivated from ground state to excited state. Meanwhile, electrons are excited from HOMO to LUMO, and then injected into semiconductor's conductive band, transported to conductive substrate, and finally "flowed" into external circuit. On the other hand, the dyes at oxidation state are reduced by redox couples in electrolyte. The circulation is completed when the redox couples accept electrons at the interface of counter electrode from external circuit.

Dye molecules attach to metal oxide substrate's surface through many ways, such as covalent attachment, electrostatic interaction, hydrogen bonding, hydrogen-phobic interaction, van der waals forces, and physical entrapment.3 In DSSCs, the most common way is the covalent bonding between the dyes and the TiO2 surface atoms in order to guarantee a strong coupling, a homogeneous dyes distribution, and device stability. While other kinds of interactions usually involve weaker bonds, leading to a quasi-reversible and unsteady adsorption.

Recently, a series of quinodimathane-based dyes were designed for DSSCs and investigated systematically, showing that the dyes with cyanoacrylic acid and carboxylic acid showing better performances compared to other dyes in this series, caused by a large twist of donor groups in its excited state which successfully blocked electron recombination.4,5 Thus, in this study, to further improve the properties of these dyes, the dimethylamino group was retained, as electrons donor, trying to maintain the property of the large ring twists in the donor group during ground-to-excited-state transition, and five anchoring groups: (1) cyanoacrylic acid (-CHC(CN)COOH), (2) carboxylic acid (-COOH), (3) pyridine (-C5H5N), (4) sulfonic acid (-SO3H), (5) phosphoric acid (-PO3H2) were investigated. Geometries at ground and excited states, frontier molecular orbitals and first excitation energies of dye 1-5 were calculated using DFT and TDDFT methods. Chemical structures of 1-5 are listed in figure 1.

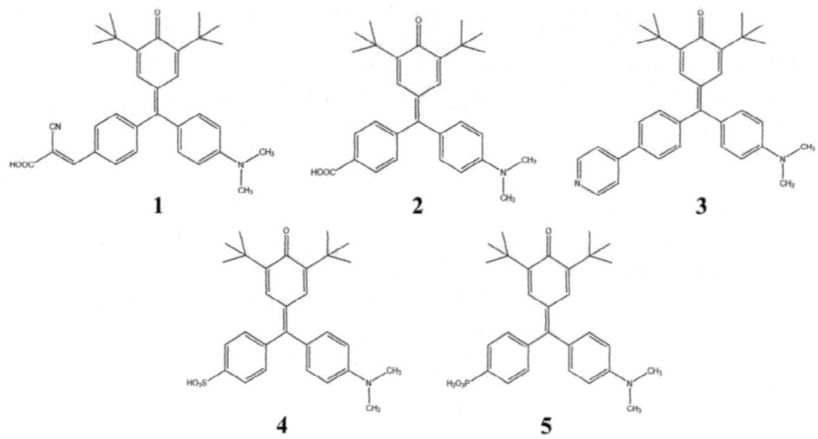

Figure 1. Chemical structures and systematic names of 1-5

2. Computational Methods

Gas-phase structures of 1−5 were firstly optimized at the B3LYP6/6-31g(d)7 level of theory using NWChem 6.6 software package8. Single-point energy calculations were performed at the B3LYP/6-31g++(d,p) level of theory. Theoretical first excited energies were calculated using time-dependent density functional theory (TD-DFT)9 in combination with the M06-2X functional10and the 6-31g (d) basis set, based on the optimized ground-state structures.

3. Results and Discussion

3.1 Optimized ground state geometries in vacuo

All frontier molecular orbitals and ground-state (GS) geometries of 1-5 were calculated and optimized as mentioned in computational methods. Selected structural parameters of the optimized structures are summarized in table 1.

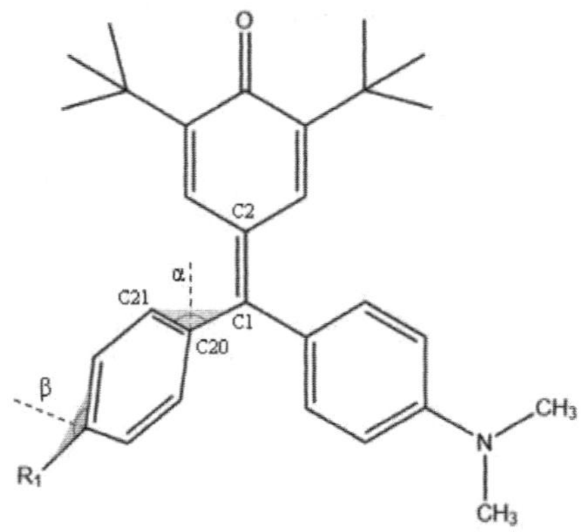

Figure 2. The chemical structure of the core part of 1-5 (R1 = -CHC(CN)COOH, -COOH, -C5H5N, -SO3H and -PO3H2 for 1-5 respectively). α is the dihedral angle of C21-C20-C1-C2, and β is the dihedral angle between the planes of anchor groups and the adjacent phenyl rings, and the bond of C-anchor refers to the bond between the anchor group and the adjacent phenyl ring.

Table 1. Selected structural parameters for 1-5.

Dye	**α** [°]	**β** [°]	**C-anchor** [Å]
1	-46.36	-0.95	1.45
2	-47.93	0.34	1.49
3	-47.20	-35.78	1.48
4	-48.72	-	1.79
5	-48.27	-	1.80

For angle α, 1-5 show similar values to each other. 4 and 5 show relatively larger α in their structures, which might be caused by stronger electron-pulling abilities of SO3H and PO3H2. For β, anchors in 1 and 2 shows excellent planar structures, which guarantee a good electron injection from dye molecules to TiO2 layers. 3 shows a much larger twist since pyridine has a larger planar ring structure, which enlarged the π conjugation, may also promote the electron injection. The bond lengths of C-anchor show that 1-3 have a typical C-C single bond, between their anchor and the adjacent benzene ring. While 4 and 5 have longer bond caused by S and P atoms which have larger electronic negativity. For structural parameters of other moities, such as the dihedral angles of the benzene rings connected to donor group and the quinodimethane rings, structural details are listed in the supporting information.

3.2 Frontier Molecular Orbitals (FMOs)

HOMO and LUMO levels are the most significant characters for dye molecules used in DSSCs. They directly exhibit the possibility of electron excitation and transportation in DSSCs. All the molecular

orbitals were calculated via DFT at the B3LYP/6-31g++(d,p) level of theory, listed in table 2 and figure 3.
The HOMO energy levels of 1-5 are lower than the redox potential of the most commonly used DSSC electrolyte (I-/I3-), which renders the regeneration of the dye molecules possible. The LUMO energy levels are sufficiently high to provide the driving force for electron injection from the excited dye molecules into the TiO2 conduction band (CB).

Table 2. DFT- and TDDFT-derived HOMO and LUMO energy levels

Dye	HOMO [eV]	LUMO [eV]	Energy gap* [eV]
1	-5.306	-2.936	2.370
2	-5.120	-2.450	2.670
3	-5.161	-2.360	2.801
4	-5.327	-2.574	2.753
5	-5.193	-2.390	2.803

*: Energy gap = LUMO – HOMO

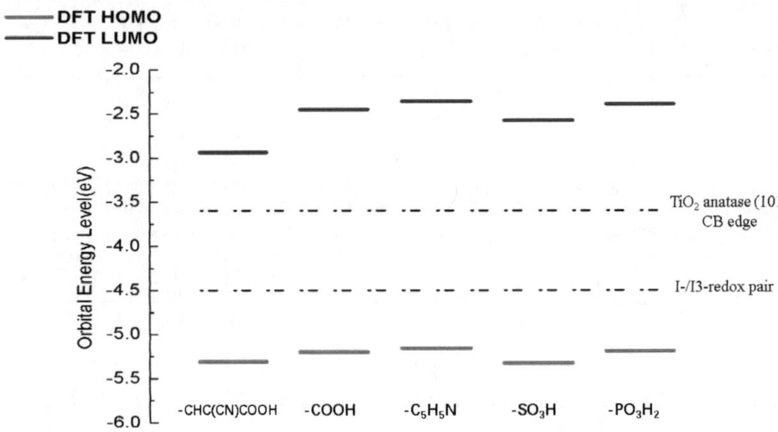

Figure 3. Calculated HOMO/LUMO energy levels of 1-5.

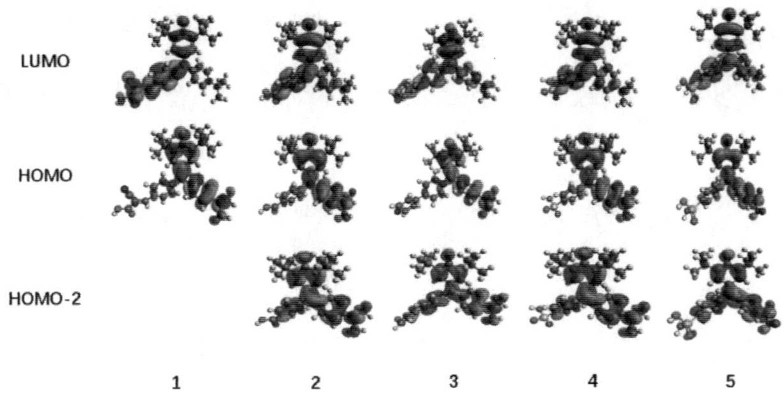

Figure 4. Calculated electron density distributions for the frontier molecular orbitals of 1-5 using DFT methods based on B3YLP/6-31++g (d,p), and shows the HOMO, LUMO, and HOMO-2 energy state

As shown in figure 4, the HOMOs in the ground state is comparable for 1-5, and the electron density is located predominantly on the donor moiety and the quinodimethane backbone. The LUMOs are located mainly on the anchor groups and the quinodimethane backbone. In addition, small parts of the LUMO of 2-5 are situated on the phenyl ring that connects the quinodimethane moiety and the donor group. Generally, 1-5 exhibit very good HOMO-to-LUMO charge redistribution from the donor to the anchor moieties, which renders the injection of excited electrons into the CB of TiO2 possible. Distribution of 1-3 are much better than 4 and 5, and 4 is slightly better than 5, especially for 5, few LUMOs is located on P atom, which may lead to a worse performance than the other four dyes.

3.3 First excitation energies

The first excitation energies of 1-5 were studied by TDDFT calculations at the M062X/6-31g (d) level of theory. The results are listed in table 3.

The first excitation energies of 1-3 are smaller indicating wider absorption band compared to that of 4-5, which are much larger and lead to difficulties of absorbing longer wavelength light.

The compositions of the excitations in terms of molecular orbital transitions were also calculated (table 3), and the results show that for dye 1, the first excitation is mainly attributed to HOMO-LUMO transition (84.12%), which is obviously an intramolecular charge transfer (ICT) process. While for 2-5, the excitations were attributed mainly to the HOMO-2-LUMO and HOMO-LUMO transitions. So in order to discuss the excitation properties of 2-5, distributions of HOMO-2 were also shown in figure 4. The results show obvious charge redistribution from donor groups to anchor groups in the HOMO-2 -> LUMO transition, indicating that ICT occurs in such transition and also the first excitation.

Table 3. Energy (eV, nm) and oscillator strength values for 1-5, as well as sizable (f > 0.1) contributions of the molecular orbitals to the transitions

Dye	1st excitation energy [eV]	[nm]	Oscillator strength (f)	Composition	
1	2.782	445.66	0.5092	84.12%	HOMO - LUMO

2	2.936	422.26	0.2663	38.26%	HOMO-2 - LUMO
				46.81%	HOMO - LUMO
3	2.959	419.02	0.2537	50.00%	HOMO-2 - LUMO
				34.65%	HOMO - LUMO
4	2.918	424.97	0.2373	49.91%	HOMO-2 - LUMO
				35.38%	HOMO - LUMO
5	2.946	420.85	0.2065	30.02%	HOMO - LUMO
				54.93%	HOMO-2 - LUMO

4. Acknowledge
For the first, I would like to gratitude my teacher Dr Yun Gong, who taught me the basic knowledge required in this project, and how to write an academic paper. And thanks to the organization Illume Research and my school Chongqing Foreign Language School, they provided the chance that I can be a part of the project; without them, I can never learn or reach about the correlation stuff.

5. Conclusion
This research investigated the effects of different anchors on quinodimethane dye molecules for dye-sensitized solar cells (DSSCs). Five anchors were selected in this research, (1) cyanoacrylic acid (-CHC(CN)COOH), (2) carboxylic acid (-COOH), (3) pyridine (-C5H5N), (4) sulfonic acid (-SO3H), (5) phosphoric acid (-PO3H2). Density functional theory (DFT) was used to calculate HOMO and LUMO levels, showing all the quinodimethane dyes are good candidates for DSSCs in terms of energy levels. The distributions of HOMOs and LUMOs show that dye 1-3 have better charge-transfer properties than 4 and 5. In TD-DFT study, the first excitation energies of 1-5 were calculated, and it is very clear that -CHC(CN)COOH is the best anchor in terms of the energy gap, since it requires the lowest first excitation energy (2.78 eV) for electrons to be excited, and also this excitation were mainly due to a intramolecular charge transfer process from HOMO to LUMO transition. Dye 3 with pyridine as its anchoring group is also an attractive one since it has a larger π conjugation, obvious intramolecular charge transfer from HOMO to LUMO transition and HOMO-2 to LUMO transition. Although the first excitation energy (2.959 eV) is slightly large, it may perform very well in shorter wavelength region and also in electron injection. And finally, as the results of 4 and 5, inorganic acid groups may not very suitable for these quinodimethane-based dyes due to their relatively worse charge transfer properties and large first excitation energies.

6. Reference
[1] O'Regan, B.; Graẗzel, M. A Low-Cost, High-Efficiency Solar Cell Based on Dye-Sensitized Colloidal TiO2 Films. Nature 1991, 353, 737–740.

[2] Kakiage, K.; Aoyama, Y.; Yano, T.; Oya, K.; Fujisawa, J.-i.; Hanaya, M. Highly-Efficient Dye-Sensitized Solar Cells with Collaborative Sensitization by Silyl-Anchor and Carboxy-Anchor Dyes. Chem. Commun. 2015, 51, 15894–15897.

[3] Kalyanasundaram, K. Applications of Functionalized Transition Metal Complexes in Photonic and Optoelectronic Devices. Coord. Chem. Rev. 1998, 177, 347–414.

[4] Yun Gong, Jaqueline M. Cole, et al, Photoexcited Phenyl Ring Twisting in Quinodimethane Dyes Enhances Photovoltaic Performance in Dye-Sensitized Solar Cells, ACS Appl. Energy Mater. 1, 3, 1127-1139.

IOP Conf. Series: Journal of Physics: Conf. Series **1298** (2019) 012025 doi:10.1088/1742-6596/1298/1/012025

[5] Lei Zhang, Jacqueline M. Cole, Anchoring Groups for Dye-Sensitized Solar Cells. ASC Appl. Mater. Interface 2015, 7, 3427-3455.

[6] Becke, A. D., Density-functional thermochemistry. III. The role of exact exchange. The Journal of Chemical Physics 1993, 98 (7), 5648-5652; (b) Lee, C.; Yang, W.; Parr, R. G., Development of the Colle-Salvetti correlation-energy formula into a functional of the electron density. Physical Review B 1988, 37 (2), 785-789.

[7] M. J. Frisch, G. W. T., H. B. Schlegel, G. E. Scuseria, M. A. Robb, J. R. Cheeseman, G. Scalmani, V. Barone, G. A. Petersson, H. Nakatsuji, X. Li, M. Caricato, A. Marenich, J. Bloino, B. G. Janesko, R. Gomperts, B. Mennucci, H. P. Hratchian, J. V. Ortiz, A. F. Izmaylov, J. L. Sonnenberg, D. Williams-Young, F. Ding, F. Lipparini, F. Egidi, J. Goings, B. Peng, A. Petrone, T. Henderson, D. Ranasinghe, V. G. Zakrzewski, J. Gao, N. Rega, G. Zheng, W. Liang, M. Hada, M. Ehara, K. Toyota, R. Fukuda, J. Hasegawa, M. Ishida, T. Nakajima, Y. Honda, O. Kitao, H. Nakai, T. Vreven, K. Throssell, J. A. Montgomery, Jr., J. E. Peralta, F. Ogliaro, M. Bearpark, J. J. Heyd, E. Brothers, K. N. Kudin, V. N. Staroverov, T. Keith, R. Kobayashi, J. Normand, K. Raghavachari, A. Rendell, J. C. Burant, S. S. Iyengar, J. Tomasi, M. Cossi, J. M. Millam, M. Klene, C. Adamo, R. Cammi, J. W. Ochterski, R. L. Martin, K. Morokuma, O. Farkas, J. B. Foresman, D. J. Fox,, Gaussian, Inc., Wallingford CT 2016, Gaussian 09, Revision A.02.

[8] Valiev, M.; Bylaska, E. J.; Govind, N.; Kowalski, K.; Straatsma, T. P.; Van Dam, H. J. J.; Wang, D.; Nieplocha, J.; Apra, E.; Windus, T. L.; de Jong, W. A., NWChem: A comprehensive and scalable open-source solution for large scale molecular simulations. Computer Physics Communications 2010, 181 (9), 1477-1489.

[9] Runge, E.; Gross, E. K. U., Density-Functional Theory for Time-Dependent Systems. Physical Review Letters 1984, 52 (12), 997-1000.

[10] Zhao, Y.; Truhlar, D. G., Density Functionals with Broad Applicability in Chemistry. Accounts of Chemical Research 2008, 41 (2), 157-167.

On the Oscillatory Convective Instability in Nanofluid Layer with Temperature Dependent Viscosity

Joginder Singh Dhiman[1] and Nivedita Sharma

Department of Mathematics, Himachal Pradesh University, Shimla 171005, India

Email[1]: jsdhiman66@gmail.com

Abstract. In the present paper, the problem of thermal convection in nanofluid layer heated from below with temperature dependent viscosity is studied mathematically for general boundary conditions, when the instability sets in as oscillatory convection. The expressions for Rayleigh numbers for exponential temperature dependent viscosity variation are derived using Galerkin method from the eigenvalue problem for each combination of rigid (the surfaces with non-slip condition) and dynamically free (the surfaces with stress free condition) boundaries. The effects of viscosity variation parameter, Lewis number, concentration Rayleigh number, modified diffusivity ratio and Prandtl number on the onset of oscillatory convection for each case of boundary conditions are investigated numerically. It is observed that the exponential variation of temperature dependent viscosity has stabilizing effect on the onset of oscillatory convection in nanofluid layer for each case of boundary conditions. However, the Lewis number and the concentration Rayleigh number have destabilizing effect on the onset of oscillatory convection.

1. Introduction

The convection in nanofluids has recently been an active field of research because of the greatly enhanced thermal properties of nanofluids. In view of this, the study of onset of thermal convection in nanofluid layer has been a subject matter of investigation for many authors in the recent years. The prominent among them are Chen *et al* [1], Tzou [2], [3], Nield and Kuznetsov [4] and Yadav *et al.* [5], who studied the problem of thermal instability of the nanofluid layer under varying assumptions of hydrodynamics. In most of the studies related to the investigation of thermal instability of ordinary fluid heated underside, much attention has been paid to the situations when the viscosity is constant. However, Rossby [6], Booker [7] and Booker and Stengel [8] have shown that for most of the classical fluids, the viscosity shows a rather pronounced variation with respect to temperature as viscosity is more sensitive to temperature than heat capacity and thermal conductivity. Therefore, it becomes important to consider the viscosity as temperature dependent in flow problems with heat transfer. Motivated by these considerations, Dhiman and Sharma [9] have recently studied the thermal convection of nanofluid layer heated underside with temperature dependent viscosity for each combination of rigid and dynamically free boundaries and established that principle of exchange of stabilities (PES) is valid for the problem when the modified diffusivity ratio $N_A \leq \frac{1}{2}$. This clearly hints that when $N_A > \frac{1}{2}$ (compliment of the condition for the existence of PES), the time-dependent oscillatory modes of instability can exist at the onset of thermal convection in nanofluids.

Motivated by the above discussion and the importance of the above discussed phenomenon in nanofluids, the aim of the present paper is to study the problem of thermal convection in a nanofluid layer heated underside for oscillatory instability and to investigate the effect of temperature dependent viscosity on the onset oscillatory convection for all cases of combinations of rigid and dynamically free boundaries. The present analysis thus extends the analysis of Dhiman and Sharma [9] for oscillatory convection in nanofluid layer heated underside with temperature dependent viscosity.

2. Physical configuration and eigenvalue problem

Consider an infinite horizontal layer of viscous, incompressible nanofluid of continuously varying density ρ and viscosity μ statically confined between two horizontal boundaries $z^* = 0$ and $z^* = d$, maintained respectively at constant temperature and volumetric fraction of nanoparticles T_0^* *and* ϕ_0^* at $z^* = 0$ and T_u^* and at $z^* = d$. A constant temperature gradient ($T_0^* > T_u^*$) is maintained across the fluid layer parallel to the body force, which is gravity here. The viscosity of the fluid is taken to be an exponential function of temperature variation given as $\mu(T) = \mu_0 e^{-\gamma(T-T_0)}$, where γ is the coefficient of viscosity variation.

Following the usual steps of linear stability theory (*cf.* Nield and Kuznetsov [4] and Chandrasekhar [10], Dhiman and Sharma [11] and [12]), the non-dimensional linearized perturbation equations governing the problem of thermal convection in nanofluids with variable viscosity as derived by Dhiman and Sharma [9] are given by;

$$f(D^2-a^2)^2w - \frac{n}{P_r}(D^2-a^2)w + 2(Df)D(D^2-a^2)w + D^2f(D^2+a^2)w = Raa^2\theta - R_N a^2\phi \quad (1)$$

$$(D^2-a^2-n)\theta + \left(\frac{2N_AN_B}{Le} - \frac{N_B}{Le}\right)D\theta + \frac{N_B}{Le}D\phi = -w \quad (2)$$

$$\left(\frac{1}{Le}(D^2-a^2) - n\right)\phi + \frac{N_A}{Le}(D^2-a^2)\theta = w \quad (3)$$

together with either of the following cases of boundary conditions;

Case 1: Both boundaries dynamically free

$$w = 0 = \theta = \phi = D^2w \qquad at\ z = 0\ and\ z = 1 \quad (4)$$

Case 2: Both boundaries rigid

$$w = 0 = \theta = \phi = Dw \qquad at\ z = 0\ and\ z = 1 \quad (5)$$

Case 3: Lower rigid and upper boundary free

$$\left.\begin{aligned} w = 0 = \theta = \phi = Dw \qquad & at\ z = 0 \\ w = 0 = \theta = \phi = D^2w \qquad & at\ z = 1 \end{aligned}\right\} \quad (6)$$

In the above equations; $D \equiv \frac{d}{dz}$ is the differential operator; $a = \left(k_x^2 + k_y^2\right)^{\frac{1}{2}}$ is a dimensionless horizontal wave number; n is a dimensionless complex growth rate (*i.e.* $n = n_r + in_i$, where n_r and n_i are real); w, θ and ϕ are the respective perturbations in the initial vertical velocity, temperature and nano particle volume fraction; $f(z)$ is the non-dimensional temperature dependent viscosity variation function; N_A is the modified diffusivity ratio; N_B is the modified particle density increment; Le is the Lewis number; P_r is the Prandtl number; Ra is the Thermal Rayleigh number and R_N is the concentration Rayleigh number.

3. Mathematical analysis

Dhiman and Sharma [9] derived the above eigenvalue problem governing the thermal convection in a nanofluid represented by equations (1)-(3) together with boundary conditions (4)-(6) and studied the case of stationary instability. Here, in the present analysis, we have extended their analysis for the case of oscillatory convection and the expressions for Rayleigh numbers for each case of boundary conditions (4)-(6) are derived using the Galerkin method (*cf.* Finlayson [13]), by taking a single term in the expansions of w, θ and ϕ. To proceed, we have considered suitably chosen trial functions; w_1, θ_1 and ϕ_1 which satisfy the respective boundary conditions given in (4)-(6), as

$$w = Aw_1(z),\ \ \theta = B\theta_1(z)\ \text{ and }\ \phi = C\phi_1(z) \quad (7)$$

Here, A, B and C are arbitrary constants.

Substituting these trial functions defined in (7) in equations (1)-(3) and multiplying the resulting equations (the *residuals*) respectively by w_1, θ_1 and ϕ_1 so that the expressions on the left-hand sides of these equations become orthogonal to the trial functions. Now, integrating these resulting equations over the vertical range of z and performing the integration by parts a suitable number of times using relevant boundary conditions (4)-(6), we obtain the following system of three linear homogeneous algebraic equations in the unknowns *A*, *B* and *C* as;

$$A\int_0^1 f[(D^2w_1)^2+2a^2(Dw_1)^2+a^4(w_1)^2]\,dz+Aa^2\int_0^1(D^2f)(w_1)^2dz+A\frac{n}{P_r}\int_0^1[(Dw_1)^2+a^2(w_1)^2]dz-Ra\,a^2B\int_0^1\theta_1w_1dz+R_Na^2C\int_0^1\phi_1w_1dz=0 \quad (8)$$

$$B\int_0^1[(D\theta_1)^2+(a^2+n)(\theta_1)^2]\,dz+B\left(\frac{(2N_A-1)N_B}{L_e}\right)\int_0^1 D\theta_1.\theta_1dz+\frac{CN_B}{L_e}\int_0^1 D\theta_1.\phi_1dz-\int_0^1\theta_1w_1dz=0 \quad (9)$$

$$C\int_0^1\frac{1}{L_e}((D\phi_1)^2+a^2(\phi_1)^2+n(\phi_1)^2)\,dz-B\frac{N_B}{L_e}\int_0^1[(D^2\phi_1.\theta_1)-a^2\phi_1.\theta_1]dz+A\int_0^1\phi_1w_1dz=0 \quad (10)$$

For the existence of non-trivial solution of these equations, the vanishing of the determinant of coefficients of equations (8)-(10) yields an *eigenvalue equation* for the system in terms of Rayleigh number Ra, for the case of oscillatory convection.

Thus, to obtain the values of the Rayleigh numbers (Ra) for each case of boundaries conditions (4)-(6), and for the exponential non-dimensional viscosity variation, we have considered the following exponential viscosity variation law (*cf.* Dhiman and Kumar [14]);

$f=e^{\delta z}$, where δ is the viscosity variation parameter. (11)

Now, proceeding in the analogous way as in the analyses of Dhiman and Sharma [9], [11] and [12], we shall derive the expressions for Rayleigh numbers for each of the cases of boundary conditions (4)-(6), separately.

Case1: Both boundaries dynamically free

For the present case of boundary conditions, choosing the suitable polynomial trial functions satisfying the boundary conditions (4) given by;

$$w=z^4-2z^3+z,\ \theta=z(z-1)\ \text{and}\ \phi=z(z-1) \quad (12)$$

Using viscosity variation law (11) and the trial functions (12) in equations (8)-(10), eliminating constants *A*, *B* and *C* amongst the resulting equations and substituting $n=in_i$ (for the case of oscillatory modes), we obtain the following dispersion relation;

$$Ra(l_{11}+in_iLe)+R_N[l_{11}(N_A+Le)+in_iLe]=\frac{1}{h_{11}}\left(M_{11}+in_i\frac{g_{11}}{P_r}\right)(l_{11}+in_iLe)(l_{11}+in_i) \quad (13)$$

Where,

$$M_{11}=\frac{1}{\delta^9}\Big[2\Big\{144\delta^4\left(-12-6\delta-\delta^2+e^\delta(12-6\delta+\delta^2)\right)+a^4\left(-20160-1080\delta-1440\delta^2+120\delta^3+48\delta^4-\delta^6+e^\delta(20160-10080\delta+1440\delta^2+120\delta^3-48\delta^4+\delta^6)\right)+2a^2\delta^2\left(-15840-7290\delta-1152\delta^2+84\delta^3+36\delta^4-\delta^6+e^\delta(15840-7290\delta+1152\delta^2+84\delta^3-36\delta^4+\delta^6)\right)\Big\}\Big] \quad (14)$$

and $l_{11}=10+a^2$; $g_{11}=\frac{(306+31a^2)}{630}$; $h_{11}=\frac{8670\,a^2}{176400}$ (15)

Now, equating the real and imaginary parts of both sides of equation (13), we obtain the following equations

$$n_i^2\left[\frac{LeM_{11}}{l_{11}}+\frac{Leg_{11}}{P_r}+\frac{g_{11}}{P_r}\right]=M_{11}l_{11}-h_{11}\left[Ra_{ff}+R_N(N_A+Le)\right] \quad (16)$$

and $n_i^2\frac{Leg_{11}}{P_r}=l_{11}M_{11}(1+Le)+\frac{l_{11}^2g_{11}}{P_r}-h_{11}\left[Le\left(Ra_{ff}+R_N\right)\right]$ (17)

If we take $n_i=0$ (non-oscillatory case) in equation (16), we get an expression for Rayleigh number

$$Ra_{ff}=\frac{M_{11}l_{11}}{h_{11}}-R_N(N_A+Le) \quad (18)$$

which is the same expression for Rayleigh number for stationary case of instability as obtained by Dhiman and Sharma [9] for the case of both dynamically free boundaries for the problem under consideration. Expression (18) also corresponds to the boundary for non-oscillatory instability.

It is clear from the expression (14) that M_{11} is a positive constant for positive δ, whereas l_{11}, g_{11} and h_{11} are clearly positive which can be verified from the expressions (15). Hence, equation (16) yields that for given n_i (real), R_N must be negative for large values of Le.
Now, eliminating ${n_i}^2$ from equations (16) and (17), we obtain the following equation;

$$\left(\frac{M_{11}}{l_{11}g_{11}}+\frac{1}{P_r}\right)Ra_{ff}+\left(\frac{M_{11}}{l_{11}g_{11}}+\frac{(1-N_A)}{LeP_r}\right)R_N=\frac{M_{11}l_{11}}{h_{11}}\left[\left(1+\frac{1}{Le}+\frac{1}{P_r}+\frac{l_{11}g_{11}}{LeM_{11}P_r}\right)\left(1+\frac{1}{LeP_r}+\frac{1}{P_r}+\frac{M_{11}}{l_{11}g_{11}}\right)-\frac{1}{LeP_r}\right] \quad (19)$$

which yields the minimum (critical value) of Rayleigh number Ra for $a^2 = 2.22$, when $\delta \to 0$, as

$$\left(Ra_{ff}\right)_c=\frac{664.525\,P_r}{(0.991865P_r+1)}\left[\left(1+\frac{1}{Le}+\frac{1.0082}{LeP_r}\right)\left(0.991865+\frac{1}{LeP_r}+\frac{1}{P_r}\right)-\frac{1}{LeP_r}\right]-\frac{(0.991865\,LeP_r+(1-N_A))}{Le(0.991865P_r+1)}R_N \quad (20)$$

This value corresponds to oscillatory instability boundary. Further, the value given in expression (20) for critical Rayleigh number when both boundaries are dynamically free is very close to the value obtained by Nield and Kuznetsov [4] for the same problem with constant viscosity.
Case 2: Both boundaries are rigid

For this case of boundary conditions, the suitable polynomial trial functions chosen are;

$$w = z^4 - 2z^3 + z^2, \quad \theta = z(z-1) \text{ and } \quad \phi = z(z-1) \quad (21)$$

Using viscosity variation law (11) and the trial functions (21) in equations (8)-(10), we obtain the following equation analogous to equation (13), with $n = in_i$;

$$Ra_{rr}(l_{22}+in_iLe)+R_N[l_{22}(N_A+Le)+in_iLe]=\frac{1}{h_{22}}\left(M_{22}+in_i\frac{g_{22}}{P_r}\right)(l_{22}+in_iLe)(l_{22}+in_i) \quad (22)$$

Where,

$$M_{22}=\frac{1}{\delta^9}\Big[4\Big\{6a^4\left(-1680-840\delta-180\delta^2-12\delta^3-\delta^4+e^\delta(1680+840\delta+180\delta^2-120\delta^3+\delta^4)\right)+\delta^4\left(-864-432\delta-96\delta^2-12\delta^3-\delta^4+e^\delta(864-432\delta+96\delta^2-12\delta^3+\delta^4)\right)+2a^2\delta^2\left(-7920-3960\delta-852\delta^2-96\delta^3-5\delta^4+e^\delta(7920-3960\delta+852\delta^2-96\delta^3+5\delta^4)\right)\Big\}\Big] \quad (23)$$

$$l_{22}=10+a^2; \quad g_{22}=\frac{1}{630}(12+a^2); \quad h_{22}=\frac{30}{19600}a^2 \quad (24)$$

Now, proceeding analogously as in *Case 1* above, we can obtain the expressions for real and imaginary parts of the equation (22) for *Case 2* also. From these obtained expressions for real and imaginary parts, eliminating ${n_i}^2$, we obtain the following dispersion relation analogous to expression (19);

$$\left(\frac{M_{22}}{l_{22}g_{22}}+\frac{1}{P_r}\right)Ra_{rr}+\left(\frac{M_{22}}{l_{22}g_{22}}+\frac{(1-N_A)}{LeP_r}\right)R_N=\frac{M_{22}l_{22}}{h_{22}}\left[\left(1+\frac{1}{Le}+\frac{1}{P_r}+\frac{l_{22}g_{22}}{LeM_{22}P_r}\right)\left(1+\frac{1}{LeP_r}+\frac{1}{P_r}+\frac{M_{22}}{l_{22}g_{22}}\right)-\frac{1}{LeP_r}\right] \quad (25)$$

Case 3: Lower rigid and upper free boundaries

For this case of boundary conditions, the suitable polynomial trial functions chosen are;

$$w = 2z^4 - 5z^3 + 3z^2, \theta = z(z-1) \text{ and } \phi = z(z-1) \quad (26)$$

Using relations (11) and (26) in equations (8)-(10), proceeding as in cases above, we have

$$Ra_{fr}(l_{33}+in_iLe)+R_N[l_{33}(N_A+Le)+in_iLe]=\frac{1}{h_{33}}\left(M_{33}+in\frac{g_{33}}{P_r}\right)(l_{33}+in_iLe)(l_{33}+in_i \quad (27)$$

Where,

$$M_{33}=\frac{1}{\delta^9}\Big[2\Big\{-18\delta^4\left(384+240\delta+66\delta^2+10\delta^3+\delta^4+-6e^\delta(64-24\delta+3\delta^2)\right)+a^4\left(-36(2240+1400\delta+370\delta^2+50\delta^3+3\delta^4)+e^\delta(80640-30240\delta+3240\delta^2+240\delta^3-72\delta^4+\delta^6)\right)+2a^2\delta^2\left(-18(3520+2200\delta+584\delta^2+80\delta^3+5\delta^4+e^\delta(63360-23760\delta+259\delta^2+168\delta^3-54\delta^4+\delta^6)\right)\Big\}-(N_A+Le)R_N\Big] \quad (28)$$

$$l_{33}=10+a^2; \quad g_{33}=\frac{1}{630}(216+19a^2); \quad h_{33}=\frac{5070}{176400}a^2 \quad (29)$$

IOP Conf. Series: Journal of Physics: Conf. Series **1298** (2019) 012026 doi:10.1088/1742-6596/1298/1/012026

Following the similar steps as in the earlier discussed two cases of boundary conditions, we obtain the following dispersion relation analogous to equation (19)

$$\left(\frac{M_{33}}{l_{33}g_{33}}+\frac{1}{P_r}\right)Ra_{fr}+\left(\frac{M_{33}}{l_{33}g_{33}}+\frac{(1-N_A)}{LeP_r}\right)R_N=\frac{M_{33}l_{33}}{h_{33}}\left[\left(1+\frac{1}{Le}+\frac{1}{P_r}+\frac{l_{33}g_{33}}{LeM_{33}P_r}\right)\left(1+\frac{1}{LeP_r}+\frac{1}{P_r}+\frac{M_{33}}{l_{33}g_{33}}\right)-\frac{1}{LeP_r}\right] \quad (30)$$

4. Results and discussion

In the present analysis, the effect of temperature dependent viscosity on the onset oscillatory thermal convection in nanofluid layer heated from below has been studied for each combination of rigid and dynamically free boundary conditions. The expressions for the Rayleigh numbers for oscillatory convection are derived using Galerkin method and the values of the critical Rayleigh numbers for various values of critical wave numbers a_c for certain fixed values of δ (exponential variation of temperature dependent viscosity) and different values of other parameters; Le, R_N, N_A and P_r are calculated numerically and the obtained results are presented in Tables 1 and 2 for *Case 1* of boundary conditions. The variation of Rayleigh number Ra_{ff} with respect to δ and Le for fixed value of $N_A = 10$, $R_N = -2$, $P_r = 5$ is presented in Figures 1(a)-1(c) and the variation of Ra_{ff} with respect to δ and R_N for fixed value of $N_A = 5$, $Le = 100$, $P_r = 5$ is shown in Figures 1(d)-1(f), for *Case 1* of boundary conditions.

From the values of Ra_{ff} presented in Tables 1 & 2 and from Figures 1(a)-1(f), we note that the exponential variation of temperature dependent viscosity (δ) has stabilizing effect on the onset of oscillatory convection in nanofluid layer heated from below. This is due to the fact that the increase in nanofluid viscosity increases the friction which results in the decrease of flow rate and consequently degrades the heat transfer performances. Also, from the values presented in Table1 and Figures 1(a), 1(b) and 1(c) for *Case 1* of the boundary conditions, we observed that for increasing values of Lewis number (Le), the values of Ra_c decreases, which implies that Le has destablizing effect on the onset of oscillatory convection. Also, it can be observed from the values presented in Table 1 and Figures 1(d), 1(e), 1(f); that the values of critical Rayleigh numbers also decrease with the increasing value of the concentration Rayleigh number R_N, hence has a destabilizing effect on the onset of oscillatory convection in nanofluids.

From the values presented in Table 2, we found that Ra_c show marginal variations with respect to increasing value of modified diffusivity ratio ($N_A > \frac{1}{2}$) and Prandtl number (P_r). This is due to the fact that the increase in N_A, which defines the ratio of thermophoretic diffusion coefficient to Brownian diffusion coefficient, increase the cross-diffusion phenomena of thermophoresis and Brownian motion, while the increase in Prandtl number, which defines the ratio of kinematic viscosity to thermal diffusivity (heat transfer rate), thus increase the viscous force which causes the retardation in the velocity and hence the heat transfer rate and therefore have marginal effect on the onset of oscillatory convection. The analogous analysis for *Case 2* and *Case 3* of boundary conditions reveals the similar behaviour of Rayleigh numbers as observed in *Case 1* boundary conditions and hence numerical values and graphical representation of the obtained results are both omitted here, for the sake of compactness.

Table 1. Variation of Ra_{ff} with respect to δ and Le for different values of a_c and fixed value of $N_A = 10$, $R_N = -2$, $P_r = 5$ and for R_N with $N_A = 5, P_r = 5$, Le =100, for *Case 1* of boundary conditions

δ	Le	a_c^2	Ra_c	R_N	a_c^2	Ra_c
0.2		4.95501	742.282		4.95503	747.483
0.5		4.93242	873.3571		4.93244	879.232
0.9	200	4.87338	1103.32	-3	4.87344	1110.37
0.2		4.9549	739.678		4.9549	753.365
0.5		4.93239	870.359		4.9324	885.257
0.9	500	4.87334	1099.64	-5	4.87337	1116.57
0.2		4.95499	739.028		4.95497	757.559
0.5		4.93239	869.61		4.93237	889.561
0.9	800	4.87333	1098.72	-10	4.87332	1121.01

Table 2. Variation of Ra_{ff} with respect to δ and N_A for different values of a_c and fixed value of $R_N = -2$, $P_r = 5$, Le =500 and for P_r with $N_A = 5$, $R_N = -3$, Le =500 for *Case 1* of boundary conditions

δ	N_A	a_c^2	Ra_c	P_r	a_c^2	Ra_c

0.2		4.95499	739.683		4.955	740.449
0.5		4.93239	870.363		4.9324	871.192
0.9	2	4.87334	1099.64	2	4.87335	1100.55
0.2		4.9551	739.681		4.95499	740.523
0.5		4.93239	870.362		4.93239	871.227
0.9	5	4.87334	1099.64	5	4.87333	1100.53
0.2		4.9551	739.678		4.95499	740.598
0.5		4.93239	870.359		4.93239	871.281
0.9	10	4.87332	1121.01	10	4.87333	871.192

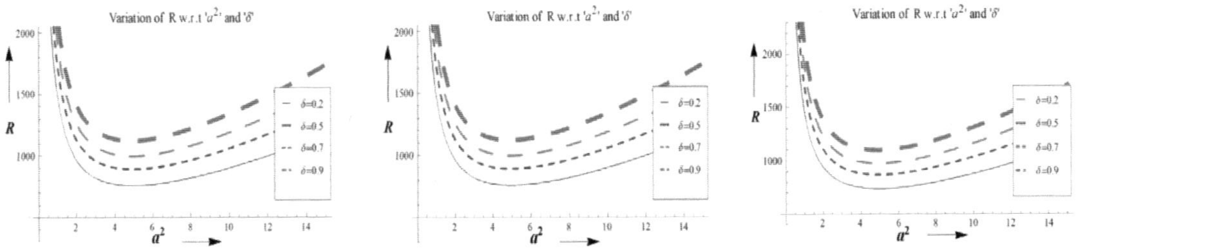

Figures 1(a)-(c): Variation of Ra_{ff} $w.\ r.\ t.\ a^2$ for $Le = 200, 500\ and\ 800$, respectively.

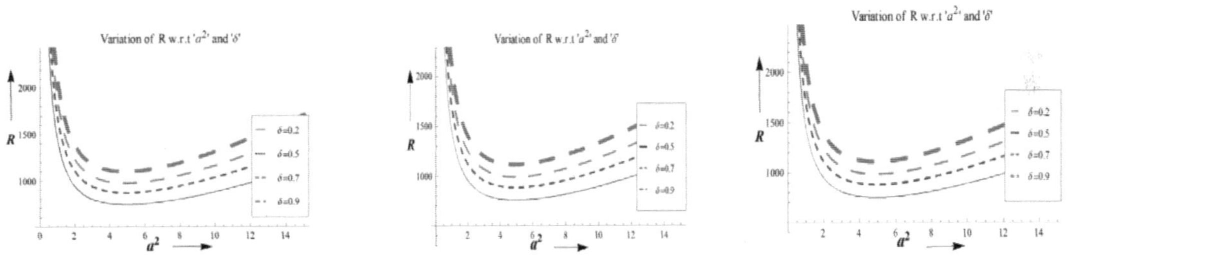

Figures 1(d)-(f): Variation of Ra_{ff} $w.r.\ t.\ a^2$ for $R_N = -10, -5\ and -\ 3$, respectively.

5. References

[1] Chen H, Ding Y and Tan C 2007 Rheological Behaviour of Nanofluids *New J. of Physics,* **9** 367(24)

[2] Tzou D Y 2008 Instability of nanofluids in Natural Convection *ASME Journal of Heat Transfer* **130**(7) 372-401

[3] Tzou D Y 2008 Thermal Instability of Nanofluids in Natural Convection *International Journal of Heat and Mass Transfer* **51**(11-12) 2967-79

[4] Nield D A and Kuznetsov A V 2010 The Onset of Convection in a Horizontal Layer of Finite Depth *European Journal of Mechanics B-Fluids* **29**(3) 217-23

[5] Yadav D Agrawal G S and Bhargava R Rayleigh 2011 Bénard Convection in Nanofluid *International Journal of Applied Maths and Mechanics* **7**(2) 61-67

[6] Rossby H T 1985 A Study of Bénard convection with and without Rotation *Journal of Fluid Mechanics* **36** 309-35

[7] Booker J R 1976 Thermal Convection with Strongly Temperature Dependent Viscosity *Journal of Fluid Mechanics* **76** 741-54

[8] Booker J R and Stengel K C 1978 Further Thoughts on Convective Heat Transport in a Variable Viscosity *Journal of Fluid Mechanics* **86** 289-291

[9] Dhiman J S and Sharma N 2015 On the Effect of Temperature Dependent Viscosity on the Thermal Convection of Nanofluid Layer *Journal of Thermophysics and Heat Transfer* **29**, 90-101

[10] Chandrasekhar S 1961 *Hydrodynamic and Hydromagnetic Stability* (London, Oxford Univ. Press)

[11] Dhiman J S and Sharma N 2018 Thermal Instability of Hot Ferrofluid Layer with Temperature Dependent Viscosity *International Journal of Fluid Mechanics Research* **45**, 389-98

[12] Dhiman J S and Sharma N 2019 Thermal Instability of a Micropolar Fluid Layer with Temperature Dependent Viscosity *Proc. National Academy of Sciences, India Section A,* https://doi.org/10.1007/s40010-018-0591-6.

[13] Finlayson B A 1972 *The Method of Weighted Residuals and Variational Principles* (New York, Academic Press)

[14] Dhiman J S and Kumar V 2013 On Stability Analysis of Rayleigh Bénard Convection with Temperature Dependent Viscosity for General Boundary Condition *International Journal of Emerging Multidisciplinary Fluid Sciences* **3,** 85-98

Design of the control law of longitudinal attitude for Tail-sitter UAV

Zhilei Ge[1], Jiabao Hou[1]

[1]School of Astronautics, Northwestern Polytechnical University, Xi'an, 710072 China

Email: 13991336193@139.com

Abstract: The longitudinal attitude control of a type of Tail-sitter UAV is studied in this paper. In order to solve the singular problem of pitch angle during the tilting process of fuselage, the double Euler angle algorithm is used for modeling. Fuzzy self-tuning PID control method and L1 adaptive control method are used to design the longitudinal attitude controller respectively to counteract the susceptibility to disturbance during the flight, especially in tilting process. The numerical simulation is carried out through MATLAB, and the simulation results show that: This controller can overcome the model parameter changes and other uncertain disturbances to the Tail-sitter UAV during the flight

1. Introduction

Fixed-wing UAV has the characteristics of high flight efficiency, high speed, simple system structure, but it needs a runway or launch and recovery units during the process of both takeoff and landing[1]. By introducing vertical takeoff and landing technology, fixed-wing UAV can have the characteristics of vertical takeoff, spot hover, vertical landing of multi-rotor UAV and high-speed level cruise of fixed-wing UAV at the same time [2]. The Tail-sitter UAV realizes the transformation between vertical flight mode and horizontal flight mode through large angle pitch maneuver and pull control. Its structure is similar to that of fixed-wing aircraft [3].

The flight speed, pitch angle, angle of attack and aerodynamic parameters of the Tail-sitter UAV will change greatly during the large angle pitch maneuver. As a result, it is difficult to control and maintain the flight attitude of the UAV stably [3].

Considering the singular problem of pitch angle in vertical flight, two SDOF dynamic models of UAV based on Euler angle are established. Fuzzy parameter self-tuning PID control method and $L1$ adaptive control method are used to design longitudinal attitude controllers in vertical and near-vertical flight phases as well as horizontal and near-horizontal flight phases respectively. Finally, the flight simulation of the designed flight control system is made in MATLAB, and the results show that the control effect is good.

2. Flight principle and dynamic model of Tail-sitter UAV

In this paper, the axisymmetric configuration mode is used for the Tail-sitter UAV. Specifically, a propeller is installed on each wingtip of four wings. Two rudder control surfaces are installed at the bottom of the wing. The aircraft is controlled by the vertical flight mode controller when the pitch angle is greater than 45 degrees, and the aircraft is controlled by the horizontal flight mode controller when the pitch angle is less than or equal to 45 degrees.

Table 1. UAV Control Strategy

Movement mode	Vertical flight mode control strategy	Horizontal flight mode control strategy
Pitching movement	Motor differential	Linkage of two rudder control surfaces
Rolling movement	Motor differential	Differential of two rudder control surfaces
Yawing movement	Motor differential	Coordinate bank-to-turn

Two reference frames are set up in this paper: geographic coordinate system (North East Down) F_{e1}, vertical geographic coordinate system (Up East North) F_{e2}, body axis system F_b and air path axis system F_a are used in a way not different from normal use. The horizontal Euler angle is determined by the relationship between F_{e1} and F_b to describe the attitude of aircraft in the horizontal and near-horizontal flight phases; the vertical Euler angle is determined by the relationship between F_{e2} and F_b to describe the attitude of aircraft in the vertical and near-vertical flight phases. The mutual transformation relationship of vertical Euler angle and horizontal Euler angle is[4]:

$$\left.\begin{array}{l} \theta = \sin^{-1}(\cos\psi_v \cos\theta_v); \phi = \sin^{-1}\left(\dfrac{\sin\psi_v \cos\theta_v}{\cos\theta}\right) \\ \psi = \sin^{-1}\left(\dfrac{\cos\psi_v \sin\theta_v \sin\phi_v + \sin\psi_v \cos\theta_v}{\cos\theta}\right) \end{array}\right\} \quad (2.1)$$

Based on the vertical Euler angle, the SDOF mathematical model of UAV in the body axis system F_b is established as follows:

$$\left.\begin{array}{l} \dot{\phi}_v = p\sec\theta_v \cos\psi_v - q\sec\theta_v \sin\psi_v \\ \dot{\theta}_v = p\sin\psi_v + q\cos\psi_v \\ \dot{\psi}_v = -p\tan\theta_v \cos\psi_v + q\tan\theta_v \sin\psi_v + r \end{array}\right\} \quad (2.2)$$

$$\left.\begin{array}{l} \dot{u} = rv - qw + g\cos\theta_v \cos\psi_v + \dfrac{1}{m}(T_{rotor} + T_{aero\_x}) \\ \dot{v} = pw - ru + g\cos\theta_v \sin\psi_v + \dfrac{T_{aero\_y}}{m} \\ \dot{w} = qu - pv - g\sin\theta_v + \dfrac{T_{aero\_z}}{m} \end{array}\right\} \quad (2.3)$$

$$\left.\begin{array}{l} \dot{p} = \dfrac{(I_y - I_z)I_z - I_{xz}^2}{I_x I_z - I_{xz}^2} qr + \dfrac{1}{I_x}(M_{rotor\_x} + M_{aero\_x}) \\ \dot{q} = \dfrac{I_z - I_x}{I_y} pr + \dfrac{1}{I_y}(M_{rotor\_y} + M_{aero\_y}) \\ \dot{r} = \dfrac{I_x(I_x - I_y) + I_{xz}^2}{I_x I_z - I_{xz}^2} pq + \dfrac{1}{I_z}(M_{rotor\_z} + M_{aero\_z}) \end{array}\right\} \quad (2.4)$$

IOP Conf. Series: Journal of Physics: Conf. Series **1298** (2019) 012027 doi:10.1088/1742-6596/1298/1/012027

Based on the horizontal Euler angle, the SDOF mathematical model of UAV in the body axis system F_b can also be derived from the equation (2.1) and equation (2.2-2.4) .

3. Design of flight controller

In the transition flight phase, the fuzzy parameter self-tuning PID method and $L1$ adaptive control method are used for the control of pitch angle respectively [5].

3.1 Design of longitudinal attitude controller for vertical flight mode

The control system is mainly composed of fuzzy parameter self-tuning controller and PID controller[6]. The inputs of the fuzzy parameter tuning part consist of the deviation e and the deviation change rate e_c , and the outputs are the corrections, $\triangle K_p$, $\triangle K_i$, $\triangle K_d$ of the PID parameters[7]. Referring to the fuzzification method in reference [8] and combining with the actual situation of the Tail-sitter UAV in this paper, a fuzzy language set {negative large, negative medium, negative small, zero, positive small, positive medium, positive large} is set up, and the universe is divided into seven parts evenly to describe. Since the pitch angle of UAV is limited to 90 degrees to - 90 degrees in radians, the universes of fuzzy variables e and e_c are both set to {-3,3}, and the fuzzy universes of $\triangle K_p$, $\triangle K_i$, $\triangle K_d$ are {- 10,10}, {- 3,3}, {- 1,1} respectively. In fuzzification, it is required to multiply the input variable by the corresponding quantization factor. In this paper, the quantization factors of the fuzzy parameter self-tuning PID controller are selected as k_e=1, k_{ec}=1, and the output scaling factors are k_{pu}=1.1, k_{iu}=0.1, k_{du}=0.1.

Using the fuzzy control rule in reference[6], the triangle membership curve is used to describe each fuzzy subset in the universe. By mamdani reasoning, each output control variable has 49 fuzzy rules. The center-of-gravity method is used to solve the fuzziness as below: $k=\sum_{i=1}^{n} u_i\lambda(u_i) \Big/ \sum_{i=1}^{n}\lambda(u_i)$. In the formula, k is the output value obtained by solving the fuzziness, u_i is the input variable, $\lambda(u_i)$ is the membership corresponding to variable u_i [6]. The initial PID parameters are obtained by adjusting by critical ratio method.The parameter self-tuning PID controller flow chart is shown in Figure 1.

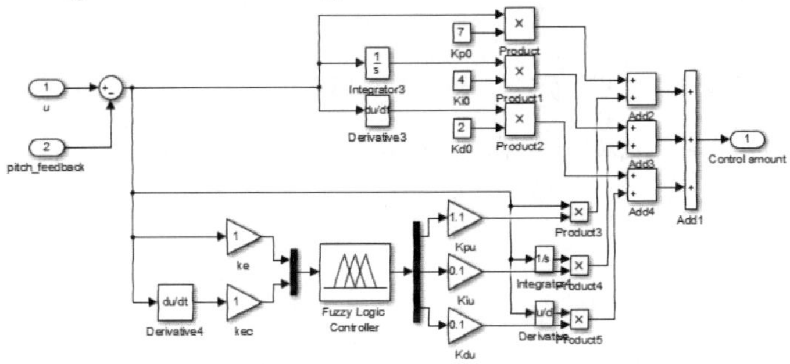

Figure 1. Flow chart of fuzzy parameter self-tuning PID controller

3.2 Design of longitudinal attitude controller in horizontal flight mode

The simplified results of the pitch channel model for the horizontal Euler angle SDOF model based on the method in reference [9] are as follows:

$$\left.\begin{aligned}
&\ddot{\theta} = -a_{\theta 1}\dot{\theta} - a_{\theta 2}\theta + a_{\theta 3}\delta_e + d_\theta \\
&a_{\theta 1} = \frac{-q_\infty s_a c C_{mq}}{2V_a I_y}; a_{\theta 2} = \frac{-q_\infty s_a c C_{ma}}{I_y}; a_{\theta 3} = \frac{q_\infty s_a c C_{m\delta_e}}{I_y} \\
&d_\theta = \frac{I_z - I_x}{I_y} pr + \frac{q_\infty s_a c}{I_y}\left(C_{m0} - C_{ma}\gamma - C_{mq}\frac{c}{2V_a} d_{\theta 1}\right) + \dot{d}_{\theta 1} \\
&d_{\theta 1} = -r\sin\phi + q(\cos\phi - 1)
\end{aligned}\right\} \tag{3.2}$$

This paper designs the $L1$ adaptive controller by Referring the $L1$ adaptive control theory in reference [10]. Taking $x_\theta = [\theta, \dot{\theta}]^T$, write the pitch channel model in formula (3.2) as the standard form of $L1$ adaptive control:

$$\left.\begin{aligned}
&\dot{x}_\theta(t) = A_\theta x_\theta(t) + b_\theta\left(w(t)u_{\delta_e}(t) + \theta^T(t)x_\theta(t) + \delta(t)\right) \\
&y_\theta(t) = c_\theta^T x_\theta(t)
\end{aligned}\right\} \tag{3.3}$$

Taking: $A_\theta = \begin{bmatrix} 0 & 1 \\ -a_{\theta 2} & -a_{\theta 1} \end{bmatrix}$, $b_\theta = \begin{bmatrix} 0 \\ a_{\theta 3} \end{bmatrix}$, $c_\theta = \begin{bmatrix} 1 \\ 0 \end{bmatrix}$ as the system matrix, input matrix and output matrix. $w(t)$ stands for input gain; $\theta^T(t)x_\theta(t)$ stands for unknown part related to state variable; $\delta(t)$ stands for unknown part independent of state variable. In this paper, the flight speed $V_a = 25m/s$ is taken as the operating point, then: $A_\theta = \begin{bmatrix} 0 & 1 \\ -299.6973 & -14.7633 \end{bmatrix}$, $b_\theta = \begin{bmatrix} 0 \\ -171.481762 \end{bmatrix}$, $c_\theta = \begin{bmatrix} 1 \\ 0 \end{bmatrix}$. According to the parameters of the Tail-sitter UAV, the projection boundary of unknown parameters are selected as follows: $w = [0.5, 1.5]$; $\delta = 4.6$; $\theta = \{\theta = [\theta_1 \ \ \theta_2]^T \in R^2 : \theta_1 = [-0.08, 0.08], \theta_2 = [-0.6, 0.6]\}$ .So: $L = \max_{\theta \in \Theta} \|\theta\|_1 = 1.3$. The input of $L1$ adaptive controller contains two parts: state feedback input and adaptive control input: $u_{\delta_e}(t) = u_m(t) + u_{ad}(t)$, $u_m(t) = -K_m^T x(t)$, where $K_m \in R^n$ makes $A_{m\theta} = A - bK_m^T$ be Hurwitz matrix. In this paper, the feedback gain matrix $K_m = [-9.3452 \ \ -7.6673]$ is obtained by using the *LQR* method in the optimal control theory, and then the closed loop system matrix is obtained:

$$A_{m\theta} = A - bK_m^T = \begin{bmatrix} 0 & 1 \\ -1902.2 & -1329.6 \end{bmatrix} \tag{3.4}$$

The pitch channel control model is converted to:

$$\left.\begin{aligned}
&\dot{x}_\theta(t) = A_{m\theta} x_\theta(t) + b_\theta\left(w(t)u_{ad}(t) + \theta^T(t)x_\theta(t) + \delta(t)\right) \\
&y_\theta(t) = c_\theta^T x_\theta(t)
\end{aligned}\right\} \tag{3.5}$$

The state estimator for pitch channel:

$$\left.\begin{aligned}
&\dot{\hat{x}}_\theta(t) = A_{m\theta}\hat{x}_\theta(t) + b_\theta\left(\hat{w}(t)u_{\delta_e}(t) + \hat{\theta}^T(t)x_\theta(t) + \hat{\delta}(t)\right) \\
&\hat{y}_\theta(t) = c_\theta^T \hat{x}_\theta(t)
\end{aligned}\right\} \tag{3.6}$$

Where: $\hat{x}_\theta(t)$, $\hat{w}(t)$, $\hat{\theta}(t)$, $\hat{\delta}(t)$ are the estimated values of $x_\theta(t)$, $w(t)$, $\theta(t)$, $\delta(t)$ respectively, and the parameter adaptive law based on projection operator is:

$$\left.\begin{array}{l} \dot{\hat{\theta}}(t)=\Gamma \operatorname{Proj}\left(\hat{\theta}(t),-\tilde{x}_\theta^T(t)Pb_\theta x_\theta(t)\right), \hat{\theta}(0)=\hat{\theta}_0 \\ \dot{\hat{\sigma}}(t)=\Gamma \operatorname{Proj}\left(\hat{\sigma}(t),-\tilde{x}_\theta^T(t)Pb_\theta\right), \hat{\sigma}(0)=\hat{\sigma}_0 \\ \dot{\hat{\omega}}(t)=\Gamma \operatorname{Proj}\left(\hat{\omega}(t),-\tilde{x}_\theta^T(t)Pb_\theta u_{ad}(t)\right), \hat{\omega}(0)=\hat{\omega}_0 \end{array}\right\} \tag{3.7}$$

Where: $\tilde{x}_\theta(t)=\hat{x}(t)-x(t)$ is the tracking error of state observer and dynamic system; $\hat{\theta}_0$, $\hat{\sigma}_0$, $\hat{\omega}_0$ are initial values of $\hat{\theta}(t)$, $\hat{\sigma}(t)$, $\hat{\omega}(t)$ respectively; $P=P^T>0$ is the solution of lyapunov equation $A_m^TP+PA_m=-Q$, where $Q=Q^T>0$;

$$\operatorname{Proj}(\theta,y)=\begin{cases} y, & f(\theta)<0 \\ y-\dfrac{\nabla f(\theta)}{\|\nabla f(\theta)\|}\left\langle\dfrac{\nabla f(\theta)}{\|\nabla f(\theta)\|},y\right\rangle f(\theta), & f(\theta)\geq 0, \nabla f(\theta)^T y>0 \\ y, & f(\theta)\geq 0, \nabla f(\theta)^T y\leq 0 \end{cases} \tag{3.8}$$

Where: $f(\theta)=\dfrac{(\varepsilon_\theta+1)\theta^T\theta-\theta_{\max}^2}{\varepsilon_\theta\theta_{\max}^2}$, $\theta_{\max}$ is the normative constraint of vector θ; ε_θ is the projection boundary; $\nabla f(\theta)$ is the gradient vector at θ. In this paper, we set Q to be the identity matrix $\begin{bmatrix}1 & 0\\ 0 & 1\end{bmatrix}$, calculate and obtain $P=\begin{bmatrix}1.0652 & 0.00002628\\ 0.00026285 & 0.00037625\end{bmatrix}$; set $\Gamma=10000$. Thus the form of adaptive control input is obtained as follows: $u_{ad}(S)=C(S)(\hat{\eta}(S)-k_g r(S))$, where:

$k_g=\dfrac{-1}{c_\theta^T A_{m\theta}^{-1}b_\theta}=-11.0927$, $\hat{\eta}(S)=\hat{\omega}(S)u_{ad}(t)+\hat{\theta}^T(S)x_\theta(S)+\hat{\sigma}(S)$, $C(S)$ is the low pass filter. Design the low pass filter as $C(S)=\dfrac{\omega KD(S)}{1+\omega KD(S)}$, $\forall\omega\in\Omega_0$, where: K is the adaptive feedback gain and greater than zero; $D(S)$ is a strictly regular transfer function. At the same time, it is required that the low pass gain of low pass filter $C(0)=1$. In order to simplify the design, take $D(S)=\dfrac{1}{S}$, then the low pass filter is $C(S)=\dfrac{\omega K}{S+\omega K}$. According to $L1$ adaptive stability theory [10], in order to ensure the stability of the system, we need to satisfy: $\lambda=\|G(S)\|_{L1}L<1$, where $G(S)=(SI-A_{m\theta})^{-1}b_\theta\dfrac{S}{S+\omega K}$. To ensure the stability of the system, here we set $K=20$. Thus the design of $L1$ adaptive controller is completed.

4. Simulation results and analysis

Initial conditions of simulation: $V_0=0m/s$, $h=0$, $\theta_{c0}=0$, $\theta_0=90°$, $\phi_{c0}=\psi_{c0}=\phi_0=\psi_0=0°$. Vertical takeoff begins at 0 seconds and transits to horizontal flight mode. At 15 seconds, the command of transformation from level flight to vertical flight mode is given.

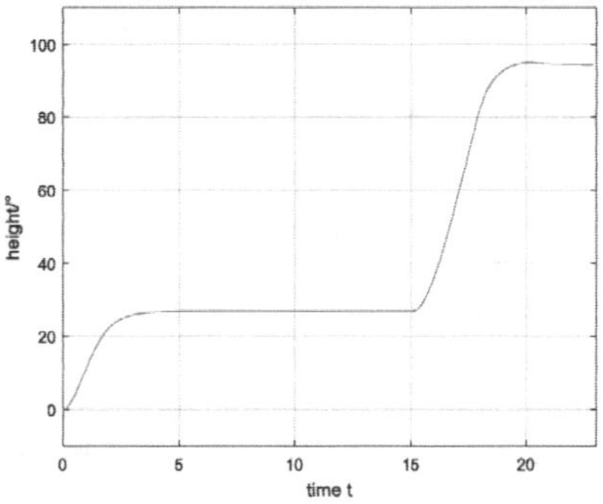

Figure 2. Altitude curve

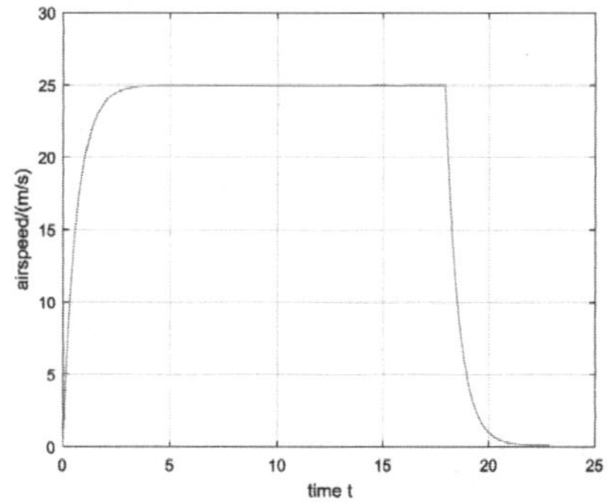

Figure 3. Airspeed curve

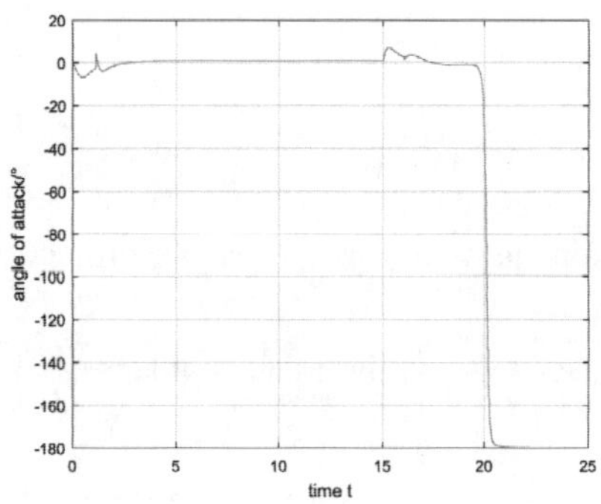

Figure 4. Curve of attack angle

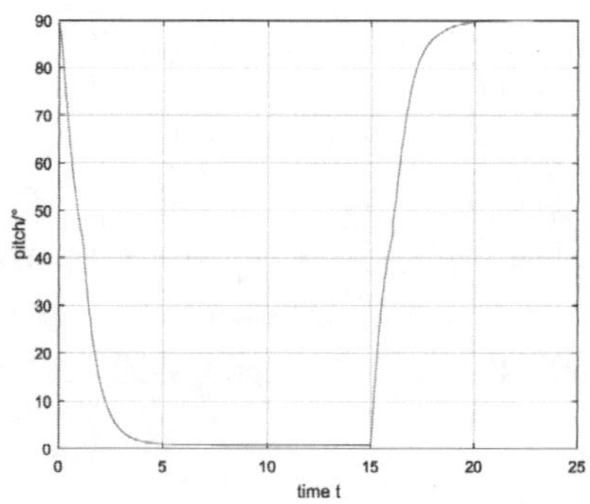

Figure 5. Pitch curve

It can be seen from the simulation results that the tail-sitter UAV completes the transition from vertical to horizontal flight in 6 seconds, and cruises at a speed of 25m/s. It completes the transition from horizontal to vertical flight in 18 seconds, after which the airspeed gradually reaches to zero. Under the change of model parameters, the change of pitch Angle, airspeed and altitude during the whole flight process is relatively stable.

5. Conclusion

In this paper, the fuzzy parameter self-tuning PID controller and $L1$ adaptive controller are designed respectively for the longitudinal attitude control of the Tail-sitter UAV in vertical and horizontal flight modes. The simulation results show that the longitudinal attitude controller designed in this paper can overcome the model parameter change problem and uncertain disturbances caused by the flight condition changes, and the control effect is good.

6. References

[1] Jiang C, Song K and Zhou H 2017 Control law design for tail-sitter UAV based on L1 adaptive control method *Ordnance Industry Automation.* 36 (08): 14-19

[2] Wang X, Chen Z and Yuan Z 2015 Modeling and control of an agile tail-sitter aircraft *Journal of the Franklin Institute.* 352(12): 5437-5472

[3] Zhang D, Chen Z and Hu Y 2018 Longitudinal decoupling control for the tail-sitter unmanned aerial vehicle during transition *Systems Engineering and Electronics.* 40(01):132-138

[4] Liu Y 2014 Flight controller design of tail-sitter UAV *Harbin Institute of Technology*

[5] Zhong J and Song B 2018 Position control of a tail-sitter MAV based on L1 adaptive controller in hover stage *Systems Engineering and Electronics.* 40 (09): 2062-2070

[6] Zhang P and Wang J 2015 Position control of a tail-sitter MAV based on L1 adaptive controller in hover stage *Computer Measurement &control.* 23(08): 2686-2688

[7] Wu D and Li X 2015 Simulation research on fuzzy parameter self-tuning PID controller *Electromechanical technology.* 1672-4801(2012)04-009-05

[8] Hu J and Hu D 2005 Design and simulation of PID parameter fuzzy self-tuning controller *JOURNAL OF NAVAL UNIVERSITY OF ENGINEERING.* 1009-3486(2005) 01-0097-04

[9] Liu Y 2005 Small fixed-wing UAV autonomous flight control law design *Harbin Institute of Technology*

[10] Xargay E, Hovakimyan N and Cao C 2010 L1 adaptive controller for multi-input multi- output systems in the presence of nonlinear unmatched uncertainties *American Control Conference (ACC).* 2010: 874-879

IOP Conf. Series: Journal of Physics: Conf. Series **1298** (2019) 012028 doi:10.1088/1742-6596/1298/1/012028

Dynamic Behaviour of a Mono-pile with Skirted Template Structure for Offshore Wind-Power System

Lee, Hsien Hua and Hsieh, H.-Y.

Department of Marine Environment and Engineering, National Sun Yat-sen University, Kaohsiung, Taiwan

hhlee@mail.nsysu.edu.tw

Abstract. The purpose of this research is to develop an offshore structural system that may sustain a stronger wave and have better stability. The object of the research is a hybrid type of offshore mono-pile-template structural system, which is more flexible and relatively easier to construct compared to the heavy mass type offshore structural system. It is expected that this hybrid type of supporting structure for an offshore wind-power generation platform will have the flexibility as presented by a typical template structural and yet, consistent properties for supporting function of the structure. Two types of structure are designed for the analytical study, namely, the mono-pile supporting structural system and the hybrid type combined with the mono-pile and the template structure as an outskirt frame system. Both annual regular waves and extreme storm waves of a 100-year return period are applied to the structural system as exerting forces.

1.Introduction

Due to the necessity for the development of offshore wind power system, it is important to find an offshore structural system that may sustain a higher wave and strong wind, and yet, have better stability during the operation of the platform. For a wind power generator installed in an offshore area, not only is the cost of installation high but the maintenance will be more complicated, especially for the safety of the structural system when it is subjected to the wave excitation, current action and mostly wind action on the generators transmitting to the supporting structure. In order to provide a safe supporting structure for an offshore wind power generator, many types of supporting system were developed, which include the gravity type, mono-pile supporting structure, multi-pile supporting structure, suction type, tri-pod supporting structure, jacket frame (template frame), floating type, and many combined types that may include two or more systems as mentioned together.

Each type of the supporting system for the offshore wind power station has its advantage and also drawback. Depending on the local environmental conditions such as seabed conditions, current and wave conditions, the applied system must be appropriately chosen to avoid the possible negative effect. The jacket frame (template frame) is a steel frame system that can be independently installed to support the offshore wind power system and it can also be combined with either a pile system or pod system. It is a flexible system in terms of its wide usage and the property of the structural system. However, according to the countable information, most construction plans proposed by the competing contractors for the possible wind-power station projects located in the late established site of wind-farm such as Taiwan Strait are mono-piles due to its economic reasons. According to the investigation that the sub-bottom sea-bed conditions for the location of most wind-power stations under plan are mostly soft soil with thick sediments of organic material, which is weak in bearing strength and shear resistance. In order to support the wind-power generator, a mono-pile with large diameter must be applied to the foundation system, such as a size larger than five meters is usually required. For mono-piles in such huge size, many difficulties under construction of infrastructure and installation for the wind-mill system will be up to another level including the mobilization, working ships like crane ship, transportation and supporting facilities around working site. Therefore, an alternative system, a combination system of mono-pile with skirted template frame is proposed and designed in this study.

The jacket frame, a template system is more flexible compared to the heavy mass type offshore structural system. In the mechanical analysis, material properties must be taken into accounts for a flexible structural system and therefore, it will take more effort for the analytical works. For the flexible type template structural system, vibration mitigation could be essentially important and to overcome this problem, some innovative measures have been proposed in as early as 1990s [1,2,3,4,5]. . If a combination between the mono-pile and template structural system can be applied, then the over-flexibility problem from the template structure or over-size problem from a mono-pile system can be solved through a more balanced hybrid structural system.

In this hybrid system, the vibration induced by either waves or current will be the main purpose of this study and moreover, the wind loading on the top structure that sustains the turbine system will also be considered. In the analysis, a prototype of mono-pile with skirted template structural system was established according to the power generation and environmental requirement of the typical west coastal offshore area, a presumed offshore wind farm. Along with the member size, material properties and joint types, two supporting systems are assumed, namely, the hybrid support and mono-pile support. When the turbine rotation was taken into accounts, the transmitting vibrations and forces to the supporting structure are also evaluated. From the analytical results, the comparison between an independent mono-pile system and a hybrid system combined with a mono-pile and template structural system can be made and conclusions are drawn.

2. Theorem Applied in the Study

The dynamic equation of motion for an offshore structural member with mass M, structural damping C, and stiffness K, subjected to the wave forces propagated in the normal direction of the structural member as shown in Figure1, can be written as

$$M\ddot{X}(t) + C\dot{X}(t) + KX(t) = P(t) \tag{1}$$

where x(t) is the displacement and the one with dot and double dot is the velocity and acceleration respectively. Taking into account of the relative motion between the structures and fluids, the wave forces P(t) exerted on the structural member,[6] can be presented as follows

$$P(t) = \rho v C_m \dot{u}(t) - \rho v C_a \ddot{X}(t) + \frac{1}{2}\rho C_d A\left|u(t) - \dot{X}(t)\right|\left(u(t) - \dot{X}(t)\right) \tag{2}$$

where C_a=C_m-1, and u and the one with dot is the velocity and acceleration of the fluid normal to the structural member resulted from the horizontal and vertical motion of the fluid, respectively. C_m and C_d are coefficients corresponding to inertia and drag effect respectively. V and A is the displaced volume and the projected front area of the structural member, respectively. The last term in the equation representing the drag force due to the relative velocity of fluid is nonlinear.

The nonlinearity of the drag term is retained through the use of the approximate relation derived by Penzien and Tseng [7] ,

$$\left|u(t) - \dot{X}(t)\right|\left(u(t) - \dot{X}(t)\right) = |u(t)|u(t) - 2 < u(t) > \dot{X}(t) \tag{3}$$

where $<u(t)>$ represent a time average value for the velocity of the fluid. Therefore, the wave force can further be rewritten into a form as the following

$$P(t) = \rho v C_m \dot{u}(t) - \rho v C_a \ddot{X}(t) + \frac{1}{2}\rho C_d A\left(|u(t)|u(t) - 2\hat{u}(t)\dot{X}(t)\right) \tag{4}$$

After substitution of the new form of wave force into the equation of motion, it becomes a form as

$$M\ddot{X}(t) + \dot{C}X(t) + KX(t) = \rho v C_m \dot{u}(t) - \rho v C_a \ddot{X}(t) + \frac{1}{2}\rho C_d A(|u(t)|u(t) - 2\hat{u}(t)\dot{X}(t)) \quad (5)$$

By grouping the unknown terms that are to be solved on one side in the equation, the equation of motion yields into the following form as

$$(M + \rho v C_a)\ddot{X}(t) + \left(C + \rho C_d A\hat{u}(t)\right)\dot{X}(t) + KX(t) = \rho v C_m \dot{u}(t) + \frac{1}{2}\rho C_d A|u(t)|u(t) \quad (6)$$

Now the normal motion for both structural displacements and fluids at the nodes of the structural member can be transformed into the global system for the whole structure, respectively, as

$$[M^*]\{\ddot{x}\} + [C^*]\{\dot{x}\} + [K^*]\{x\} = \{F^*\} \quad (7)$$

When we have the equations of motion and the forces exerted on the structural system ready, the analysis can be carried out by using the step-by-step integration schemes for the nonlinear structural system such as Newmark method [8] , Wilson's method [9] etc.. In this study the Newmark method was adopted due to its stability advantage.

3. Analytical Model

An offshore template structure located in the sea can be in various forms, shapes and sizes. A typical one will be like a wedge shape or part of a pyramid of which the lower part is wider. However, a square shape like a box of template structure with vertical column-piles is also popular due to its easier to be installed. Therefore, in this study, for the purpose of enhancement of a mono-pile foundation system, a square box type of hybrid foundation of mono-pile and template infrastructural system was designed for an analytical study.

The study applied here is a case utilized by Tai-power to be applied to an offshore wind-power generator in the west coast of Taiwan, of which sizes of most members except for the vertical ones are adopted. For the purpose of this study to compare the responses of the infrastructure with or without an out-skirt frame system under various wave conditions, the thickness of the vertical members is reduced in order to have a compatible stiffness when combined with a mono-pile system.

The related dimensional data for each category of members are illustrated in Table 1. A list of geometrical data for the mono-pile is also presented in Table 2.

Table 1 Dimensional data for the template structure

Member	OD (mm)	ID (mm)
vertical	2200	2180
horizontal	800	765
diagonal	900	865

Table 2 Geometrical data for the mono-pile

Member	OD (mm)	ID (mm)
Foundation	7000	6940
Tower/ lower	7000	6940
Tower /upper	5500	5440

During the analysis, as mentioned in the previous section, a finite element method was applied. In addition, a lumped mass method was also adapted for the dynamic analysis. Shown in Figure 1 is the analytical model for the numerical analysis of mono-pile system, where as illustrated in the figure, a mono-pile type of supporting structural system with wedge-shape in the upper part is presented. There are two sections in the pile, namely, the upper section with a wedge shape and the lower section with an uniform area. The cross section of the upper part with the wedge shape begins with a diameter of 5.5 m up to 7.0 m in a length of 30 meters, while the cross section of the lower part of the pile is uniform with diameter of 7.0 m all the way to the seabed. The analytical model was further simplified into a lumped model of three degree of freedom system for dynamic analysis as shown in the left hand side of Figure 1. Numbers in the circle represent a node along with lumped mass accumulated in the related section (as indicated in a square box) for the dynamic analysis.

For the comparison, a second infrastructure of hybrid type combined with a mono-pile and out-skirt template (frame) structural system is presented in the right hand side of Figure 1. The out-skirt template structure is proposed according to an equivalent manners for a compatible mono-pile system. The span of the template structure is 16 m and the height is 21 m while a two-layer of cross bracing were installed to reinforce the lateral resistant capability. The whole structure is made of steel and therefore, the property for a typical steel material was applied to the study.

In order to simulate an on-situ environment, a typical environment of offshore water was applied, where the water depth is around 32 m and wave conditions: H=6.6 m, L=168 m are obtained from a 100 years return period data base. For the annual average waves, the wave height H=1.5 m, the wave length L=75.75 m.

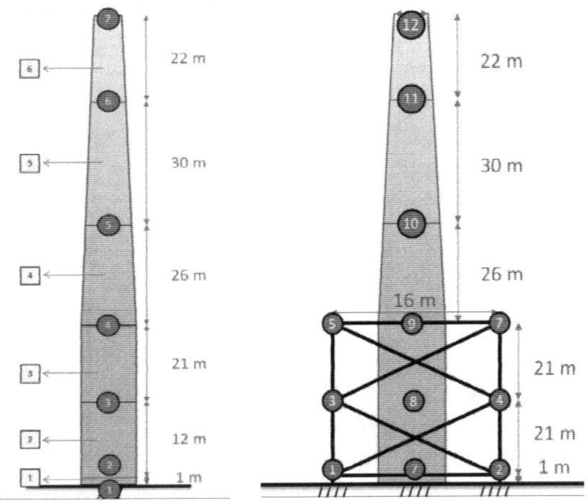

Figure 1 2-D model for the analytical study for the two types of structure

4. Analytical Results

The major responses to be observed from the dynamic analysis will include mainly the displacement, velocity and acceleration of the structure while the exerting forces will include both the wave of annual average and the maximum wave of 100 years return period.

Since the structure is symmetric at each direction, a three dimensional analysis can be simplified into a two dimensional one. The wave force calculated through the Airy theorem for small amplitude linear wave is applied. The maximum wave force can reach as high as 1500 kN on the top level of the template supporting structure near the water surface level while the other level submerged under the water may reach 500 kN.

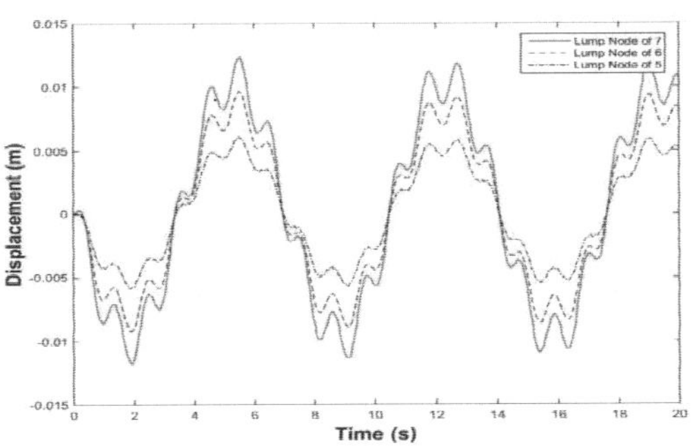

Figure 2 Displacement response for the mono-pile

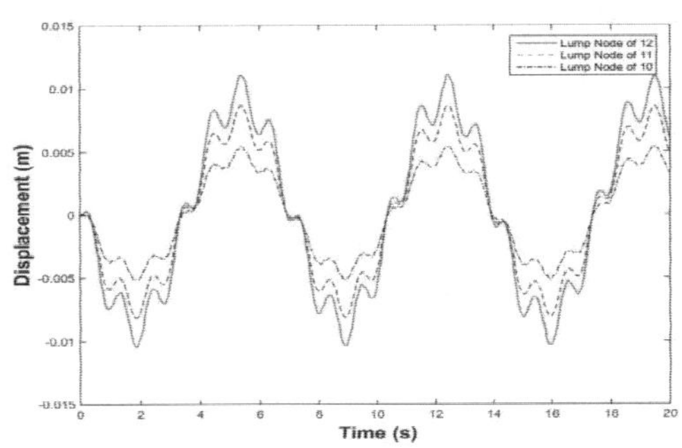

Figure 3 Displacement response for the hybrid mono-pile-template structure

Figure 2 shows the displacement response for the upper three levels (as node 7, 6, node 5) of the mono-pile structure. The movement of the structure seems to be minor when subjected to the wave force as presumed and calculated through Airy theorem. The maximum displacement occurs at the top level, where the value is about 1.23 cm that is not significant, probably due to the large rigidity of the vertical piles. The displacement for the lower second level of the platform is about 0.96 cm.

Figure 3 also shows the displacement response for the upper three levels of the hybrid mono-pile-template structural system. The displacement of the hybrid structure subjected to the waves seems to be very similar to the mono-pile system except for a smaller value at each corresponding levels was

IOP Conf. Series: Journal of Physics: Conf. Series **1298** (2019) 012028 doi:10.1088/1742-6596/1298/1/012028

found. The maximum displacement occurs at the top level, where the value is about 1.1 cm that is not significant either. The displacement for the lower second level of the platform is about 0.88 cm.

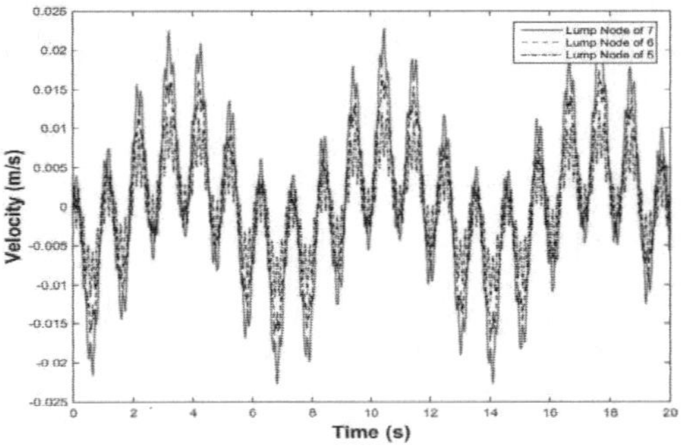

Figure 4 velocity response for the mono-pile

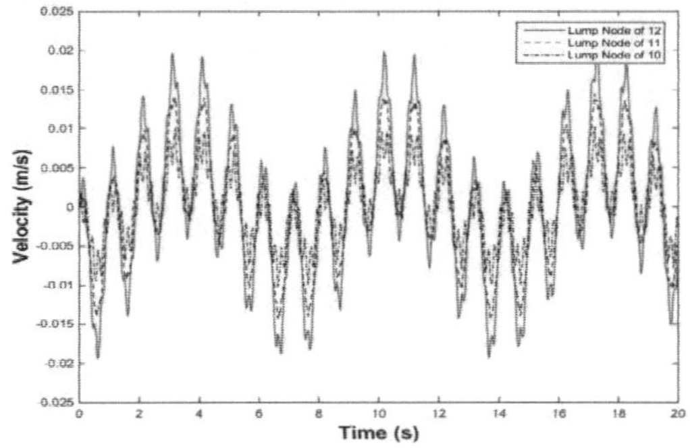

Figure 5 Velocity response for the hybrid mono-pile-template structure

Similarly, the time history of velocity for the upper part of various levels of both structures, namely, 4 mono-pile and the hybrid type with mono-pile and template structure, is also presented in Figure 4 and Figure 5. Again the maximum response occurs at the top level of both supporting structures, which is about 2.284 m/s and 1.985 m/s corresponding to the mono-pile type and the hybrid type supporting structures, while the velocity response at the second level of the platform is not significantly reduced for the mono-pile system compared to the hybrid type of supporting system.

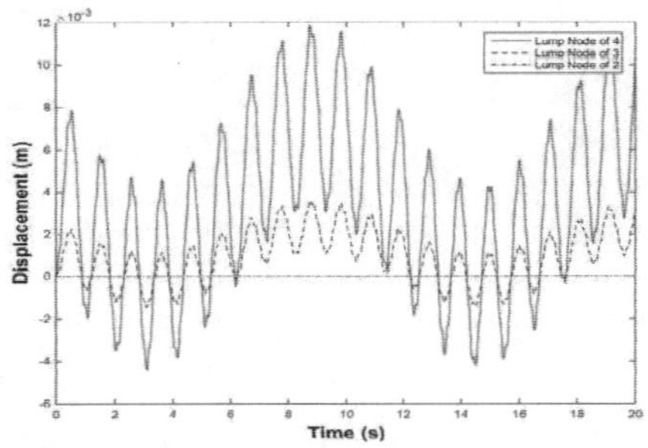

Figure 6 Displcement responses for various levels of the mono-pile

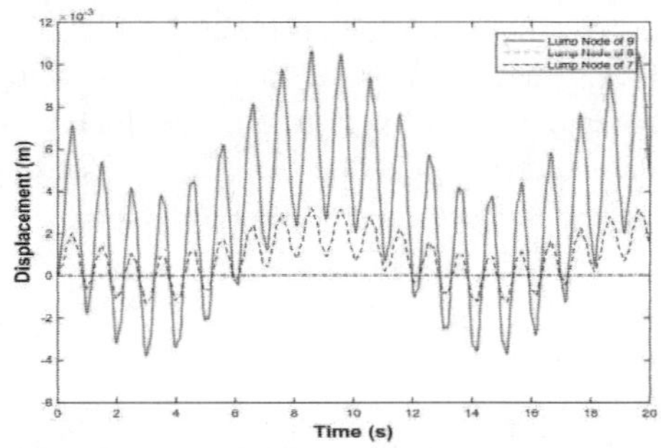

Figure 7 Displacement response for various levels of the hybrid mono-pile-template structure

A design wave of 100 years return period is applied in the second analysis. The displacement responses corresponding to the lower level (close to the wave motions) of structures for both structures are presented in Figure 6 and Figure 7 respectively. For the levels under analysis most close to the wave motions, namely node 4 for the mono-pile and node 9 for hybrid type of mono-pile-template structure respectively, we found that the displacement is 1.184 cm and 1.064 cm correspondingly.

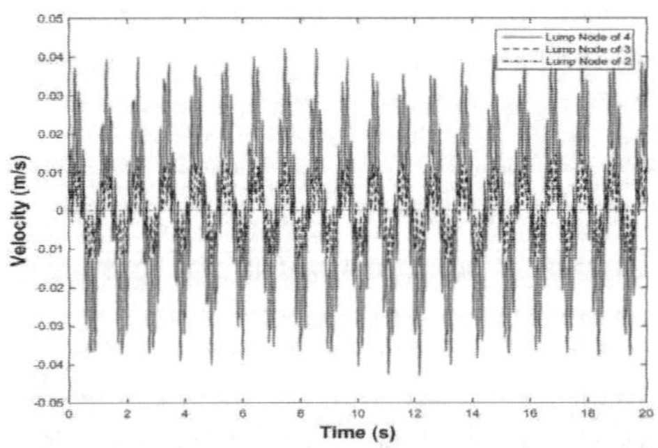

Figure 8 velocity responses for various levels of the mono-pile

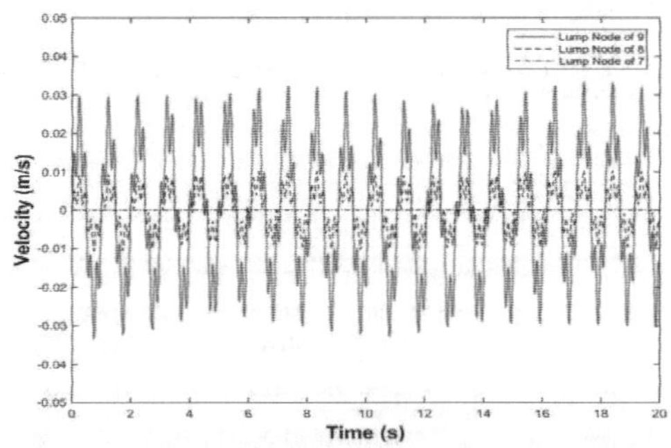

Figure 9 Velocity responses for various levels of the hybrid mono-pile-template structure

IOP Conf. Series: Journal of Physics: Conf. Series **1298** (2019) 012028 doi:10.1088/1742-6596/1298/1/012028

A phenomenon of significant directional motion for the displacement is observed in this analysis. The motion of the supporting structure near the water surface moves along with the wave and tends to have larger amplitude in one direction than the other.
In the velocity response, a comparison between two types of structural system is also presented in Figure 8 and Figure 9. The velocity response of positions at the structure is same as the displacement responses. For the mono-pile structure the velocity at the near water level is 4.22 m/s and 3.338 m/s for the hybrid mono-pile-template structure. It is also found that the fluctuations of the velocity in hybrid system is much smaller and better mitigated, especially for the large frequency motions, compared to the mono-pile system.

5. Conclusions

During the comparison for two systems subjected to two storm conditions, namely, annual regular storm and a storm of 100-year return period, some phenomena were found as follows:

1. Both systems subjected to regular annual waves have very small responses in both displacement and velocity. However, out-skirted template structure of the hybrid system may reduce the responses by 10%.
2. For the designed waves of 100-year return period, the displacement response may reach as large as five times. The reduction for the displacement response from the outskirt frames is more or less than 10%, similar to the case subjected to an annual regular wave condition.
3. For the velocity response, the amplitude can be reduced by 21% for the hybrid system, which is larger than the reduction in displacement response. Nevertheless, a subtler vibration mode in velocity was also found in the hybrid system such that the fluctuations are mitigated and high frequency motion is reduced.

Therefore, it is concluded from the results of this study that the installation of an outskirt template structure can enhance the stability of a mono-pile supporting structural system. The enhancement will be benefit in the high frequency motion and the motion close to resonant frequency of supporting structure.

6. References

[1] Lee, H.H. and Wu, R.-J. 1996 Vibration Mitigation of Structures in the Marine Environment. *Ocean Engineering - An Int. Journal*, Vol. 23 No.8 pp.741-759.

[2] Lee, H.H. 1997 Stochastic Analysis for Offshore Structures with Added Mechanical Dampers. *Ocean Engineering -An Int. Journal*, Vol. 24, No.9, pp.817-834.

[3] Lee, H.H. 1998 Seismic and Vibration Mitigation for the A-type Offshore Template Platform System. *Structural Engineering and Mechanics-An Int. Journal* Vo. 6, No. 3, pp.347-362

[4] Lee, H. H. and Juang, H.-H. 2012 Experimental Study on the Vibration Mitigation of Offshore Tension Leg Platform System with UWTLCD, *Smart Structures and Systems—An Int. J.* Vol.9, No.1, pp.71-104.

[5] Cheng, C.-W. Lee H. H. and Luo Y.-Z. 2015 Experimental Study of Controllable MR-TLCD Applied to the Mitigation of Structure Vibration, *Smart Structures and Systems—An Int. J.* Vol. 15, No. 6, pp.1481-1501

[6] Isaacson, M. 1979 Nonlinear inertia forces on bodies. *J. Waterways Div. ASCE*, No.WW3, pp.213-227.

[7] Penzien, J. and Tseng, S. 1978 Three-Dimensional Dynamic Analysis of Fixed Offshore Platforms. In *Numerical Methods in Offshore Engineering*, John Wiley and Sons, New York, pp.221-243.

[8] Newmark, N.M. 1962 A method of computation for structural dynamics. *Trans. ASCE*, Vol.127, pt.1, pp.1406-1435.

[9] Bathe K.J. and Wilson E.L. 1976 *Numerical Methods In Finite Element Analysis*. Prentice-Hall, Englewood Cliffs, NJ

AUTHOR INDEX